General Theory of
Algebraic Equations

THÉORIE
GÉNÉRALE
DES ÉQUATIONS
ALGÉBRIQUES;

Par M. BÉZOUT, de l'Académie Royale des Sciences & de celle de la Marine; Examinateur des Gardes du Pavillon & de la Marine, des Aspirans-Gardes de la Marine, des Eleves & Aspirans au Corps Royal de l'Artillerie; Censeur Royal.

A PARIS,

De l'Imprimerie de Ph.-D. PIERRES, rue S. Jacques.

M. DCC. LXXIX.

AVEC APPROBATION ET PRIVILEGE DU ROI.

Original title-page to the 1779 edition. The text beneath the title reads: "By M. Bézout, of the Royal Academy of Sciences and the Navy; Examiner of the Guards of the Pavilion and the Navy, Midshipman-Candidates of the Navy, of the Students and Midshipmen in the Royal Artillery Corps; Royal Censor."

General Theory of
Algebraic Equations

Etienne Bézout

Translated by Eric Feron

PRINCETON UNIVERSITY PRESS

PRINCETON AND OXFORD

Published by Princeton University Press,
41 William Street, Princeton, New Jersey 08540
In the United Kingdom: Princeton University Press,
3 Market Place, Woodstock, Oxfordshire OX20 1SY

Library of Congress Cataloging-in-Publication Data

Bézout, Etienne, 1730–1783.
[Théorie générale des équations algébriques. English]
General theory of algebraic equations / Etienne Bézout; translated by Eric Feron.
p. cm.
ISBN-13: 978-0-691-11432-3 (cloth : alk. paper)
ISBN-10: 0-691-11432-3 (cloth: alk. paper)
1. Equations, Theory of. I. Feron, Eric, 1967– II. Title.
QA211.B49 2006
519.9'4—dc22   2005054518

British Library Cataloging-in-Publication Data is available

The publisher would like to acknowledge the author of this
volume for providing the camera-ready copy from which this book was printed.

This book has been composed in Times Roman

Printed on acid-free paper. ∞

pup.princeton.edu

Printed in the United States of America

1 3 5 7 9 10 8 6 4 2

Contents

Section III

Book Two
*In which we give a process for reaching the final equation
resulting from an arbitrary number of equations in the
same number of unknowns, and in which we present many
general properties of algebraic quantities and equations*   137

Translator's Foreword

This document is a literal translation of Bézout's seminal work on the theory of algebraic equations in several unknowns. The notation of this translation strictly follows that of the original manuscript. Bézout's purpose was to provide an in-depth analysis of systems of algebraic equations. His main push was devoted to determining the degree of the final equation in one unknown, resulting from the original set of polynomial equations.

Translating Bézout's research centerpiece became necessary to me after attending an illuminating presentation made by Pablo Parrilo at MIT sometime around 2002. His presentation was devoted to polynomially constrained polynomial optimization via sum-of-square arguments. It was illuminating because much of sum-of-square optimization methods rely on (i) using *polynomial multipliers*, and (ii) considering the various monomials appearing in the polynomial expressions as *independent variables*, resulting in interesting algorithmic simplifications. Such was also Bézout's approach when dealing with systems of polynomial equations. I decided I needed to investigate the matter in more detail, by reading Bézout's work and writing the present translation.

A secondary motivation for performing this translation is the great success Bézout's textbooks encountered 200 years ago, not only in his home country, but also in the United States of America. The clarity of style and the informal presentation of advanced mathematical material resulted in early translations of Bézout's textbooks, most notably in Cambridge, Massachusetts. I just could not resist the temptation of re-enacting this activity in the same town, nearly 200 years later, with the goal of bringing Bézout's research masterpiece within reach of the same audience.

The document is several hundred pages long, and I would like to point out several elements I believe are prominent in Bézout's work:

- The preface is an excellent summary of the whole work and contains Bézout's essential ideas about polynomial multipliers.

- Paragraph (47) states Bézout's well-known theorem about the number of solutions to a set of algebraic equations.

- Many of the paragraphs between (47) and (193) are devoted to determining cases where the degree of the final equation is lower than the general case, yet can still be expressed in closed form.

- Paragraphs (194)–(223) contain an elegant, self-contained approach to solving simultaneous linear equations. They demonstrate Bézout's interest not only in algebraic results related to systems of algebraic equations in many unknowns, but also in efficient numerical approaches to obtain the final equation.

- Paragraphs (338) and following deal with the solution of equations whose number exceeds the number of unknowns by one. Likewise, paragraphs (440) and following deal with the solution of equations whose number exceeds the number of unknowns by two. These paragraphs are further generalized in paragraphs (502) and following. In addition to the role they play within Bézout's central purpose, these paragraphs can also easily lead the reader to *Bézout's identity* relative to characterizing coprime polynomials in one and, to some extent, *several* variables.

The strenuous process of translating 18th century French into English is not error-free. It probably even contains counter-interpretations. The reader is kindly invited to communicate any mistakes to the author of this translation and these will be promptly addressed. This work would probably never have been completed if it were not for a few key individuals:

I wish first to thank Alain le Pourhiet, who made me aware of Bézout's work and was kind enough to put an original copy of it under my custody.

Christine-Marie Feron and Jacques Pedron have offered me constant support and understanding during the many evenings I spent on this work. I also thank Lauren Clark, Angela Olsen, Lisa Gaumond, Animesh Chakravarthy, Emily Craparo, Sommer Gentry, Masha Ishutkina, Gregory Marks, Glenn Tournier and Mario Valenti for sharing constructive criticisms and good laughs about my Frenglish. Many thanks also to my colleagues Erik Demaine, Pablo Parrilo, and Bruce Reznick for so many useful discussions.

I finally wish to thank the Air Force Office of Scientific Research (MURI award F49620-01-1-0361), the Defense Advanced Research Projects Agency (MURI award F49620-02-1-0325), the National Science Foundation (Grant CNS-0451865), the Office of Naval Research (Grant N00014-03-1-0171) and the Office National d'Etudes et Recherches Aérospatiales, Centre d'Etudes et Recherches de Toulouse, for their financial support.

Dedication from the 1779 edition

To Monseigneur de Sartine
Minister and Secretary of State, Department of the Navy

MONSEIGNEUR,

Many administrations in the capital remember your active, wise and enlightened leadership.

Since then, the King gave you a much wider ranging assignment: This important mission proves, Monseigneur, that the same spirit can make significant contributions to very different missions.

The multiple and urgent tasks imposed upon you today do not, nevertheless, deviate your sight from the future. Keeping your eyes opened to everything that may improve Naval forces and the current glory of the King, you simultaneously keep in mind its long-term future. You make sure that useful knowledge be transmitted to new and enlightened officers, thereby compensating for the unavoidable human losses associated with the intensity by which the Marine Corps enforces respect for the French Flag. These considerations, Monseigneur, urged me to dedicate this work to you. Its goal is to push and improve a sub-part of the mathematical sciences, to be used by all other sciences for their own advancement. Therefore, it may contribute to advancing knowledge that will be useful to the Navy. I cannot determine whether this work will fulfill its goals, but I have spared no efforts to make it worth the general public's approval, and thereby to be dedicated to you.

Respectfully yours,

Monseigneur,
Your very humble and faithful servant,
BÉZOUT.

Preface to the 1779 edition

Algebraic analysis is applied to various mathematical questions, essentially with the help of equations. Many efforts have been made to develop a theory for the latter, to perfect the methods to obtain general and specific statements that the solutions to these equations may provide, and to provide the surest, simplest and fastest means to solve them.

However, when the variables involved in answering a particular question become somewhat numerous, and when at the same time, the various constraints connecting these variables become more complex, the art of dealing with the whole system using general and simple rules requires sustained attention. This attention is all the more easily lost when the unlimited field of mathematical investigations continuously offers more interesting objects, which can be rapidly enjoyed, and where one's own sagacity finds intense satisfaction.

This probably is one of the reasons which almost led researchers to abandon the analysis of finite quantities when infinitesimal analysis was discovered. This happened in spite of the fact that investigations hardly touched upon or examined the difficulties remaining to be solved in that context, or observed the value, the rigor and the extent of the methods that one thought to possess to solve questions related to this problem.

Infinitesimal analysis was considered so attractive and important because of its numerous and useful applications; as such it attracted upon itself all research attention and efforts. Concurrently, algebraic analysis appeared to be a field where nothing remained to be done, or where whatever remained to be done would have only been worthless speculation.

The modest growth of algebraic analysis may not be attributed to this reason alone. Looking back at our work, we dare to believe that the matter was very difficult to tackle with the appropriate level of generality. Starting from known methods, analysts may have been discouraged from making any significant progress in this area. This impression does not stem from a bias towards our work: We candidly acknowledge that we have long thought the same way and worked without success, when we tried to deal with the matters contained in the present work using known methods only.

Nevertheless, the major contributors to infinitesimal analysis are well aware of the need to improve algebraic analysis: Their own progress depends upon it. Among many very distinguished analysts, the famous MM. Euler and de la Grange wrote memoirs on this topic that are no less deep nor sagacious than the other works of these illustrious analysts. Nevertheless, all research efforts towards general methods dealing with equations reduce (with the exception of first-order equations) to methods to obtain the simplest solution to a system of two equations in two unknowns. And those methods essentially all reduce to recasting the system under study to equations involving only one unknown. But nowhere do we see any systematic and general method to deal with this very restricted class of equations, let alone with the case of an arbitrary number of equations and unknowns.

Recognizing that, among all problems involving an infinite number of equations and unknowns, we only know how to solve systems involving two equations in two unknowns; recognizing, I repeat, that only that case could be solved, making sure that it not be perturbed in any way, we can easily acknowledge that everything remained to be done on this topic. Let us now pause and look at the state of the art when we started the work presented here.

Any early attempt at combining higher degree equations with each other naturally began with a very small number of such equations. The lack of maturity of the methods led to very complex computations, and the overall effort did not go very far. Research therefore started with low-degree equations in two unknowns. Combining these equations following various approaches, the values of the consecutive powers of the unknown to be eliminated were determined, from the highest power to the lowest. Substituting this variable and its powers into one of the equations finally resulted in one equation in one unknown. Once this first step was made, it was concluded that the same procedure could be followed for three equations in three unknowns: Combining, for example, the first and second equations leads to an equation with only two unknowns. Using the same procedure with the first and third equation or the second and the third, similarly leads to an equation with the same two unknowns. Having finally recast the three initial equations into two equations in two unknowns, it was concluded that in general, the same procedure could be used and always end up with a unique equation in one unknown from an arbitrary set of equations with an equal number of unknowns.

Even if this procedure had not suffered from the essential defects we will describe later, it would still merely be a means to bring the question back to a problem involving only one unknown, and it would still be insufficient to develop a General Theory of Equations.

Indeed, with a limited knowledge of calculus and attention to this method, we see that the proposed equations will, depending on the way the calculations are performed, contribute in many possible and different ways to the formulation of the last equation: Depending upon the way and the order in which calculations are made, we can obtain very different expressions for that last equation. This equation must, however, be unique. What insight can such a method provide about the general properties of the proposed equations? What useful knowledge can a General Theory of Equations gain from such a method, when it seems to be masking and hiding the general properties of the system instead, possibly more so than the original problem formulation? There is therefore a long way to go before claiming that such a method is useful towards a General Theory of Equations.

Let us consider this process, in the light of its expected use, to concentrate all proposed equations into one and to ascertain the actual number of solutions and the actual solutions to the question.

Our previous observation showed that the final equation obtained through this process can vary, depending upon the order in which the process is applied. However, intuition suggests that only one such final equation exists, which must be completely insensitive to the way it was obtained. We must therefore conclude that the final equation obtained through all these successive eliminations is not the actual final equation, but only contains it, multiplied by quantities that do not relate to the original question. Thus such a method (which is not practical anyway, given the very large number of computations it requires, even in simple cases) leads to useless computations. It is also deceptive regarding the true degree of the final equation. Also, it provides no means to determine the number of solutions, or which ones among them are true solutions, distinguishing them from those irrelevant to the initial question. One should know at least what the true degree of the final equation is, and available methods are still far from accessing this information. Even if this information had been available, there would still have been a need for analytic tools to extract the one or many irrelevant factors. But the currently available analysis tools fall far short of solving this difficult problem. Whether the extent of these difficulties was perceived or not, they could already be felt in the case of equations involving two unknowns. The enormous complexity of the calculations necessary for successive variable eliminations is probably one of the reasons for which no general result may be found in the written works of the analysts, when dealing with equations involving more than two unknowns, except in the case of linear equations. Distinguished analysts, realizing the difficulties of this problem, all turned their attention to equations in two unknowns.

M. Euler gave means to reach the final equation devoid of any superfluous factor and at the same time determined the actual degree of the final equation, in the special case when these equations are complete or when the missing terms are those in the highest powers of one or the other unknown.

M. Cramer gave a very elegant and simple method to address the same problem in his excellent analysis of curved lines. Various other very distinguished Analysts have dealt with this problem since then, but they focused their attention on simplifying computations and making their results more insightful with respect to the general properties of these kinds of equations.

I do not intend to downplay the merit of this work, but I must observe that, although these methods are very useful for two equations in two unknowns, they cannot be applied to a larger number of equations and unknowns without raising the same difficulties as those of the initial method.

Applying these methods to a larger number of equations and unknowns requires, as we have shown earlier, combining these equations two by two. However, although the results of these combinations have no superfluous factor, they remain more complicated than necessary. The ensuing eliminations that must be made are not only much more demanding than necessary, but lead to ever more complicated expressions, which grow very fast as the number of required eliminations becomes larger. In addition, there is nothing available to recognize superfluous factors, which appear only in the last equation.

Thus, and in spite of the degree of perfection reached in solving equations in two unknowns, analysis still lacked tools for systems involving large numbers of equations and unknowns.

I had the opportunity to think about this problem during various research projects, and tried to partially solve it. One of the main reasons for the complications came from the need to combine the equations two by two in the methods of both M. Euler and M. Cramer.

It appeared to me that using pairs of equations for successive elimination was introducing irrelevant information during the elimination process; I then concluded that simpler results could be obtained by combining larger numbers of equations all at once. This intuition led me to produce a paper, now part of a 1764 Memoir of the Academy of Sciences.

Although the methods proposed in this Memoir indeed always led to a simpler final equation than the previous methods, they still did not lead to the simplest possible final equation. And although the superfluous factor appearing in the final equation had a lower degree than those generated by other procedures, its complexity usually still grew with the complexity of the initial equations.

As a result of these analyses, I felt even more how imperfect the analysis was, and I thought it would be useful to develop a method that would be free of these defects. I did not attempt to analyze the global impact of such a method on analysis, but I thought it was well worth spending much research effort towards the sole purpose of finding with confidence the final equation with the lowest degree resulting from a system of equations.

I had suspected for a long time that the general reason for the defects in existing methods was the sequential elimination of unknowns. Following much reflection upon this topic, I became convinced about it.

I therefore felt it was not worth thinking about using known methods to make progress on this problem, and that I had to use new methods.

I had many opportunities to think about the idea of multiplying the proposed equations by functions of all involved unknowns, to sum up all these products, and to assume that in this sum all terms containing the unknowns to be eliminated must vanish. This idea probably came to others as well. But what must these functions be to provide a satisfactory outcome? These functions could provide fewer, as many, or more coefficients than are necessary to remove the terms to be eliminated. What usage could be made of the redundant coefficients? Which ones were they and how many of them were there? And if it is possible to use fewer coefficients than the number of terms to cancel (as we will see in many cases towards the end of this work), how should we behave to avoid generating condition equations?

These questions are precisely the core difficulty of the problem. By being completely unaware of the degree of the final equation, we also did not know the degree of the polynomial multipliers, and therefore the total number of available coefficients. We did not even know how many useless coefficients there were. We would have made a significant mistake if, taking random degrees for the polynomial multipliers, the number of arbitrary coefficients was computed as the difference between the total number of available coefficients and the number of terms to eliminate.

In one word, the idea of performing variable elimination by multiplying the proposed equations was still a sterile idea as long as these questions remained unanswered.

I first decided I had to find a general expression for the number of coefficients which are not necessary in the elimination process.

The question reduced then to the following: Given an arbitrary polynomial, with a given number of unknown coefficients in its terms; given also an arbitrary number of equations involving these coefficients, how many independent coefficients are there in this polynomial?

It is clear that if we use these equations for arbitrary purposes, then the number of useful coefficients is not the the total number of terms in the polynomial, but rather the difference between the number of terms and the number of terms constrained by the arbitrary equations.

In my approach to solving this question, I did not first consider, as may be understood, all possible equation forms. Rather, I proposed to solve it for an arbitrary number of complete equations, that is, equations with no missing terms.

The solution to this first question gave me insight as to how I should approach the problem of finding the degree of the final equation resulting from an arbitrary number of complete equations, of arbitrary degrees and having the same number of unknowns.

Indeed, assume we multiply any of these equations by a complete polynomial of undetermined order; in addition, assume that we use all the remaining equations to cancel some terms in the polynomial multiplier; then, by the same token, we can cancel some terms in the product equation. By performing this operation, we can express the contents of all other equations in the last equation.

Therefore, since we can truly express all the original questions via this process, the terms that may still contain the unknowns to be eliminated must disappear by themselves. The polynomial multiplier must therefore have introduced a sufficient number of coefficients in the product equation, so that these terms disappear; that is, excluding the arbitrary coefficients, there remains enough coefficients to cancel the terms that must disappear in the product equation.

Once these fundamental ideas were established, I had to apply them. This required two things: The first is a general expression for the number of terms in a complete polynomial; this is an easy task. The second is an expression for the number of terms remaining in an arbitrary complete polynomial, after canceling all terms that can be eliminated with a given number of equations. This required, as I hope one will see, some attention and cleverness to quickly obtain the very simple result it leads to. It would have been quite difficult to untangle this problem, if not for all the attention that we spent linking all these different expressions to finite differences.

We wanted to spare the reader the need to look elsewhere for what we mean by the expression of the number of terms in a polynomial, as well as concepts about finite differences, and sums of a few finite quantities; thus we have begun this work with an introduction containing all these useful notions.

By applying these means and these ideas to complete equations, we found the following general theorem: *The degree of the final equation resulting from an arbitrary number of complete equations, containing an equal number of unknowns, and of arbitrary degrees, is equal to the product of the powers of the degrees of these equations.* This theorem was known and proven for two equations only.

No matter how general this theorem is, and irrespective of the usefulness it may have in research, it left many obscure points behind. Given the little information available about equations in two unknowns, where the highest powers of these unknowns are missing, we felt there were many equations leading to a final equation with lower degree, because some of their terms are missing. This class of equations is infinitely larger than the former, although that one is also infinite.

This class is infinite with respect to the arbitrary number of unknowns it can contain and also with respect to the many different ways for terms to be missing; these missing terms do influence the degree of the final equation.

To proceed in an orderly fashion, I first decided to determine the degree of the final equation resulting from an arbitrary number of incomplete equations with the same number of unknowns, satisfying the following conditions:

1. Assuming the total number of unknowns to be n, the combination of n variables results in an arbitrary, different degree for each equation.

2. The combination of $n-1$ unknowns results in terms of arbitrary degrees, not only for each equation, but also for each of these combinations.

3. The combination of $n-2$ unknowns results in terms of arbitrary degrees, not only for each equation, but also for each of these combinations, and so on.

But since it was not possible to attack this problem up front, I decided to work on it backwards, that is, by assuming first the absence of the highest dimensions in single variables alone, then the absence of highest dimensions in combinations of two of these variables, etc., first with a few restrictions, to help me understand the method without loss of generality, however.

In order to succeed with this new class of equations, I opted for a different path from the one I followed for complete equations; I could have chosen the same path, but I would have been led to further developments and would have had to account for details that the new method did not need; I was happier to choose this other method because it is also applicable to complete as well as incomplete equations.

This new process, like the first, requires an expression for the number of terms of the product equation, of the polynomial multiplier, and of all different polynomials that contribute to the expression of the number of unnecessary coefficients of the polynomial multiplier. I therefore give the means to calculate the number of terms of the polynomials I use, and of the different expressions that must contribute to the degree of the final equation: Only through practice can one understand the need for these expressions. Since the complete equations and a few classes of incomplete equations that I first dealt with gave me only one form of polynomial multiplier and thus a unique expression for the degree of the final equation, I was quite amazed to find many different expressions for the degree of the final equation when I extended my work to more extended objects. After thinking about it for a while, I realized this apparent drawback was getting worse as the object of this study expanded.

Admittedly, I quickly started suspecting that these different expressions did arise from the various relative orders of the given exponents, which can influence the degree of the final equation. It took me a lot of concentration on this matter to find a way to determine the criteria to identify the only valid expression for the degree of the final polynomial, when these expressions disagree. One would be mistaken to think that it is sufficient to pick, among these different expressions, the one that gives the lowest final equation degree. Very different considerations enter into determining which expression is the right one.

After having provided much insight to determine the general expression for the degree of the final equation for any of these incomplete equations, we have considered incomplete equations with higher order: We refer the reader to the book itself to get an idea about these. We do not deal with these equations as before. The form of the polynomial multiplier is not nearly as easy to determine: Depending upon the relative magnitude of the known exponents, it can be a polynomial of higher or lower order; also, our considerations expressing all factors that contribute to the expression of the degree of the final equation are not sufficient to make it a function of the known exponents of the proposed equations only; that is our goal, however.

Beyond these equations of various orders, which include all those that I will call regular equations later, we must still deal with equations of so-called irregular form to claim we can determine the degree of the final equation for all equation forms. The considerations used to determine this degree for equations of irregular form are the same that must be used for incomplete equations of various orders; we have postponed the treatment of both types of equations to the second part of this book, because several of the tools introduced in the second part are useful for dealing with them.

The aim of the second part of this work, or the Second Book, is to give the method to reach the final equation, and, more generally, to discover the general properties of the equations.

In the first book, we needed only one polynomial multiplier, as long as we were only concerned with determining the degree of the final equation. But when considering computational issues, either to obtain the final equation or to obtain any function depending upon the conditions given by the initial equations, we must multiply each proposed equation by a polynomial and add the products to obtain what we call the sum equation. If we seek the final equation, we first set to zero all the useless coefficients of the polynomial multipliers; we then set to zero the total coefficient of each term of the sum equation that contains one or more of the unknowns to be eliminated; this leads to a system of first-degree equations in the undetermined coefficients of the polynomial multipliers. Substituting the values of these coefficients in the remaining terms of the final equation leads to the final equation.

It therefore appears that nothing remains to be done once the degree of the final equation is determined, since it then seems to reduce to the elimination process for first-order equations: We hope that the second part of this work will lead the reader to think otherwise. To give a small idea of what remained to be done to perfect the theory of equations, we observe:

1. That it is necessary to determine at least the form of each polynomial multiplier.

2. That it is no less necessary to know the number of useless coefficients for each of these polynomials, and that it is even more necessary to examine and to determine whether these arbitrary coefficients are arbitrary in an unlimited fashion or whether they are subject to certain conditions; computing the number of useless coefficients following what is prescribed or what results from what is prescribed in the first book, can we arbitrarily decide which coefficients are arbitrary?

3. Is it really justified to say that the problem is solved when the question is reduced to the elimination of first-degree unknowns? Aren't the two following questions important to Analysis?

   Are the methods available so far to solve first-degree equations as efficient and perfect as possible? When applied to several cases, and in particular to elimination in equations of higher degree, don't these require many more useless calculations than necessary? Wouldn't it be possible to have a method that would compute only what is necessary, especially when, as in the case of interest in this work, there is such a large number of unknowns to compute? Finally, and that is a point of interest here, could there exist a method that gives all unknowns, or a given number of them, all at once? This question is really important to analysis, and we believe we have found a simple, general and useful solution.

   The second question is the following: Wouldn't it be possible that, independently from the number of useless coefficients (which we will

call useless because it is always possible to make them disappear from
the various polynomial multipliers), the requirement to cancel terms
from the sum equation leads to the disappearance of several other coefficients? And wouldn't means to locate them before proceeding to the
computations exist? We will see that the solution to this question still
considerably reduces the number of coefficients and consequently considerably simplifies the computations. After having thus perfected the
elimination method for first-order equations, without which computations would have been immediately impractical, several new questions
did arise from this effort.

We first deal with equations in all their generality: This is the only way to
get all the information related to the proposed equations, and we do not have
to fear the prospect of reaching a final equation whose degree is too high or
whose roots are unrelated to the question. However, the different terms in
this equation have one or many common factors that are a function of the
known coefficients of the proposed equations. What do these factors mean?
This question became more important to solve, because its solution is closely
related to the following other important question: Which relations between
the coefficients of the proposed equations can lead to a final equation with
lower degree?

To succeed in solving those different objects, I had to improve the elimination method for first-order equations, but this would not have been sufficient.
We had to rely on new methods that may be very useful in analysis, to recognize the factors in the final equation: These are ways to find functions of
an arbitrary number of quantities, that are identically zero. At this point
we will not expand any further on this and other topics we had to deal with
when considering general equations in the Second Book.

By recasting the n proposed equations to n equations in $n-1$ unknowns,
computations are immensely reduced; but this approach may hide the possibility to reduce the degree of the final equation when particular relations
exist between known coefficients and when there are more than two unknowns. Indeed, these relations can lead to additional factors. Fortunately,
those factors do not complicate the degree of the final equation when these
equations are complete, and we give means to recognize them. But it would
have been better to avoid them to speed up computations. We believe this
is not possible and we think we have an argument supporting this claim. By
reading this book, the reader will see that a rigorous application of analysis
never yields any useless information. We will see that the factors of interest
here are not unrelated to the question; computations are accelerated when
they can be avoided, but at the expense of hiding a part of the knowledge
that may be gained about the proposed equations.

In a work whose object is the General Theory of Equations, we also had to
deal with equations that contain more or fewer unknowns than their number:
Both led a large amount of research and remarks that we think will be useful
to the analysis; we think that one can appreciate this only by reading this

work.

Finally, we complete this work by presenting the way to determine the degree of the final equation for regular or irregular equations, that is, equations for which we may or may not have an algebraic expression for the number of their terms: By doing so, we believe there is no kind of algebraic equations for which we have not given a way to find the lowest degree of the final equation even when there exists a relation between the coefficients that may lead to lowering the degree of the final equation. We also believe we have given a large number of new and very general properties of arbitrary numbers of equations; we have also developed methods that find more than one useful application in analysis. We hope that this work will be the opportunity for significant progress in analysis by turning the talent and attention of today's analysts towards this important subject. We will consider ourselves satisfied if, considering the point where we started from, and the point where we arrived, the reader agrees that we have done our part of the duty that any human owes to society.

Introduction

Theory of differences and sums of quantities

Definitions and preliminary notions

(1.) A function of a given variable is defined as any arithmetic expression involving this variable, irrespective of how it appears in it.

Thus x, $a + bx$, $(c - 3dx^3 + fx^4)^5$, $(a + fx^p + gx^q)^r$ etc. are functions of x.

Consider X an arbitrary function of x, and define X' as what becomes of X when x is replaced by $x+k$; then $X'-X$ represents the variation of X when x increases by k. $X' - X$ is called *the difference of X*. Thus, although strictly speaking, one may not talk about the difference of one quantity, we will adopt this commonly used expression; it means the difference between this quantity, considered in an arbitrary state, and the same quantity, considered in another arbitrary state.

We use the letter d to represent the difference of an arbitrary quantity or function. It will not be used for any other purpose to avoid any confusion. Thus, instead of $X' - X$, we write dX or $d(X)$.

And to express, at the same time, the amount by which the quantity x varies, we thus write $d(X) \ldots \left(\begin{array}{c} x \\ k \end{array} \right)$ to express the *difference of X when x varies by an amount of k*.

We consider increasing quantities here; we will see later what happens when considering decreasing quantities.

Assume the function whose variation or difference under consideration is a function of several variables, x, y or z, whose respective variations are k, l, m; denoting this function by P, we write its difference as $d(P) \left(\begin{array}{ccc} x & : & y & : & z \\ k & : & l & : & m \end{array} \right)$, which means the *difference of P when x varies by an amount of k, y by an amount of l, and z by an amount of m*.

Applying to $X' - X$ the same ideas as above, assume that x is replaced by $x + k'$ in $X' - X$. Then X' becomes X''', and X becomes X''. Then $(X''' - X'') - (X' - X)$ is known as the *second difference of X*, because it is the difference between two successive differences of X.

The second difference will be denoted $dd(X) \ldots \left(\begin{array}{c} x \\ k, k' \end{array} \right)$, which means the *second difference of X, when x varies first by k and then by k'*.

(2.) We will very soon give the rules to determine first differences. But we show right now that the second differences are determined by applying to first differences the same rules as those that generate them.

Indeed, the quantity $(X''' - X'') - (X' - X)$ can also be written as follows, $(X''' - X') - (X'' - X)$. Since by assumption X''' is what becomes of X' when substituting $x + k'$ for x and, likewise, X'' is what becomes of X, we therefore obtain $X''' - X' = d(X') \ldots \begin{pmatrix} x \\ k' \end{pmatrix}$ and $X'' - X = d(X) \ldots \begin{pmatrix} x \\ k' \end{pmatrix}$; so $(X''' - X') - (X'' - X)$ or $(X''' - X'') - (X' - X) = d(X') \ldots \begin{pmatrix} x \\ k' \end{pmatrix} - d(X) \ldots \begin{pmatrix} x \\ k' \end{pmatrix} = d(X' - X) \ldots \begin{pmatrix} x \\ k' \end{pmatrix}$. However, $X' - X = d(X) \ldots \begin{pmatrix} x \\ k \end{pmatrix}$, therefore $(X''' - X'') - (X' - X)$ or

$$dd(X) \ldots \begin{pmatrix} x \\ k, k' \end{pmatrix} = d \left(d(X) \ldots \begin{pmatrix} x \\ k \end{pmatrix} \right) \ldots \begin{pmatrix} x \\ k' \end{pmatrix}.$$

That is, we must first compute $d(X) \ldots \begin{pmatrix} x \\ k \end{pmatrix}$ to obtain $dd(X) \ldots \begin{pmatrix} x \\ k, k' \end{pmatrix}$: We must first take the difference of x, when x varies by an amount of k; we then take the difference of the resulting expression, when x varies by an amount of k'.

(3.) The order of the variation of x (whether x varies by an amount of k in the first difference and k' in the second, or vice versa) makes no difference. Indeed, $(X''' - X'') - (X' - X)$ contains $X''' - X'' = d(X'') \ldots \begin{pmatrix} x \\ k \end{pmatrix}$; it also contains $X' - X = d(X) \ldots \begin{pmatrix} x \\ k \end{pmatrix}$. Therefore $(X''' - X'') - (X' - X)$ or $dd(X) \ldots \begin{pmatrix} x \\ k, k' \end{pmatrix} = d(X'') \ldots \begin{pmatrix} x \\ k \end{pmatrix} - d(X) \ldots \begin{pmatrix} x \\ k \end{pmatrix} = d(X'' - X) \ldots \begin{pmatrix} x \\ k \end{pmatrix}$. But, by definition, $X'' - X = d(X) \ldots \begin{pmatrix} x \\ k' \end{pmatrix}$. Thus

$$d(X'' - X) \ldots \begin{pmatrix} x \\ k \end{pmatrix} = d(X'') \ldots \begin{pmatrix} x \\ k \end{pmatrix} - d(X) \ldots \begin{pmatrix} x \\ k \end{pmatrix}$$
$$= d \left(d(X \ldots \begin{pmatrix} x \\ k' \end{pmatrix}) \right) \ldots \begin{pmatrix} x \\ k \end{pmatrix}.$$

Thus $dd(X) \ldots \begin{pmatrix} x \\ k, k' \end{pmatrix} = d(d(X) \ldots \begin{pmatrix} x \\ k' \end{pmatrix}) \ldots \begin{pmatrix} x \\ k \end{pmatrix}$, but we also just saw that $dd(X) \ldots \begin{pmatrix} x \\ k, k' \end{pmatrix} = d \left(d(X) \ldots \begin{pmatrix} x \\ k \end{pmatrix} \right) \ldots \begin{pmatrix} x \\ k' \end{pmatrix}$; thus

$$d \left(d(X) \ldots \begin{pmatrix} k \\ x \end{pmatrix} \right) \ldots \begin{pmatrix} x \\ k' \end{pmatrix} = d \left(d(X) \ldots \begin{pmatrix} x \\ k' \end{pmatrix} \right) \ldots \begin{pmatrix} x \\ k \end{pmatrix}.$$

Assume the function under consideration contains several variables x, y, z, etc., whose first variation is k, l, m, etc., respectively; we call the second difference of this function (whose name I assume to be P)

$$dd(P)\dots\left(\begin{array}{ccc} x & y & z \\ k,k' & : & l,l' & : & m,m' \end{array}\text{ etc.}\right).$$

(4.) To have an idea of the third difference, imagine that x is replaced by $x + k''$ in $(X''' - X'') - (X' - X)$. Then if $X^{\text{VII}}, X^{\text{VI}}, X^{\text{V}}, X^{\text{IV}}$ are what becomes of X''', X'', X' and X with this substitution, the quantity $((X^{\text{VII}} - X^{\text{VI}}) - (X^{\text{V}} - X^{\text{IV}})) - ((X''' - X'') - (X' - X))$ is what is called the *third difference* of X, because it is the difference of two second differences. If k, k', k'' are the successive variations of x, the third difference is written $d^3(X)\dots\left(\begin{array}{c} x \\ k,k',k'' \end{array}\right)$. It is easy to see how this extends to the definition of the fourth, fifth and further differences.

About the way to compute the differences of quantities

(5.) Once the algebraic expression of a quantity is given, it is very easy to compute its difference. For example, assume we want to compute the difference of x^3 when x varies by k; we just have to evaluate $(x + k)^3$ and subtract x^3. This difference is $3kx^2 + 3k^2x + k^3$. Computing the difference of a quantity is known as *differentiating this quantity*.

(6.) The differentiation rules are simply the common rules provided by algebra to compute the power of a binomial expression. But to ease and speed up this computation, we give the following rule, already known for other purposes. It is known that the expansion of the binomial $(x + k)$ to the mth power, is $x^m + mx^{m-1}k + m\frac{m-1}{2}x^{m-2}k^2 + m\frac{m-1}{2}\frac{m-2}{3}k^3 + $ etc.

Paying attention to the rules by which those terms are derived from one another, we see that their construction can be performed by using the following rule:

Write on the first line	x^m
Under this line, write	m
Multiply by this exponent, and, diminishing the exponent of x by one unit, replace the factor x that currently misses by the factor k, and get in the second line	$mx^{m-1}k$
Under this line, write one half of the current exponent of x; that is,	$\frac{m-1}{2}$
Multiply by the latter, and, diminishing the current exponent of x by one unit, replace the new missing x factor by a new k factor, and get in the third line	$m\frac{m-1}{2}x^{m-2}k^2$
Under this line, write the third of the current exponent of x; that is,	$\frac{m-2}{3}$
Multiply by the latter, and, diminishing the x exponent by one unit, replace the x factor that is missing again by a new factor k, and get in the fourth line	$m\frac{m-1}{2}\frac{m-2}{3}x^{m-3}k^3$

Keep multiplying according to the same process, successively by one fourth, one fifth, etc. of the exponent of x, and keep lowering the exponent of x by one unit. Replace the missing x factor by a k factor. Then the value of $(x+k)^m$ is the sum of the first, second, third, fourth etc. lines, until the line where the exponent of x becomes 0 which is obvious by comparison with the first formula.

(7.) Therefore it is sufficient to omit the first line in the result from the preceding rule to obtain the difference of x^m where x varies by an amount of k, that is, to obtain the value of $(x + k)^m - x^m$.

(8.) Since the polynomial $Ax^p + Bx^q + Cx^r$ only consists of terms of the form x^m, computing the difference of such a polynomial can be done by simply applying the rule above given for x^m.

Thus, to obtain the difference of $x^3 - 5x^2 + 3x - 6$, where x varies by an amount of k, I write as follows:

First line	x^3	$-$	$5x^2$	$+$	$3x$	$-$	6
Exponent of x	3		2		1		0
Second line	$3x^2k$	$-$	$10xk$	$+$	$3k$		
Half of exponents of x	$\frac{2}{2}$		$\frac{1}{2}$		$\frac{0}{2}$		
Third line	$3xk^2$	$-$	$5k^2$				
Third of exponents of x	$\frac{1}{3}$		$\frac{0}{3}$				
Fourth line	k^3						

Thus $d(x^3 - 5x^2 + 3x - 6) \ldots \left(\dfrac{x}{k} \right) = 3x^2k + 3xk^2 - 10xk + k^3 - 5k^2 + 3k,$
which is the sum of lines 2, 3 and 4.

(9.) We can use the same rule to differentiate quantities involving several variables. Thus, we can compute $d(x^3y^2) \ldots \left(\dfrac{x}{k} : \dfrac{y}{l} \right)$ using the method below, by writing successively under each variable its exponent, then one half of its exponent, one third of it, etc. of its according to the line number being computed.

First line	x^3 y^2								
	3 2								
Second line	$3x^2$ y^2k	$+$	$2x^3$ yl						
	$\frac{2}{2}$ $\frac{2}{2}$		$\frac{3}{2}$ $\frac{1}{2}$						
Third line	$3x$ y^2k^2	$+$	$3x^2$ ykl	$+$	$3x^2$ ykl	$+$	x^3l^2	
or	$3x$ y^2k^2	$+$	$6x^2$ ykl	$+$	x^3l^2			
	$\frac{1}{3}$ $\frac{2}{3}$		$\frac{2}{3}$ $\frac{1}{3}$		$\frac{3}{3}$			
Fourth line	y^2k^3	$+$	$2x$ yk^2l	$+$	$4x$ $yk^2l + 2x^2kl^2 + x^2kl^2$			
or	y^2k^3	$+$	$6x$ yk^2l	$+$	$3x^2kl^2$			
	$\frac{2}{4}$		$\frac{1}{4}$ $\frac{1}{4}$		$\frac{2}{4}$			

Fifth line $\frac{1}{2}yk^3l \;+\; \frac{3}{2}yk^3l \;+\; \frac{3}{2}xk^2l^2 \;+\; \frac{3}{2}xk^2l^2$

or $2yk^3l \;+\; 3xk^2l^2$

$$\frac{1}{5} \qquad\qquad \frac{1}{5}$$

Sixth line $\frac{2}{5}k^3l^2 \;+\; \frac{3}{5}k^3l^2$

or k^3l^2

Thus $d(x^3, y^2) \ldots \left(\begin{matrix} x & : & y \\ k & & l \end{matrix} \right) = 3x^2y^2k + 2x^3yl + 3xy^2k^2 + 6x^2ykl + x^3l^2 +$ $y^2k^3 + 6xyk^2l + 3x^2kl^2 + 2yk^3l + 3xk^2l^2 + k^3l^2$.

(10.) The same rule applies to functions of two variables: Simply compare the result of $(x+k)^m \times (y+l)^n$ found with this rule, with the result of the expansion of this quantity using ordinary rules of algebra. These indeed lead to

$$x^my^n + mx^{m-1}y^nk + m.\tfrac{m-1}{2}x^{m-2}y^nk^2 + m.\tfrac{m-1}{2}.\tfrac{m-2}{3}.x^{m-3}y^nk^3, \text{ etc.}$$
$$+nx^my^{n-1}l + mnx^{m-1}y^{n-1}kl + mn.\tfrac{m-1}{2}.x^{m-2}y^{n-1}k^2l, \text{ etc.}$$
$$+n.\tfrac{n-1}{2}.x^my^{n-2}l^2 + mn.\tfrac{n-1}{2}.x^{m-1}y^{n-2}kl^2, \text{ etc.}$$
$$+n.\tfrac{n-1}{2}.\tfrac{n-2}{3}.x^my^{n-3}l^3, \text{ etc.}$$

By applying our rule, we find as follows:

First line. $x^m \quad y^n,$

          $m \quad\; n$

Second line. $mx^{m-1} \quad y^nk \;+\; nx^m \quad y^{n-1}l,$

             $\tfrac{m-1}{2} \quad \tfrac{n}{2} \qquad\quad \tfrac{m}{2} \quad \tfrac{n-1}{2}$

Third line. $m.\tfrac{m-1}{2}.x^{m-2} \quad y^nk^2 + \quad \tfrac{mn}{2}.x^{m-1} \quad y^{n-1}kl + \quad \tfrac{mn}{2}.x^{m-1} \quad y^{n-1}kl$

                      $+ \qquad\qquad\; n.\tfrac{n-1}{2}.x^m \quad y^{n-2}l^2$

or $m.\tfrac{m-1}{2}.x^{m-2} \quad y^nk^2 + \quad mnx^{m-1} \quad y^{n-1}kl + \quad n.\tfrac{n-1}{2}.x^m \quad y^{n-2}l^2,$

              $\tfrac{m-2}{3} \quad \tfrac{n}{3} \qquad\qquad \tfrac{m-1}{3} \quad \tfrac{n-1}{3} \qquad\qquad \tfrac{m}{3} \quad \tfrac{n-2}{3}$

Fourth line $m.\tfrac{m-1}{2}.\tfrac{m-2}{3}.x^{m-3}y^nk^3 + \tfrac{mn}{3}.\tfrac{m-1}{2}.x^{m-2}y^{n-1}k^2l$

          $+mn.\tfrac{m-1}{3}.x^{m-2}y^{n-1}k^2l + mn.\tfrac{n-1}{3}.x^{m-1}y^{n-2}kl^2$

          $+\tfrac{mn}{3}.\tfrac{n-1}{2}.x^{m-1}y^{n-2}kl^2 + n.\tfrac{n-1}{2}.\tfrac{n-2}{3}x^my^{n-3}l^3;$

or $m.\tfrac{m-1}{2}.\tfrac{m-2}{3}.x^{m-3}y^nk^3 + mn.\tfrac{m-1}{2}.x^{m-2}y^{n-1}k^2l$

          $+mn.\tfrac{n-1}{2}.x^{m-1}y^{n-2}kl^2 + n.\tfrac{n-1}{2}.\tfrac{n-2}{3}.x^my^{n-3}l^3, \text{ etc.}$

We therefore see that the sum of the first, second, third and fourth lines gives exactly the same result.

(11.) We can use the same method to show that the same rule can be applied to an arbitrary number of variables.

We have demonstrated in (2) that it is enough to apply the same rules to first differences to obtain second differences, and that this also holds true for third, fourth, etc. differences; thus the method to compute arbitrary differences reduces to the only rule given in (4). Consider for example the computation of second differences: We want to compute the value of $dd(x^3 + 2x^2y - 3xy + 2xy^2 - 2x + 3y + 6)\ldots \begin{pmatrix} x & y \\ k, k' & l, l' \end{pmatrix}$. I write as follows:

First line	x^3+	$2x^2$	$y-$	$3x$	$y+$	$2y^2-$	$2x+$	$3y+$	6
	3	2	1	1	1	2	1	1	0

Second line	$3x^2k+$	$4x$	$yk+$	$2x^2l-$	$3yk-$	$3xl+$	$4yl-$	$2k+$	$3l$
	$\frac{2}{2}$	$\frac{1}{2}$	$\frac{1}{2}$	$\frac{2}{2}$	$\frac{1}{2}$	$\frac{1}{2}$	$\frac{1}{2}$	$\frac{0}{2}$	$\frac{0}{2}$

Third line	$3xk^2+$	$2yk^2+$	$2xkl+$	$2xkl-$	$\frac{3}{2}kl-$	$\frac{3}{2}kl+$	$2l^2$

or	$3xk^2+$	$2yk^2+$	$4xkl-$	$3kl+$	$2l^2$
	$\frac{1}{3}$	$\frac{1}{3}$	$\frac{1}{3}$	$\frac{0}{3}$	$\frac{0}{3}$

Fourth line	k^3	$+$	$\frac{2}{3}k^2l$	$+$	$\frac{4}{3}k^2l$
or	k^3	$+$	$2k^2l$		

Therefore

$$d(x^3 + 2x^2y - 3xy + 2y^2 - 2x + 3y + 6)\ldots \begin{pmatrix} x & y \\ k & l \end{pmatrix}$$

$$= 3x^2k \quad +4x \quad yk \quad +2x^2l \quad -3yk \quad -3xl \quad -2k$$
$$+4yl \quad +3xk^2 \quad +3l$$
$$+2yk^2 \quad +4xkl \quad -3kl$$
$$+2l^2$$
$$+2k^2l$$
$$+k^3$$

$\left.\right\}$ First line for the second difference

bottom: $2 \quad 1 \quad 1 \quad 2 \quad 1 \quad 1 \quad 0$

Second line	$6xkk'$	$+4ykk'$	$+4xkl'$	$-3kl'$	$-3lk'$
			$+4xk'l$	$+4ll'$	$+3k^2k'$
			$+2k^2l'$	$+4kk'l$	
	$\frac{1}{2}$	$\frac{1}{2}$	$\frac{1}{2}$	$\frac{0}{2}$	$\frac{0}{2}$

Third line	$3kk'^2$	$+$	$2kk'l'$	$+$	$2kk'l'$
				$+$	$2k'^2l$
or	$3kk'^2$	$+$	$4kk'l'$	$+$	$2k'^2l.$

Therefore

$$dd(x^3 + 2x^2y - 3xy + 2y^2 - 2x + 3y + 6)\ldots \begin{pmatrix} x & y \\ k, k' & l, l' \end{pmatrix}$$
$$= 6xkk' + 4ykk' + 4xkl' + 4xk'l + 3kk'^2 + 2k'^2l$$
$$+2k^2l' + 4kk'l + 4kk'l' + 3k^2k' + 4ll' - 3kl' - 3lk'.$$

A general and fundamental remark

(12.) Whatever the number of variables entering in the quantity to be differentiated, and whatever the dimension these variables can reach, either alone or together, we can generally observe that:

1. If T is the highest dimension reached by these variables, either alone or together, then $T - 1$ is the highest dimension these variables reach in the first difference, since the rule prescribes to reduce the exponent of the variable of interest by one unit.

   Consequently, $T - 2$ is the highest degree of the variables in the second difference; $T - 3$ is the highest degree of the variables in the third difference; and in general, $T - n$ is the highest dimension of the variables in the difference of order n. Thus, if the order of the difference has the same exponent as that of the highest dimension of the variables, the degree of the variables in the difference is zero; that is, the difference contains no more variables and is only a function of their respective variations.

   For example, $d(ax + by + c) \ldots \left(\begin{array}{ccc} x & : & y \\ k & & l \end{array} \right) = ak + bl$; we see that x and y do not enter in the difference, but their respective variations k and l do.

   Likewise, the above rule yields

   $$dd(ax^2 + bxy + cy^2 + ex + fy + g) \ldots \left(\begin{array}{ccc} x & : & y \\ k, k' & & l, l' \end{array} \right)$$
   $$= 2akk' + bkl' + bk'l + 2cll',$$

   where we see that x and y have vanished and only their respective variations k, k' and l, l' remain.

2. If there are constant quantities in the function to be differentiated, that is, if there are terms where no variables are present, these terms will not be found in the first derivative, and therefore not in the subsequent differentials either; indeed, the rule prescribes to multiply them by the exponent of the variable, which is zero in that case.

3. The terms where the variables do not exceed, either together or separately, the first dimension, are not to be found in the second difference, since they all become constant by the process of the first differentiation; consequently they will disappear in the second differentiation. For example, assume we must differentiate the quantity $ax^2 + bxy + cy^2 + ex + fy + g$ twice; the quantity g is not present in the first difference, which is $2axk + byk + bxl + 2cyl + ek + fl + ak^2 + bkl + cl^2$. Likewise, the terms ex and fy do not appear in the second difference, which is $2akk' + bkl' + bk'l + 2cll'$. Indeed, during the first differentiation, these terms become ck and fl. Since these terms are constant, they cannot be found in the following difference.

Likewise, the terms where the variables do not exceed, either together or separately, the dimension 2 do not appear in the third difference; in general, the terms where the degree of the variables does not exceed, either together or separately, the dimension $n - 1$, disappear in the difference of order n.

The differentiations we have to perform later are all, or almost all, of the order of the total dimensions of the quantities involved; thus it is appropriate to present here the simplifications that the observations we just made can bring to the usage of the differentiation method.

Reductions that may apply to the general rule to differentiate quantities when several differentiations must be made

(13.) The terms where the variables do not exceed, either together or alone, the dimension $n - 1$ cannot be found in the differential of order n; thus the calculations can be considerably simplified, if we strictly follow the general rule we first gave.

This simplification is about rejecting all terms of all dimensions from 0 to $n - 1$ included before any computations are performed; n is the planned number of differentiations.

Thus, if we must differentiate twice the quantity $ax^2 + bxy + cy^2 + ex + fy + g$, the question reduces to differentiating twice the quantity $ax^2 + bxy + cy^2$.

If we must differentiate twice the quantity $ax^3 + bx^2y + cx^2z + exy^2 + fxyz + gxz^2 + ky^3 + ly^2z + myz^2 + nz^3 + px^2 + qxy + rxz + a'y^2 + b'yz + c'z^2 + e'x + f'y + g'z + h'$, the question reduces to differentiating twice the quantity $ax^3 + bx^2y + cx^2z + exy^2 + fxyz + gxz^2 + ky^3 + ly^2z + myz^2 + nz^3 + px^2 + qxy + rxz + a'y^2 + b'yz + c'z^2$.

And if the differentiation were to be performed three times, the question would reduce to differentiating three times the quantity $ax^3 + bx^2y + cx^2z + exy^2 + fxyz + gxz^2 + ky^3 + ly^2z + myz^2 + nz^3$.

(14.) This simplification is not the only one resulting from the previous observations. After rejecting the various terms that cannot enter the differential, we proceed with the differentiation of the remaining terms; we observe that when computing the various parts that we have called *lines*, it is superfluous to perform computations beyond the line number $T - n + 2$, where T is the total dimension of the quantity that we want to differentiate and n is the number of differentiations to be performed.

Indeed, the total dimension diminishes by one unit in every line starting from the second line; when reaching the line $T - n + 2$, the dimension is $n - 1$; so it is clear that the lines computed beyond this point disappear through successive differentiations, because their dimension is less than $n - 1$. It is therefore useless to consider them.

So, if the degree of the differential equals that of the total dimension of the quantity to be differentiated, (i) we must keep only the terms with the highest dimension, and (ii) we need not go beyond the second line for each differentiation.

For example, assume we need to differentiate three times the quantity $x^3 - 3xyz + 2y^3 - x^2 + 2xz - y + 2z - 2$:

1. We reject the dimensions 2, 1 and 0, which reduce this quantity to $x^3 - 3xyz + 2y^3$.

2. We take, in the first difference, the second line only, yielding $3x^2k - 3yzk - 3xzl - 3xym + 6y^2l$.

3. We take, in the second difference, the second line only, yielding $6xkk' - 3zkl' - 3ykm' - 3zlk' - 3xlm' - 3ymk' - 3xml' + 12yll'$.

4. We take, in the third difference, the second line only, yielding $6kk'k'' - 3kl'm'' - 3km'l'' - 3lk'm'' - 3lm'k'' - 3mk'l'' - 3ml'k'' + 12ll'l''$ as the third difference.

Remarks about the differences of decreasing quantities

(15.) Until now we have assumed that each variable was increasing. If, conversely, they were all decreasing, it would not be necessary to establish different rules, but simply to make a slight change in the computed lines.

Indeed, if x becomes $x - k$ instead of $x + k$, there is no other difference between these expressions than k becoming $-k$.

Concerning the differential, there is another change, because we must differentiate x^n, for example; in the first case, we must expand $(x+k)^n - x^n$ and in the second case we must expand $x^n - (x - k)^n$.

If we had to expand $(x - k)^n - x^n$, we would clearly have nothing else to do than to differentiate x^n according to the preceding rules, and making x vary by the quantity $-k$ instead of k.

Thus, in the case of $x^n - (x - k)^n$, we should differentiate x^n, making x vary by the amount $-k$; we should then change the sign of all lines of the result, or, alternatively, we should write along each part of the result, with the sign of one line opposite to that yielded by the differentiation obtained by making x vary by an amount $-k$.

(16.) Thus we see that, in general, the differential of a function is different when its variables are increasing quantities from the same differential when all variables are decreasing. There are, however, two cases when these differentials are the same. The first case is when the variations of the variables are infinitesimally small. The second is when the quantity must be differentiated as many times as the size of its exponent of the highest dimension.

This last case is the only one of interest to this work: Thus, in the differentiations we will perform later, we will not need to examine whether the variables are increasing or decreasing. We will differentiate following the rules we have first given.

About certain quantities that must be differentiated through a simpler
process than that resulting from the general rule

(17.) The principles that we just elicited are general and could even, with slight changes, be applied to fractional and irrational quantities. They can be applied to convert functions of several variables into series, and to many other objects. But our goal is not to discuss these applications. We will only consider rational quantities that can be differentiated faster than through the general rule: We consider only those that will be useful to us later on.

Assume we must differentiate a quantity such as $(x+a).(x+a+b).(x+a+2b).(x+a+3b)\ldots(x+a+(n-1)b)$, where n is the number of factors and x varies by a quantity b; the differential is $nb.(x+a+b).(x+a+2b).(x+a+3b)\ldots(x+a+(n-1)b)$, where $n-1$ is the number of factors in arithmetic progression.

But if the variation is $-b$, the differential is $nb(x+a).(x+a+b).(x+a+2b)\ldots(x+a+(n-2)b)$, where $n-1$ is the number of factors in arithmetic progression.

Indeed,

$$d[(x+a).(x+a+b).(x+a+2b)\ldots(x+a+(n-1)b)]\ldots\left(\begin{array}{c}x\\b\end{array}\right)$$

$$= \quad (x+a+b).(x+a+2b)(x+a+3b)\ldots(x+a+nb)$$
$$\quad -(x+a)(x+a+b)(x+a+2b)\ldots(x+a+(n-1)b)$$
$$= \quad [(x+a+b)(x+a+2b)(x+a+3b)\ldots(x+a+(n-1)b)]$$
$$\quad \times(x+a+nb-x-a)$$
$$= \quad nb(x+a+b)(x+a+2b)(x+a+3b)\ldots(x+a+(n-1)b).$$

Likewise,

$$d[(x+a)(x+a+b)(x+a+2b)\ldots(x+a+(n-1)b)]\ldots\left(\begin{array}{c}x\\-b\end{array}\right)$$

$$= \quad (x+a)(x+a+b)(x+a+2b)\ldots(x+a+(n-1)b)$$
$$\quad -(x+a-b)(x+a)(x+a+b)\ldots(x+a+(n-2)b)$$
$$= \quad [(x+a)(x+a+b)(x+a+2b)\ldots(x+a+(n-2)b)]$$
$$\quad \times(x+a+(n-1)b-x-a+b)$$
$$= \quad nb(x+a)(x+a+b)(x+a+2b)\ldots(x+a+(n-2)b).$$

About sums of quantities

(18.) Imagine that P is an arbitrary function of one or many variables x, y, z, etc., and that, giving successively to each of these variables the values k, l, m, etc., k', l', m', etc., k'', l'', m'', etc. respectively, the quantity P becomes successively P', P'', P''', etc. the sum $P+P'+P''+P'''+$ etc. is what we will call *sum of P*, and we will write it as $\int P$.

We will not attempt, by far, to deal with this matter to the whole extent that it deserves. Our purposes only require a very specific branch of this theory and we will restrict ourselves to it.

We therefore consider rational functions of a single variable, with no variable divider.

We will also assume that the variable increases or decreases by equal amounts.

About sums of quantities whose factors grow arithmetically

(19.) Those products are usually represented by

$$(x + a)(x + a + b)(x + a + 2b) \ldots (x + a + (n - 1)b),$$

where n is the number of factors.

Substituting for x the quantities $(x - b)$, $(x - 2b)$, $(x - 3b)$, etc., the quantities of interest become

$$(x + a)(x + a + b)(x + a + 2b) \ldots (x + a + (n - 1)b),$$
$$(x + a - b)(x + a)(x + a + b) \ldots (x + a + (n - 2)b),$$
$$(x + a - 2b)(x + a - b)(x + a) \ldots (x + a + (n - 3)b),$$
$$(x + a - 3b)(x + a - 2b)(x + a - b) \ldots (x + a + (n - 4)b),$$

etc.

Let P be the sum of all these products, and P' the sum of all these products, except the first. We have $P - P' = (x+a)(x+a+b)(x+a+2b)\ldots(x+a+(n-1)b)$. But $P - P' = d(P) \ldots \left(\dfrac{x}{-b} \right)$. Thus $d(P) \ldots \left(\dfrac{x}{-b} \right) = (x + a)(x + a + b)(x + a + 2b) \ldots (x + a + (n - 1)b)$.

Finding P therefore reduces to *finding the function whose difference is* $(x + a)(x + a + b)(x + a + 2b) \ldots (x + a + (n - 1)b)$, *when x varies by* $-b$.

Considering what was said in (17), it is easy to see that this function is

$$\frac{1}{(n + 1)b}(x + a)(x + a + b)(x + a + 2b) \ldots (x + a + nb),$$

where $n + 1$ is the number of factors.[1]

Thus $P = \frac{1}{(n+1)b}(x + a)(x + a + b)(x + a + 2b) \ldots (x + a + nb)$.

Remarks

(20.) First, we have supposed that the variation of x was precisely equal to the difference b present in the progression of the factors. We will soon see how to determine this sum when this variation is any other quantity.

(21.) Second, from (12) the constant terms present in a quantity to be differentiated vanish in the difference; thus it follows that a constant must always be added to the quantity to be summed. From a computational standpoint, this constant is arbitrary, since the differential is always the same. But in each question, this constant has a specific value, which is easily found by the conditions of the problem.

---

[1] One must be careful, when comparing with what was said in (17), that what was denoted n in (17) is now $n + 1$.

From now on, we write this constant as C. Thus the value of P we have found is, more generally,

$$P = \frac{1}{(n+1)b}(x+a)(x+a+b)(x+a+2b)\ldots(x+a+nb) + C.$$

To give an example about the way to determine this constant C, assume we need to compute the sum of the products $2\times4\times6$, $4\times6\times8$, $6\times8\times10$, $8\times10\times12$ until $14\times16\times18$; we therefore have $(x+a).(x+a+b).(x+a+2b) = 14\times16\times18$ and $n = 3$.

Assume $a = b = 2$; we will have $x = 12$. Thus $P = \frac{1}{4\times2}14\times16\times18\times20+C$.

But we want the sum only from $2 \times 4 \times 6$; comparing this product with $(x+a).(x+a+b).(x+a+2b)$, we have $x = 0$; thus when $x = 0$, the sum P becomes $2 \times 4 \times 6$; we therefore have $2 \times 4 \times 6 = \frac{1}{4.2} \times 2 \times 4 \times 6 \times 8 + C$, or $C = 48 - 48 = 0$. The sum is therefore simply $\frac{1}{4.2} \times 14 \times 16 \times 18 \times 20$, that is, 10080. It is easy to check this result by adding the products together.

If instead of assuming that $a = 2$ we had assumed that $a = 0$, then we would have had $x = 14$ as the final value of x, and $x = 2$ as its initial value; the sum would then be $P = \frac{1}{4.2} \times 14 \times 16 \times 18 \times 20 + C$. To determine the constant C, we could rely on the condition that the sum P must become $2 \times 4 \times 6 = \frac{1}{4.2} \times 2 \times 4 \times 6 \times 8 + C$ when $x = 2$. Thus $C = 0$, and P is again 10,080 as expected.

About sums of rational quantities with no variable divider

(22.) For the sake of clarity, assume first that we must sum a simple quantity, such as x^3 or mx^3. The question asked is ill-posed, because we must know by which amounts x increases or decreases. Assume therefore that x decreases by equal amounts of amplitude b.

Then the true meaning of the question is the following: Assuming that x becomes $x-b$, $x-2b$, $x-3b$, successively, compute the sum of the quantities mx^3, $m(x-b)^3$, $m(x-2b)^3$, $m(x-3b)^3$, etc.

To answer this question, I reduce it to the one solved in (19), by bringing mx^3 back to the form $(x+b).(x+2b).(x+3b)$, etc.

I therefore write $mx^3 = A(x+b).(x+2b).(x+3b) + B(x+b).(x+2b) + C(x+b) + D$. I then obtain:

$$\begin{aligned} mx^3 = \quad & Ax^3 + 6Abx^2 + 11Ab^2x + 6Ab^3 \\ & +Bx^2 + 3Bbx + 2Bb^2 \\ & +Cx + Cb \\ & +D. \end{aligned}$$

Since this equality must hold true for any value of x, I conclude that $A = m$, $6Ab + B = 0$, $11Ab^2 + 3Bb + C = 0$, $6Ab^3 + 2Bb^2 + Cb + D = 0$; that is, $A = m$, $B = -6mb$, $C = +7mb^2$, $D = -mb^3$; thus

$$mx^3 = m(x+b).(x+2b).(x+3b) - 6mb.(x+b).(x+2b) + 7mb^2(x+b) - mb^3.$$

The value of mx^3 is therefore composed of four parts, each of which is of the form we know to sum, from (19). We therefore easily find, through what was already established in (19), that

$$\int mx^3 = \frac{m}{4b}.(x+b).(x+2b).(x+3b).(x+4b)$$
$$-2m(x+b).(x+2b).(x+3b)$$
$$+\frac{7mb}{2}.(x+b).(x+2b) - mb^2(x+b) + C,$$

where C is a constant (21).

(23.) Consider the quantity $mx^3 + nx^2 + px + q$: We see that each term reduces to the form $(x+b).(x+2b).(x+3b)$, etc., as we have seen with mx^3. Thus the total can also be reduced to that form. Therefore, I can sum a quantity such as $mx^3 + nx^2 + px + q$ by writing

$$
\begin{aligned}
mx^3 + nx^2 + px + q \;=\;& A(x+b).(x+2b).(x+3b) \\
+\;& B(x+b).(x+2b) \\
+\;& C(x+b) + D,
\end{aligned}
$$

by determining the coefficients A, B, C, D, and by equating the coefficients of the same powers of x in the left- and right-hand side of the equations. Thus I only have to compute the sum of the quantity $A.(x+b).(x+2b).(x+3b) + B.(x+b).(x+2b) + C.(x+b) + D$, which is easy to do from what was said in (19), and which is

$$\frac{A}{4b}.(x+b).(x+2b).(x+3b).(x+4b)$$

$$+ \quad \frac{B}{3b}.(x+b).(x+2b).(x+3b)$$

$$+ \quad \frac{C}{2b}.(x+b).(x+2b)$$

$$+ \quad \frac{D}{b}.(x+b) + C,$$

where A, B, C and D must be replaced by their values.

(24.) Looking at the form of the sum in this example and the previous one, we see that computing these sums can be made simpler. We do not need to bring the proposed quantity back to the form $(x+b).(x+2b).(x+3b).$, etc., since the sum is of the same form; we can immediately determine the coefficients of the sum as follows. Let us go back to the example of mx^3.

(25.) I assume

$$
\begin{aligned}
\int mx^3 \;=\;& A.(x+b).(x+2b).(x+3b).(x+4b) \\
+\;& B.(x+b).(x+2b).(x+3b) \\
+\;& C.(x+b).(x+2b) \\
+\;& D.(x+b) + C
\end{aligned}
$$

right away; to obtain the coefficients, I differentiate each one of the terms (17) and I get

$$
\begin{aligned}
mx^3 \;=\;& 4Ab.(x+b).(x+2b).(x+3b) \\
=\;& 3Bb.(x+b).(x+2b) \\
=\;& 2Cb.(x+b) + Db,
\end{aligned}
$$

that is,

$$
\begin{aligned}
mx^3 = \ & 4Abx^3 + 24Ab^2x^2 + 44Ab^3x + 24Ab^4 \\
& +3Bbx^2 + 9Bb^2x + 6Bb^3 \\
& +2Cbx + 2Cb^2 \\
& +Db.
\end{aligned}
$$

I therefore obtain

$$
4Ab = m, \quad 24Ab^2 + 3Bb = 0,
$$
$$
44Ab^3 + 9Bb^2 + 2Cb = 0,
$$
$$
24Ab^4 + 6Bb^3 + 2Cb^2 + Db = 0;
$$

thus $A = m/4b$, $B = -2m$, $C = 7mb/2$, $D = -mb^2$; this leads to the same expression as computed earlier for $\int mx^3$.

(26.) In general we see that in order to integrate a rational polynomial without variable divider, such as $ax^p + bx^q + cx^r +$ etc., we first write

$$
\begin{aligned}
& \int (ax^p + bx^q + cx^r +, \text{etc.}) \\
= \ & A.(x + b).(x + 2b).(x + 3b) \ldots (x + (p + 1).b) \\
+ \ & B.(x + b).(x + 2b).(x + 3b) \ldots (x + pb) \\
+ \ & C.(x + b).(x + 2b).(x + 3b) \ldots (x + (p - 1).b) \\
+ \ & D.(x + b).(x + 2b).(x + 3b) \ldots (x + (p - 2).b) + \cdots \\
+ \ & P.(x + b).(x + 2b) + Q.(x + b) + C,
\end{aligned}
$$

where we assume that p is the largest of the exponents p, q, r, etc.; we then compute the coefficients as we have discussed earlier.

If we had $(ax^p + bx^q + cx^r + \text{etc.})^k$, we would recast the problem to the preceding problem by expanding this power of a polynomial.

(27.) We now see, as promised in (20), how we can compute the sum of

$$
(x + a).(x + a + b).(x + a + 2b) \ldots (x + a + (n - 1)b)
$$

when x increases or decreases by steps other than b. For example, if k is the stepsize by which x grows, we write

$$
\begin{aligned}
& \int (x + a).(x + a + b).(x + a + 2b) \ldots (x + a + (n - 1)b) \\
= \ & A.(x + k).(x + 2k).(x + 3k)(x + (n + 1)k) \\
& +B.(x + k).(x + 2k).(x + 3k) \ldots (x + n.k) \\
& +C.(x + k).(x + 2k) \ldots (x + (n - 1)k) + \cdots \\
& +Q.(x + k) + C.
\end{aligned}
$$

(28.) If we wanted to compute the value of $\int Ax^m$ when $m = 0$, it follows from what we have just said that this value would be $A(x + b)$. Indeed, since m is zero, the question reduces to computing the sum of A from a given value of x to another value of x. Thus if $x + b$ represents the range over which we compute the sum of A, the sum is $A.(x + b)$.

Book One

SECTION I
About complete polynomials and complete equations

(29.) Any polynomial that contains only one unknown x can be represented in general by $ax^T + bx^{T-1} + cx^{T-2} \cdots + s$, where T is the highest degree of x and a, b, c, etc. are arbitrary coefficients.

Likewise, any equation in one unknown can generally be represented by $ax^T + bx^{T-1} + cx^{T-2} + \cdots + s = 0$.

However, the large number of terms that can enter polynomials and equations, as their degree and the number of unknowns increases, requires us to represent these as compactly as possible. We first present the various notations that we propose to use.

(30.) We represent any polynomial in one unknown by the abbreviated expression $(x)^T$, which means *a polynomial in one unknown, with degree T*.

Likewise, we represent any equation in one unknown x by the abbreviated expression $(x)^T = 0$.

The number of terms of such a polynomial or equation is written $N(x)^T$.

(31.) A *complete polynomial* is defined as a polynomial that contains all combinations of x, y, z, etc. allowed by its degree.

For example, any complete polynomial of degree 3 in two variables must contain all the following terms (where we have omitted to write the nonzero coefficients):

$$x^3, \ x^2y, \ xy^2, \ y^3$$
$$x^2, \ xy, \ y^2$$
$$x, \ y$$
$$1$$

Any complete polynomial in three unknowns x, y, z must contain all the following terms up to degree 3:

$$x^3, \ x^2y, \ xy^2, \ x^2z, \ xyz, \ xz^2, \ y^3, \ y^2z, \ yz^2, \ z^3$$
$$x^2, \ xy, \ xz, \ y^2, \ yz, \ z^2$$
$$x, \ y, \ z$$
$$1$$

In general, a complete polynomial must contain all possible products of the unknowns, from the lowest dimension (degree 0) up to the highest dimension T; the same holds for a complete equation.

(32.) We write $(u \ldots 2)^T$ to represent a complete polynomial in two un-knowns; for an equation, we write $(u \ldots 2)^T = 0$; to indicate the number of terms of this polynomial or that equation, we write $N(u \ldots 2)^T$.

(33.) In general, we write $(u \ldots n)^T$ to denote a polynomial in n unknowns. The corresponding equation is written $(u \ldots n)^T = 0$, and the number of terms is written $N(u \ldots n)^T$.

About the number of terms in complete polynomials

(34.) Determining the number of terms of a polynomial is of fundamental importance to the theory we are about to present. We first consider the number of terms of complete polynomials only.

PROBLEM I

(35.) *Compute the value of $N(u \ldots n)^T$.*

First, it is obvious that $N(u \ldots 1)^T = T + 1$.

(36.) Assume all terms of the polynomial $(u \ldots 1)^T$ are made homogenous by introducing a new unknown, yielding the following terms: $u^T, u^{T-1}x$, $u^{T-2}x^2, u^{T-3}x^3, u^{T-4}x^4 \ldots \ldots u^2 x^{T-2}, ux^{T-1}, x^T$. It is clear these are the terms of dimension T of the polynomial $(u \ldots 2)^T$, and their number is $T+1$.

Imagine that we successively replace T by the quantities T, $T-1$, $T-2$, $T-3$, etc. in the expression $T+1$. We see the number of terms of dimension $T, T-1, T-2, T-3$ is $T+1, T, T-1, T-2$ for the polynomial $(u \ldots 2)^T$.

Thus, following the ideas we gave in (18) to compute the sum of quantities, we can obtain $N(u \ldots 2)^T$ by simply computing the sum of $T + 1$, where T varies by increments of -1 from T to zero, included. From (19), that sum is $\frac{(T+1)(T+2)}{2}$.

Thus $N(u \ldots 2)^T = \frac{(T+1)(T+2)}{2}$.

(37.) Likewise, assume we create homogenous terms of order T from all those composing the polynomial $(u \ldots 2)^T$, by introducing a new variable y.

Doing so, we generate all terms that can be part of the dimension T of the polynomial $(u \ldots 3)^T$.

For example, assume we construct homogenous terms of degree 3 from all terms of the polynomial $(u \ldots 2)^T$, that is, all the following terms:

$$
\begin{array}{cccc}
u^3 & u^2x & ux^2 & x^3 \\
u^2 & ux & x^2 & \\
u & x & & \\
1 & & &
\end{array}
$$

using the unknown y. We obtain the terms

$$u^3 \; u^2x \; ux^2 \; x^3 \; u^2y \; uxy \; x^2y \; uy^2 \; xy^2 \; y^3,$$

which are all those present in the third dimension of the polynomial $(u \ldots 3)^3$.

The number of these terms is therefore the number of terms of the poly-nomial $(u \ldots 2)^T$, that is, $\frac{(T+1).(T+2)}{2}$; thus we only need to replace T by

$T - 1$ in $\frac{(T+1).(T+2)}{2}$ to get the number of terms in the dimensions $T - 1$, $T - 2$, $T - 3$, etc., of the polynomial $(u \ldots 3)^T$. Thus, we only have to sum up $\frac{(T+1).(T+2)}{2}$, where T varies from T to 0 included in decrements of -1, to obtain the total number of terms of all dimensions. From (19) this sum is $\frac{(T+1).(T+2).(T+3)}{1.2.3}$.

Thus $N(u \ldots 3)^T = \frac{(T+1).(T+2).(T+3)}{1.2.3}$.

(38.) Following the same argument for $(u \ldots 4)^T$, we see that we must compute the sum of $N(u \ldots 3)^T$, where T varies from T down to 0 included in decrements of -1 to compute $N(u \ldots 4)^T$. Thus

$$N(u \ldots 4)^T = \frac{(T + 1).(T + 2).(T + 3).(T + 4)}{1.2.3.4}.$$

(39.) Thus in general we have

$$N(u \ldots n)^T = \frac{(T + 1).(T + 2).(T + 3).(T + 4) \ldots (T + n)}{1.2.3.4 \ldots \ldots n}.$$

About the number of terms of a complete polynomial that can be divided by certain monomials composed of one or more of the unknowns present in this polynomial

WARNING

(40.) We frequently use the signs $>$ and $<$, which usually mean to indicate the inequality of two quantities; the quantity to the left of $>$ or to the right of $<$ is the larger quantity, and the quantity to the right of $>$ or to the left of $<$ is the smaller quantity. We warn the reader that we will always use this inequality sign to include the possibility of equality between quantities. Thus, when we write $a < b$, this means that b is greater than or equal to a. This must be remembered throughout this work.

PROBLEM II

(41.) *Given a complete polynomial in an arbitrary number of unknowns u, x, y, z, etc. we ask how many terms can be divided by u^P; in addition to these terms, how many terms can be divided by x^Q, and, in addition to those that can be divided by u^P and x^Q, how many can be divided by y^R; and how many, in addition to the previous ones, may be divided by z^S, etc.? We assume that $P + Q + R + S + $ etc. $< T$, where T is the exponent of the dimension of the polynomial.*

Let us assume that we have brought together all terms that can be divided by u^P and that, having factored out u^P, the total number of the terms multiplied by this factor is a polynomial of the form $(u \ldots n)^K$; all terms that can be divided by u^P are therefore included in the generic expression $(u \ldots n)^K \times u^P$. It is obvious, however, that we must have $K + P = T$ for this expression to contain them all; therefore $K = T - P$; the number of terms that can be divided by u^P is therefore $N(u \ldots n)^{T-P}$, and consequently it is easy to express as a function of $T - P$ from (39).

Likewise, the number of terms that can be divided by x^Q is $N(u\ldots n)^{T-Q}$. But we do not simply ask for the number of terms that can be divided by x^Q; rather we ask for how many of these terms exist, beyond those that may be divided by u^P. The number of terms that may be divided by both u^P and x^Q is $N(u\ldots n)^{T-P-Q}$. Thus we must subtract this amount from $N(u\ldots n)^{T-Q}$.

Therefore, the number of terms that may be divided by x^Q, beyond those that may also be divided by u^P, is expressed as $N(u\ldots n)^{T-Q} - N(u\ldots n)^{T-P-Q}$. This can also be written $d[N(u\ldots n)^{T-Q}]\ldots \begin{pmatrix} T-Q \\ -P \end{pmatrix}$.

The number of terms that can be divided by y^R is $N(u\ldots n)^{T-R}$. But among the terms that may be divided by u^P, there are some that may also be divided by y^R, a number expressed as $N(u\ldots n)^{T-P-R}$, and among the terms that may be divided by x^Q beyond those that may be divided by u^P, the number of those that may be divided by y^R is $N(u\ldots n)^{T-Q-R} - N(u\ldots n)^{T-P-Q-R}$. Thus, in addition to the terms that may be divided by u^P and those that may be divided by x^Q, the number of terms that may be divided by y^R is

$$N(u\ldots n)^{T-R} - N(u\ldots n)^{T-P-R} - N(u\ldots n)^{T-Q-R} + N(u\ldots n)^{T-P-Q-R},$$

that is,

$$d[N(u\ldots n)^{T_R}]\ldots \begin{pmatrix} T-R \\ -P \end{pmatrix} - dN(u\ldots n)^{T-Q-R}\ldots \begin{pmatrix} T-Q-R \\ -P \end{pmatrix}$$

or, in other terms, $dd[N(u\ldots n)^{T-R}]\ldots \begin{pmatrix} T-R \\ -P,-Q \end{pmatrix}$.

The number of terms that may be divided by z^S is $N(u\ldots n)^{T-S}$, but among the terms that may be divided by u^P, there is a number $N(u\ldots n)^{T-P-S}$ of them that may be divided by z^S. And among those terms, the number of those that may also be divided by u^P is $N(u\ldots n)^{T-Q-S} - N(u\ldots n)^{T-P-Q-S}$. And among these, the number of those that may also be divided by y^R is $N(u\ldots n)^{T-R-S} - N(u\ldots n)^{T-P-R-S} - N(u\ldots n)^{T-Q-R-S} + N(u\ldots n)^{T-P-Q-R-S}$. Thus, the number of terms that may be divided by z^S but not by u^P, nor x^Q, nor y^R is $N(u\ldots n)^{T-S} - N(u\ldots n)^{T-P-S} - N(u\ldots n)^{T-Q-S} + N(u\ldots n)^{T-P-Q-S} - N(u\ldots n)^{T-R-S} + N(u\ldots n)^{T-P-R-S} + N(u\ldots)^{T-Q-R-S} - N(u\ldots n)^{T-P-Q-R-S}$, that is,

$$dd[N(u\ldots n)^{T-S}]\ldots \begin{pmatrix} T-S \\ -P,-Q \end{pmatrix}$$
$$-dd[N(u\ldots n)^{T-R-S}]\ldots \begin{pmatrix} T-R-S \\ -P,-Q \end{pmatrix}$$
$$= d^3[N(u\ldots n)^{T-S}]\ldots \begin{pmatrix} T-S \\ -P,-Q,-R \end{pmatrix}.$$

It is is now easy to see that if there is a fifth unknown r, and if we ask for the number of terms that may be divided by r^M standing beyond those that may already be divided by u^P, x^Q, y^R and z^S, this number is expressed as

$$d^4 N(u \ldots n)^{T-M} \ldots \begin{pmatrix} T-M \\ -P, -Q, -R, -S \end{pmatrix}.$$

In general, we see clearly what this expression should be for an arbitrary number of unknowns.

Remark

(42.) Thus we have found an expression for the terms of interest when $T > P+Q+R+S+$ etc. This is the only case we need for complete equations. This expression would not be valid if $T < P + Q + R + S+$ etc. Only when we deal with incomplete equations shall we examine relevant expressions in this case.

PROBLEM III

(43.) *Assume that we exclude all the terms that may be divided by u^P, or x^Q, or y^R, or z^S, etc. from the polynomial $(u \ldots n)^T$. What is the number of remaining terms?*

From the preceding problem, it is clear that if we only exclude the terms that may be divided by x^Q, the number of remaining terms is $N(u \ldots n)^T -$

$N(u \ldots n)^{T-P}$ or $dN(u \ldots n)^T \ldots \begin{pmatrix} T \\ -P \end{pmatrix}.$

Excluding the terms that may be divided by u^P and those that may be divided by x^Q, the number of remaining terms is

$$d[N(u \ldots n)^T] \ldots \begin{pmatrix} T \\ -P \end{pmatrix} - d[N(u \ldots n)^{T-Q}] \ldots \begin{pmatrix} T-Q \\ -P \end{pmatrix},$$

that is, $dd[N(u \ldots n)^T] \ldots \begin{pmatrix} T \\ -P, -Q \end{pmatrix}.$

Excluding the terms that may be divided by u^P, x^Q or y^R, the number of remaining terms is

$$dd[N(u \ldots n)^T] \ldots \ldots \begin{pmatrix} T \\ -P, -Q \end{pmatrix} - dd[N(u \ldots n)^{T-R}] \ldots \begin{pmatrix} T-R \\ -P, -Q \end{pmatrix}$$

$$= d^3[N(u \ldots n)^T] \ldots \begin{pmatrix} T \\ -P, -Q, -R \end{pmatrix}.$$

Excluding the terms that may be divided by u^P, x^Q, y^R or z^S, the number of remaining terms is

$$d^3[N(u \ldots n)^T] \ldots \begin{pmatrix} T \\ -P, -Q, -R \end{pmatrix}$$

$$-d^3[N(u \ldots n)^{T-S}] \ldots \begin{pmatrix} T-S \\ -P, -Q, -R \end{pmatrix}$$

$$= d^4[N(u \ldots n)^T] \ldots \begin{pmatrix} T \\ -P, -Q, -R, -S \end{pmatrix},$$

and so on.

Remark

(44.) The expression we have found for the number of terms of interest is not the easiest to handle if we really want to know this number of terms. Should this be the case, we must write this expression explicitly, as we will see in the following example.

But, unless I am mistaken, this expression is the most perfect for the purpose we have assigned later.

Assume for example that we want to know the number of terms remaining in the polynomial $(u \ldots 3)^5$ if we exclude the terms that may be divided by u^3, x^2 or y.

All the terms of this polynomial are

u^6	u^5x	u^5y	u^4x^2	u^4xy	u^4y^2	u^3x^3	u^3x^2y	u^3xy^2	u^3y^3
u^2x^4	u^2x^3y	$u^2x^2y^2$	u^2xy^3	u^2y^4	ux^5	ux^4y	ux^3y^2	ux^2y^3	uxy^4
uy^5	x^6	x^5y	x^4y^2	x^3y^3	x^2y^4	xy^5y^6			
u^5	u^4x	u^4y	u^3x^2	u^3xy	u^3y^2	u^2x^3	u^2x^2y	u^2xy^2	u^2y^3
ux^4	ux^3y	ux^2y^2	uxy^3	uy^4	x^5	x^4y	x^3y^2	x^2y^3	xy^4
y^5									
u^4	u^3x	u^3y	u^2x^2	u^2xy	u^2y^2	ux^3	ux^2y	uxy^2	uy^3
x^4	x^3y	x^2y^2	xy^3	y^4					
u^3	u^2x	u^2y	ux^2	uxy	uy^2	x^3	x^2y	xy^2	y^3
u^2	ux	uy	x^2	xy	y^2				
u	x	y							
1									

From (37) the total number of terms is

$$N(u \ldots 3)^6 = \frac{7 \times 8 \times 9}{2 \times 3} = 84.$$

The number of terms that may be divided by u^3 is

$$N(u \ldots 3)^{6-3} = \frac{4 \times 5 \times 6}{2 \times 3} = 20.$$

The number of terms dividable by x^2, excluding the terms that may be divided by u^3, is

$$N(u \ldots 3)^{6-2} - N(u \ldots 3)^{6-5} = N(u \ldots 3)^4 - N(u \ldots 3)^1 = 35 - 4 = 31.$$

The number of terms that may be divided by y, excluding the terms that may be divided by u^3 and those that may be divided by x^2 is

$$N(u \ldots 3)^{6-1} - N(u \ldots 3)^{6-3-1} - N(u \ldots 3)^{(6-2-1)} + N(u \ldots 3)^{6-3-2-1}$$
$$= 56 - 10 - 20 + 1 = 27.$$

Thus the number of remaining terms is 6. Indeed the remaining terms are

$$u^2 x$$
$$u^2 \quad ux$$
$$u \quad x$$
$$1$$

Initial considerations about computing the degree of the final equation resulting from an arbitrary number of complete equations with the same number of unknowns

(45.) Assume that we have an arbitrary number n of complete equations, containing the same number of unknowns. We write these equations $(u\ldots n)^t = 0$, $(u\ldots n)^{t'} = 0$, $(u\ldots n)^{t''} = 0$, $(u\ldots n)^{t'''} = 0$, etc.

Assume that we are able to determine the value of $x^{t'}$, $y^{t''}$, $z^{t'''}$, etc., with the help of the $n-1$ last equations; this is always easy to do when the equations are as general as possible, and we will show how to do this later; for now, we simply need to assume it is possible. These equations can only give the values of these quantities or their multiples (which adds no information); thus we can substitute only the value of the terms x^t, $y^{t'}$, $z^{t''}$, etc., that is, the terms that may be divided by x^t, those that may be divided by $y^{t'}$, those that may be divided by $z^{t''}$, etc., in the first equation. However, we feel this is not enough to eliminate the other multiples of x, y, z, etc., and thereby to give the equation as a function of u alone, except by accident and under special circumstances, for certain values of the coefficients in these equations. This is impossible in our case since we consider arbitrary equations. Thus we see that the final equation can be neither of degree t, nor of a lower degree than t. However, imagine that we multiply the equation $(u\ldots n)^t$ by a complete polynomial with degree T in the same number of unknowns. Moreover, assume that we replace x^t, $y^{t'}$, $z^{t''}$, etc., by their values in the resulting equation $(u\ldots n)^{T=t} = 0$ (we call this equation the *product equation*), wherever possible. The polynomial multiplier will have introduced as many different coefficients as there are terms in the product equation; after these substitutions, the number of terms multiples of x, y, z, etc. will have been reduced by as many terms as those that it was possible to eliminate by adjusting the coefficients of the polynomial multiplier.

We see not only that this can happen, but also that this must happen; that is, there must exist a polynomial multiplier that provides the necessary coefficients to completely destroy the terms affected by x, y, z, etc., after eliminating the terms that can be divided by x^t, $y^{t'}$, $z^{t''}$, etc. by substituting these quantities by their values.

Indeed, we can reach the equation in u only if we use the values that $n-1$ of these equations can yield and substitute these values in the nth equation, or in a function of the nth equation. The $n-1$ last equations, for example, can give nothing more than the value of $x^{t'}$, $y^{t''}$, $z^{t'''}$, etc. Thus these values, when they are substituted in a certain function of the first equation, must be enough to express all the information contained by these equations; thus, we must be able to destroy all terms containing x, y, z, etc., after these substitutions, since the question must reduce to an equation in u at the end.

However, the most general polynomial form in which this substitution can be made is a complete polynomial: It must therefore be the product of one of the proposed equations by a complete polynomial. There must therefore exist a complete polynomial whose coefficients must be able to satisfactorily destroy all terms containing x, y, z, etc. after substitution by $x^{t'}$, $y^{t''}$, $z^{t'''}$, etc.

We would, however, make a big mistake by thinking all coefficients of this polynomial can be useful for that purpose.

Indeed, we can always use the values of $x^{t'}$, $y^{t''}$, $z^{t'''}$, etc., to cancel all terms that may be divided by $x^{t'}$, $y^{t''}$, $z^{t'''}$. Therefore, since those terms can disappear at will, the value of their corresponding coefficients does not matter: In short, since we can always cancel these terms, the solution must be independent from these coefficients. We must therefore omit them for the sake of simplicity.

Another important consideration that will conclude our investigation of the qualities the polynomial multiplier must have to annihilate all terms except those in u is that the degree of this polynomial cannot be less than the sum of the exponents $t' + t'' + t''' + $ etc. of the $n-1$ equations used for the substitutions.

Indeed the polynomial multiplier must be as general as possible; thus we must be able to replace $x^{t'}$, $y^{t''}$, etc. by all their possible values. It must therefore contain all possible combinations of $x^{t'}$, $y^{t''}$, $z^{t'''}$, etc. Its degree must therefore not be less than $t' + t'' + t''' + $ etc.

Following these considerations, we can now proceed with computing the degree of the final equation.

Determination of the degree of the final equation resulting from an arbitrary number of complete equations containing the same number of unknowns

(46.) Assume that the proposed equations are given by $(u \ldots n)^t = 0$, $(u \ldots n)^{t'} = 0$, $(u \ldots n)^{t''} = 0$, etc. Assume that we replace $x^{t'}$, $y^{t''}$, $z^{t'''}$, etc., by their values resulting from the $n-1$ other equations in the *product equation* $(u \ldots n)^{T+t} = 0$, obtained by multiplying the first equation by the complete polynomial $(u \ldots n)^T$. We see that this substitution will cancel all terms that may be divided by $x^{t'}$, $y^{t''}$, $z^{t'''}$, etc., in the product equation. Thus by (43) the number of terms remaining in the product equation after all these substitutions is $d^{n-1}[N(u, \ldots n)^{T+t}] \ldots \left(\begin{array}{c} T+t \\ -t', -t'', -t''', \text{etc.} \end{array} \right)$.

Let D be the degree of the final equation. $D+1$ is therefore the number of its terms and consequently also the number of terms where only powers of u are present. Thus the number of terms that contain x, y, z, etc., is $d^{n-1}[N(u \ldots n)^{T+t}] \ldots \left(\begin{array}{c} T+t \\ -t', -t'', -t''', \text{etc.} \end{array} \right) - D - 1$.

Let us assume that we replace $x^{t'}$, $y^{t''}$, $z^{t'''}$, etc. by their values in the polynomial multiplier. These substitutions cancel all terms that may be divided by $x^{t'}$, by $y^{t''}$, by $z^{t'''}$, etc.; they consequently reduce the number of terms of this polynomial to $d^{n-1}[N(u \ldots n)^T] \ldots \left(\begin{matrix} T \\ -t', -t'', -t''', \text{etc.} \end{matrix} \right)$.

The polynomial is therefore only able to produce that number of coefficients to eliminate the terms in x, y, z, etc., after the substitutions.

In fact, we must even reduce this number by 1, because we can always assume one of the coefficients of one of the terms to be one or any other quantity in the product equation.

This being said, we must use one undetermined coefficient of the polynomial multiplier to destroy each term affected by x, y, z, etc., in the product equation. Thus the following equation must hold:

$$d^{n-1}[N(u \ldots n)^T] \ldots \left(\begin{matrix} T \\ -t', -t'', -t''', \text{etc.} \end{matrix} \right) - 1$$

$$= d^{n-1}[N(u \ldots n)^{T+1}] \ldots \left(\begin{matrix} T+t \\ -t', -t'', -t''', \text{etc.} \end{matrix} \right) - D - 1.$$

Thus we conclude that

$$\begin{aligned} D \;=\; & d^{n-1}[N(u \ldots n)^{T+t}] \ldots \left(\begin{matrix} T+t \\ -t', -t'', -t''', \text{etc.} \end{matrix} \right) \\ & -d^{n-1}[N(u \ldots n)^T] \ldots \left(\begin{matrix} T \\ -t', -t'', -t''', \text{etc.} \end{matrix} \right), \end{aligned}$$

that is,

$$D = d^n[N(u \ldots n)^{T+t}] \ldots \left(\begin{matrix} T+t \\ -t, -t', -t'', -t''', \text{ etc.} \end{matrix} \right).$$

Remembering that:

1. $N(u \ldots n)^{T+t} = \dfrac{(T+t+1).(T+t+2).(T+t+3)\ldots\ldots(T+t+n)}{1.2.3\ldots\ldots.n}$,

2. Since from (12) we can omit certain terms when computing the nth difference,

the value of D can first be reduced to

$$D = \frac{d^n(T+t)^n \ldots \ldots \left(\begin{matrix} T+t \\ -t, -t', -t'', -t''', \text{etc.} \end{matrix} \right)}{1.2.3\ldots.n}.$$

Finally, remembering from (14) that we can simplify the computation of the successive differences leading to the difference of degree n, and from (15) that we need not distinguish whether quantities are increasing or decreasing when computing the difference of an expression whose degree equals the number of successive differenciations, we get

$$D = \frac{d^n(T+t)^n \ldots \ldots \left(\begin{matrix} T+t \\ t, t', t'', t''', \text{etc.} \end{matrix} \right)}{1.2.3\ldots\ldots.n};$$

that is,

$$D = tt't''t''', \text{ etc.}$$

We therefore have the following general theorem:

(47.) *The degree of the final equation resulting from an arbitrary number of complete equations containing the same number of unknowns and with arbitrary degrees is equal to the product of the exponents of the degrees of these equations.*

Remarks

(48.)

1. Assume we have two equations in two unknowns $(u \ldots 2)^t = 0$ and $(u \ldots 2)^{t'} = 0$, and the degree of the final equation is tt'; that is, the product of the exponents of the degrees of these two equations. All that we have derived up to now regarding the elimination in complete equations reduces to that statement.

2. Assume $t'' = t''' = t^{IV} = \text{etc.} = 1$; then we have $D = tt'$, that is, the degree of the final equation is the same as if we only had two equations in two unknowns, one of degree t and the other of degree t'. It is easy to see that this must be true, since by using the $n-2$ first-order equations, we feel we can eliminate $n - 2$ unknowns without changing the degree of the two equations $(u \ldots n)^t = 0$ and $(u \ldots n)^{t'} = 0$; these become two equations of the form $(u \ldots 2)^t = 0$, $(u \ldots 2)^{t'} = 0$ via appropriate eliminations. However, the method we are about to give to reach the final equation will not require such partial eliminations. We will see this in detail in the Second Book. Our question here is only about computing the degree of the final equation.

3. We know, using geometry and algebra, that two curves drawn on a plane can meet only as many times as the product of the exponents of the degrees of their equations. This is a very simple consequence of what was said in the first remark.

   We also know, using geometry, that surfaces can be expressed by equations in three unknowns. Thus if three surfaces can be expressed by three algebraic equations, the following general theorem in geometry results immediately from our general theorem (47):

   The number of intersection points of three surfaces expressed by algebraic equations is not greater than the product of the three exponents of the degrees of these equations.

   Thus, we can say in passing that three cylinders, three spheres, three cones, three ellipsoids, three paraboloids, three hyperboloids, cannot intersect in more than eight points. This is true irrespective of their location and orientation. The same holds with the intersection of a cylinder, a sphere and an ellipsoid and, in general, the intersection of any three of the surfaces we have just introduced. This is because these three surfaces can be expressed by three second-order equations in three unknowns.

4. We can get an idea of the large number of useless roots produced by the successive elimination procedure in the final equation: Indeed, consider for example four equations, all of them of degree t. Comparing one of the equations to the other three, we generate three equations, each of them with degree t^2.

Likewise, comparing one of these three equations with the other two yields two equations of degree t^4.

Finally, comparing one of the equations to the other yields a final equation of degree t^8. But we just saw that the final equation can only have degree t^4.

For example, considering four equations of degree 2 only, the successive elimination method gives a final equation of degree 256, whereas it is only of degree 16.

If the four equations were of third order, the final equation obtained with the elimination method would be of degree 6561, but it is in fact of degree 81 only.

It is, however, true that if we proceed by eliminating variables according to the method we gave in the 1764 Academy of Sciences Memoir, we would avoid many of these useless roots. But there would still remain a large number of them; this number was unknown up until now.

Thus the reader can see how right we were to say, several years ago,[2] that all useless roots could be eliminated only when we would have found a method to eliminate all unknowns at once, except one of them.

---

[2]See the *Cours de Mathématiques à l'usage des Gardes du Pavillon et de la Marine, troisième partie*, pages 209 and 210.

SECTION II

About incomplete polynomials and first-order incomplete equations

(49.) We will not insist on the broad impact of the general theorem (47) about complete equations. We will simply remark that it not gives only a precise expression for the degree of the final equation resulting from an arbitrary number of complete equations with all possible terms and coefficients being present, but also an upper bound on the final degree of any equation, complete or incomplete, which may or may not be reducible due to either the absence of some terms or existing relations among their coefficients.

(50.) As useful as this upper bound may already be, it is far more useful to tighten it even more and even to determine the precise degree of the final equation for all possible cases, even when the equations may be reduced because of specific relations among their coefficients.

(51.) This subject is so broad that the reader will undoubtedly not expect us to explore it completely. What may be reasonable to expect, however, is to know a generic method to reach this goal in any case: This is what we will attempt to do.

(52.) Whatever idea the reader may already have about the extent of the matter we intend to deal with, we believe it will be surpassed by later developments. We must therefore proceed methodically and be first very general so as to prepare the reader to deal with broader subjects.

We will therefore introduce the different types of incomplete polynomials and incomplete equations only as we encounter them. But before beginning our investigations, we must point the reader's attention towards the following observations.

(53.) Any equation where a term is missing is, in general, called an *incomplete equation*. But all the terms that may be missing from a complete equation are not as important as far as lowering the degree of the final equation.

If the exponents of several unknowns in the missing terms are less than the highest exponent of the same unknowns in the remaining terms, and if at the same time, they belong to lower dimensions than those where they can be found, their absence does not influence the degree of the final equation: It will be the same as if the equations were complete. Occurrences of a lower degree will only be accidental and the consequence of a coincidental relation between coefficients.

For example, consider the two equations $ax^2 + bxy + cy^2 + g = 0$, $a'x^2 + b'xy + c'y^2 + g' = 0$, where the first-order terms are missing from both equations, but the second order terms are not. They lead to a final equation with order 4, the same as for two complete equations $ax^2 + bxy + cy^2 + ex + fy + g = 0$, $a'x^2 + b'xy + c'y^2 + e'x + f'y + g' = 0$. The main difference is that the coefficients of the final equation are simpler in the first case than the second.

(54.) The incomplete equations that we will consider are those whose missing terms can influence the degree of the final equation. Although they are incomplete, that is, they have fewer terms than a complete equation of the same degree, they cover a much wider range of problems than that covered by complete equations, which they contain as a very particular case.

We could have skipped the specific treatment of complete equations. However, not only would we have had to resort to a less intuitive presentation, but also we believe that the idea based on substitutions, upon which most of our reasoning was based, brings our problem and methodologies most closely to those used to solve equations of the first degree.

Although we can apply the same ideas to incomplete equations, we will present things from a different standpoint, which remains generally applicable and always in the same way: Instead, the principle of substitution requires modifications and a particular care that we have found useful to detail, by the way.

(55.) Let us assume for more simplicity that we have only three complete equations and three unknowns, and all three have degree t. Assume I use two of these equations to compute the value of y^t and z^t; since these two quantities have no common divider, the two equations which generated these values cannot give anything more than these two values and their multiples. Thus the question must be resolved solely using substitutions of y^t and z^t in the proper expression: The product equation.

But assume that the equations are incomplete: Assume, for example, that y has degree A and z has degree $\underset{\prime}{A}$. Then we can obtain the value of the term $y^A z^{t-A}$ from one equation and the value of the term $y^{\underset{\prime}{t-A}} z^{\underset{\prime}{A}}$ from the other; this is what must be done because these are the terms which have the smallest common divider in the highest dimension. However, these two equations can give not only these two values, but also others that are not simply multiples of these. For example, assume the two equations have degree 4 and y and z do not reach a degree larger than 3 in both equations; then we can get the values of $y^2 z$ and $y z^2$ using these two equations. However, these are not the only values that we can get from these equations. We can also obtain the value of $x^2 y^2 z^2$, or $x^2 y^4$, or $x^2 z^4$. This observation is made easily with the shorthand notation for $y^2 z$ and $y z^2$, represented as $y^2 z = M$ and $y z^2 = M$. Dividing one equation by the other, we obtain $y^2/z^2 = M/N$ or $N y^2 = M z^2$, a sixth-order equation that provides any one of the values that we just described. And since $x^2 y^2 z^2$, for example, is neither a multiple of $y^3 z$ nor $y z^3$, it is clear that in addition to the terms that may be divided by $y^3 z$ and by $y z^3$, we will also be able to suppress, by substituting, all terms or at least some of the terms that may be divided by $x^2 y^2 z^2$. Thus by substituting only the value of $y^3 z$ and the value of $y z^3$ provided by two equations, we clearly do not express all the initial conditions of the question, and we would not get out of these equations all the information they can and must give. We also have to substitute the value of $x^2 y^2 z^2$.

Thus there are usually more possible substitutions to be made in higher order or different incomplete equations.

(56.) We therefore see that the question becomes more complex and that talent and know-how are necessary to persist and use the substitution principle. But we believe we are making a useful remark for analysis by observing that when the quantities whose value is determined through a number of equations have a common divider among them, these values do not summarize all the information provided by these equations.

We will not pursue this observation further for now. We will come back to it when we come to the process for elimination. It is sufficient to know that we have justified the necessity or, at least, the usefulness of using another method to determine the degree of the final equation.

(57.) The first kind of incomplete equation whose final degree we are looking for is that where the degree of each unknown does not exceed a given maximum degree (different for each unknown); however, we will assume that the unknowns, when combined pairwise, three-by-three, etc., reach the total dimension of the equation.

About incomplete polynomials and incomplete equations in which each unknown does not exceed a given degree for each unknown. And where the unknowns, combined two-by-two, three-by-three, four-by-four, etc., all reach the total dimension of the polynomial or the equation

(58.) Let us denote by A, $A_{\prime}$, $A_{\prime\prime}$, $A_{\prime\prime\prime}$ the degrees that can be reached by each unknown, and by T the degree of the polynomial or the equation. We represent the polynomial of interest as $(u^A \ldots n)^T$, and an equation by $(u^A \ldots n)^T = 0$.

Problem IV

(59.) *We ask for the number of terms of the polynomial $(u^A \ldots n)^T$, or the value of $N(u^A \ldots n)^T$.*

The solution to this problem is very easy, following what was said in (41). Indeed, the degree of u, for example, must not exceed A; thus it follows that all terms that may be divided by u^{A+1} are missing, and that number, in a complete polynomial, is from (41) $N(u \ldots n)^{T-A-1}$.

Since the degree of x must not exceed $A_{\prime}$, all terms that may be divided by $x_{\prime}^{A+1}$ are missing, and the number of these terms is $N(u \ldots n)^{T-A-1}_{\prime}$. But since the dimension of u and x togeth must reach T, we must have $A + A_{\prime} > T$; thus eliminating the terms that may be divided by u^{A+1} does not imply elimination of terms that may be divided by $x_{\prime}^{A+1}$, since the only such term of lowest degree is $u^{A+1}x_{\prime}^{A+1}$, which exceeds the dimension T.

Thus, even after eliminating the terms that may be divided by u^{A+1}, eliminating the terms that may be divided by $x^{A'+1}$ removes a number of terms equal to $N(u \ldots n)^{T-A-1}$.

Since y must not exceed the degree $\underset{''}{A}$, the complete polynomial misses all terms that may be divided by $y''^{\underset{}{A}+1}$, and since we assume that the combination of u and y, and that of x and y, must reach the dimension T in the proposed polynomial, we have $A + \underset{''}{A} > T$, $\underset{'}{A} + \underset{''}{A} > T$; thus eliminating terms that may be divided by u^{A+1} and those that may be divided by $x'^{\underset{}{A}+1}$ does not eliminate any of the terms that may be divided by $y''^{\underset{}{A}+1}$. Thus the number of these terms is $N(u\ldots n)^{T-\underset{''}{A}-1}$.

By reasoning along the same lines, we see that the missing number of terms in z is $N(u\ldots n)^{T-\underset{'''}{A}-1}$, and so on.

Thus $N(u^A\ldots n)^T = N(u\ldots n)^T - N(u\ldots n)^{T-A-1} - N(u\ldots n)^{T-\underset{'}{A}-1} - N(u\ldots n)^{T-\underset{''}{A}-1} - N(u\ldots n)^{T-\underset{'''}{A}-1} -$, etc.

PROBLEM V

(60.) *Consider* $(u^{a'}\ldots n)^{t'} = 0$, $(u^{a''}\ldots n)^{t''} = 0$, $(u^{a'''}\ldots n)^{t'''} = 0$, *etc. an arbitrary number of equations containing* n *unknowns. Let* $(u^A\ldots n)^T$ *be a polynomial such that*

$$\left.\begin{array}{l} A - a' - a'' - a''', \text{ etc.}\\ + \underset{'}{A} - \underset{'}{a}' - \underset{'}{a}'' - \underset{'}{a}''', \text{ etc.}\end{array}\right\} > T - t' - t'' - t''', \text{ etc.}$$

$$\left.\begin{array}{l} A - a' - a'' - a''', \text{ etc.}\\ + \underset{''}{A} - \underset{''}{a}' - \underset{''}{a}'' - \underset{''}{a}''', \text{ etc.}\end{array}\right\} > T - t' - t'' - t''', \text{ etc.}$$

$$\left.\begin{array}{l} A - a' - a'' - a''', \text{ etc.}\\ + \underset{'''}{A} - \underset{'''}{a}' - \underset{'''}{a}'' - \underset{'''}{a}''', \text{ etc.}\end{array}\right\} > T - t' - t'' - t''', \text{ etc.}$$

$$\left.\begin{array}{l} \underset{'}{A} - \underset{'}{a}' - \underset{'}{a}'' - \underset{'}{a}''', \text{ etc.}\\ + \underset{''}{A} - \underset{''}{a}' - \underset{''}{a}'' - \underset{''}{a}''', \text{ etc.}\end{array}\right\} > T - t' - t'' - t''', \text{ etc.}$$

$$\left.\begin{array}{l} \underset{'}{A} - \underset{'}{a}' - \underset{'}{a}'' - \underset{'}{a}''', \text{ etc.}\\ + \underset{'''}{A} - \underset{'''}{a}' - \underset{'''}{a}'' - \underset{'''}{a}''', \text{ etc.}\end{array}\right\} > T - t' - t'' - t''', \text{ etc.}$$

$$\left.\begin{array}{l} \underset{''}{A} - \underset{''}{a}' - \underset{''}{a}'' - \underset{''}{a}''', \text{ etc.}\\ + \underset{'''}{A} - \underset{'''}{a}' - \underset{'''}{a}'' - \underset{'''}{a}''', \text{ etc.}\end{array}\right\} > T - t' - t'' - t''', \text{ etc.}$$

and so on.

We ask how many terms these equations can help eliminate, without introducing new terms.

Let us first assume we have one equation only; after multiplying it by a polynomial $(u^{A'}\ldots n)^{T'}$, assume we add the product $(u^{A'+a'}\ldots n)^{T'+t'}$ to the proposed polynomial: It is clear that

1. this addition does not change the value of the proposed polynomial,

2. assuming the polynomial multiplier has the appropriate characteristics so as to not introduce new terms, we are able to cancel several terms in the proposed polynomial. The number of these is the number of terms in the polynomial multiplier, since each of these provides a coefficient.

3. in order for this polynomial multiplier to cancel as many terms as possible without introducing new ones, we must have

$$T' + t' = T; \ A' + a' = A; \ A' + a' = A, \ A' + a' = A, \ A' + a' = A;$$

and so on. We therefore have

$$T' = T - t'; \ A' = A - a'; \ A' = A - a'; \ A' = A - a'; \ A' = A - a';$$

and so on.

But it so happens from the conditions presented in the problem formulation that

$$A - a' + A - a' > T - t'; \qquad A - a' + A - a' > T - t';$$

$$A - a' + A - a' > T - t'; \quad A - a' + A - a' > T - t', \text{ etc.}$$

Thus the polynomial $(u^{A'} \ldots n)^{T'}$, which becomes $(u^{A-a'} \ldots n)^{T-t'}$, is of the same kind as the proposed polynomial and equations.

The number of terms that may be cancelled using the first equation alone is therefore $N(u^{A-a'} \ldots n)^{T-t'}$; consequently, it is easy to express this number as a function of $T - t'$ and $A - a'$, from what was said in (39).

Assume now that we have two equations.

Assume we multiply, as we have done before, the first equation by the polynomial $(u^{A'} \ldots n)^{T'}$. The number of terms that may be cancelled is not $N(u^{A'} \ldots n)^{T'}$ anymore. Indeed, since there exists a second equation, we are always able to use this equation to cancel $N(u^{A-a''} \ldots n)^{T'-t''}$ terms in the polynomial $(u^{A'} \ldots n)^{T'}$; thus the polynomial $(u^{A'} \ldots n)^{T'}$ can only provide $N(u^{A'} \ldots n)^{T'} - N(u^{A'-a''} \ldots n)^{T'-t''}$ coefficients, that is, replacing A' and T' by their values, $N(u^{A-a'} \ldots n)^{T-t'} - N(u^{A-a'-a''} \ldots n)^{T-t'-t''}$ coefficients. The first equation can therefore cancel the same number of terms only. Concerning the second equation, if there is no third equation, there is nothing that can justify reducing the number of coefficients of the possible polynomial multiplier, whose product with the second equation is to be added to the proposed polynomial. Following the same argument as the one we used for the case involving a single equation, we can show that this second equation can be used to cancel $N(u^{A-a''} \ldots n)^{T-t''}$ terms.

Thus we can use the two equations to cancel $N(u^{A-a'} \ldots n)^{T-t'}$ $+ N(u^{A-a''} \ldots n)^{T-t''} - N(u^{A-a'-a''} \ldots n)^{T-t'-t''}$ terms.

According to the conditions

$$A - a' - a'' + A - a' - a'' > T - t' - t'',$$

etc., we see that the polynomials $(u^{A-a'}\dots n)^{T-t'}$, $(u^{A-a''}\dots n)^{T-t''}$, $(u^{A-a'-a''}\dots n)^{T-t'-t''}$ are similar in nature to the proposed polynomial and equations.

Let us now consider the case of three equations.

Assume we multiply the first equation by the polynomial $(u^{A'}\dots n)^{T'}$, the second, by the polynomial $(u^{A''}\dots n)^{T''}$, the third, by the polynomial $(u^{A'''}\dots n)^{T'''}$. Assume we determine the largest possible values for A', A'', A''', etc., T', T'', T''', etc. without adding new terms; as earlier we find

$$T' = T - t'; \qquad A' = A - a'; \qquad \underset{\prime}{A}' = \underset{\prime}{A} - \underset{\prime}{a}', \text{ etc.,}$$

$$T'' = T - t''; \qquad A'' = A - A''; \qquad \underset{\prime}{A}'' = \underset{\prime}{A} - \underset{\prime}{a}'', \text{ etc.,}$$

$$T''' = T - t'''; \qquad A''' = A - a'''; \qquad \underset{\prime}{A}''' = \underset{\prime\prime\prime}{A} - \underset{\prime}{a}''', \text{ etc.}$$

From what was said about two equations, we can always use the last two equations to cancel $N(u^{A'-a''}\dots n)^{T'-t''} + N(u^{A'-a'''}\dots n)^{T'-t'''}$ $-N(u^{A-a''-a'''}\dots n)^{T'-t'-t'''}$ terms in the polynomial multiplier $(u^{A'}\dots n)^{T'}$; consequently this polynomial can provide only $N(u^{A'}\dots n)^{T'}$ $-N(u^{A'-a''}\dots n)^{T'-t''} - N(u^{A'-a'''}\dots n)^{T'-t'''} + N(u^{A'-a''-a'''}\dots n)^{T'-t''-t'''}$ coefficients; replacing A' and T' by their values, this expression becomes $N(u^{A-a'}\dots n)^{T-t'} - N(u^{A-a'-a''}\dots n)^{T-t'-t''} - N(u^{A-a-a'''}\dots n)^{T-t'-t'''}$ $+ N(u^{A-a'-a''-a'''}\dots n)^{T-t'-t''-t'''}$. Thus, using the first equation, only that number of terms can be cancelled in the proposed polynomial.

Likewise, using the third equation, we can always cancel $N(u^{A''-a'''}\dots n)^{T''-t'''}$ terms in the polynomial multiplier $(u^{A''}\dots n)^{T''}$ of the second one. This polynomial can therefore provide only $N(u^{A''}\dots n)^{T''} - N(u^{A''-a'''}\dots n)^{T'''-t'''}$ coefficients, or, replacing A'' and T'' by their values, $N(u^{A-a''}\dots n)^{T-t''} - N(u^{A-a''-a'''}\dots n)^{T-t'-t'''}$ coefficients.

Thus we can use the second equation to cancel only $N(u^{A-a''}\dots n)^{T-t''} - N(u^{A-a''-a'''}\dots n)^{T-t''-t'''}$ terms. The third equation can be used to cancel $N(u^{A'''}\dots n)^{T'''}$ terms, that is, replacing A''' and T''' by their values, $N(u^{A-a'''}\dots n)^{T-t'''}$ such terms.

Thus finally, the number of terms that we can cancel using all three equations is

$$N(u^{A-a'}\dots n)^{T-t'} - N(u^{A-a'-a''}\dots n)^{T-t'-t''}$$

$$- N(u^{A-a'-a'''}\dots n)^{T-t'-t'''} + N(u^{A-a'-a''-a'''}\dots n)^{T-t'-t''-t'''}$$

$$+ N(u^{A-a''}\dots n)^{T-t''} - N(u^{A-a''-a'''}\dots n)^{T-t''-t'''}$$

$$+ N(u^{A-a'''}\dots n)^{T-t'''}.$$

Remembering the hypotheses $A - a' - a'' - a''' + \underset{\prime}{A} - \underset{\prime}{a}' - \underset{\prime}{a}'' - \underset{\prime}{a}''' >$ $T-t'-t''-t'''$, etc. we can show, as we have done earlier, that all polynomials entering this expression are of the same type as the proposed polynomial and equations.

It is now easy to see what the expression for the number sought is, given a larger number of equations.

Problem VI

(61.) *We ask for the number of terms remaining in the polynomial*
$(u^A \ldots n)^T$ *when, using a given number of equations of the same nature as
that polynomial, we have cancelled all terms that can possibly be cancelled*

It is easy to see, from the preceding problem, that if there is only one equation, the number of remaining terms is $N(u^A \ldots n)^T - N(u^{A-a'} \ldots n)^{T-t'}$,

which reduces to $d[N(u^A \ldots n)^T] \ldots \left(\begin{array}{ccc} T & A & \overset{A}{,} \\ -t' & : & -a' & : & - \underset{,}{a'} \end{array} : \text{etc.} \right)$. As we

have shown before, this reduction holds when the two polynomials $(u^A \ldots n)^T$,
$(u^{A-a'} \ldots n)^{T-t'}$ share the same characteristics, but not otherwise.

If there are two equations, the number of remaining terms is $N(u^A \ldots n)^T - N(u^{A-a'} \ldots n)^{T-t'} - N(u^{A-a''} \ldots n)^{T-t''} + N(u^{A-a'-a''} \ldots n)^{T-t'-t''}$. This
number reduces to

$$
dd[N(u^A \ldots n)^T] \ldots \left(\begin{array}{ccc} T & A & \overset{A}{,} \\ -t', -t'' & : & -a', -a'' & : & - \underset{,}{a'}, - \underset{,}{a''} \end{array} : \text{etc.} \right)
$$

because the four polynomials are of the same nature.

Likewise, in the case of three equations, the number of remaining terms
is

$$
d^3[N(u^A \ldots n)^T] \ldots \left(\begin{array}{cc} T & A \\ -t', -t'', -t''' & : & -a', -a'', -a''' \end{array} : \\ \overset{A}{,} \\ - \underset{,}{a'}, - \underset{,}{a''}, - \underset{,}{a'''} \quad , \text{etc.} \right),
$$

and so on.

Problem VII

(62.) *We ask for the degree of the final equation resulting from an arbitrary number n of equations of the form* $(u^a \ldots n)^t$ *in the same number of
unknowns*

Assume the equations are $(u^a \ldots n)^t = 0$, $(u^{a'} \ldots n)^{t'} = 0$,
$(u^{a''} \ldots n)^{t''} = 0$, $(u^{a'''}, \ldots n)^{t''''} = 0$, etc. After having multiplied the first
equation by a polynomial $(u^A \ldots n)^{T'}$ that meets the conditions expressed
in the formulation of Problem V, assume we then cancel all terms that may
be cancelled using the $n - 1$ remaining equations without introducing new
terms. Then by (61) there remain

$$
d^{n-1}[N(u^{A+a} \ldots n)^{T+t}] \ldots \left(\begin{array}{cc} T+t & A+a \\ -t', -t'', -t''', \text{etc.} & : & -a', -a'', -a''', \text{etc.} \end{array} : \\ \overset{A+a}{,} \quad \overset{A+a}{,,} \\ - \underset{,}{a'}, - \underset{,}{a''}, - \underset{,}{a'''}, \text{etc.} \quad : \quad - \underset{,,}{a'}, - \underset{,,}{a''}, - \underset{,,}{a'''}, \text{etc.} \quad : \text{etc.} \right)
$$

terms in the product equation $(u^{A+a}\ldots n)^{T+t} = 0$. Let D be the degree of the final equation; $D+t$ is the number of terms it contains; then the number of terms that remains to cancel is

$$d^{n-1}[N(u^{A+a}\ldots n)^{T+t}]\ldots \left(\begin{matrix} T+t \\ -t', -t'', -t''', \text{etc.} \end{matrix} \quad\vdots\quad \begin{matrix} A+a \\ -a', -a'', -a''', \text{etc.} \end{matrix} \quad\vdots\quad \right.$$
$$\left.\begin{matrix} A+\underset{\prime}{a} \\ -\underset{\prime}{a}', -\underset{\prime}{a}'', -\underset{\prime}{a}''', \text{etc.} \end{matrix} \quad : \text{etc.}\right) - D - 1.$$

The polynomial multiplier must provide as many coefficients plus one to cancel these terms. Looking at the number of terms that may be cancelled in the polynomial multiplier using the $n-1$ last equations, this polynomial can only provide a number of coefficients equal to

$$d^{n-1}[N(u^{A}\ldots n)^{T}]\ldots \left(\begin{matrix} T \\ -t', -t'', -t''', \text{etc.} \end{matrix} \quad\vdots\quad \begin{matrix} A \\ -a', -a'', -a''', \text{etc.} \end{matrix}\right.$$
$$\left.\vdots\quad \begin{matrix} A \\ -\underset{\prime}{a}', -\underset{\prime}{a}'', -\underset{\prime}{a}''', \text{etc.} \end{matrix} \quad : \text{etc.}\right),$$

We therefore obtain the following equation:

$$d^{n-1}[N(u^{A}\ldots n)^{T}]\ldots \left(\begin{matrix} T \\ -t', -t'', -t''', \text{etc.} \end{matrix} \quad\vdots\quad \begin{matrix} A+a \\ -a', -a'', -a''', \text{etc.} \end{matrix} \quad\vdots\quad\right.$$
$$\left.\begin{matrix} A+\underset{\prime}{a} \\ -\underset{\prime}{a}', -\underset{\prime}{a}'', -\underset{\prime}{a}''', \text{etc.} \end{matrix} \quad : \text{etc.}\right) - D - 1$$
$$= d^{n-1}(N(u^{A}\ldots n)^{T}]\ldots \left(\begin{matrix} T \\ -t'-t'', -t''', \text{etc.} \end{matrix} \quad\vdots\quad \begin{matrix} A \\ -a', -a'', -a''', \text{etc.} \end{matrix} \quad\vdots\quad\right.$$
$$\left.\begin{matrix} A \\ -a', -a'', -a''', \text{etc.} \end{matrix} \quad\text{etc.}\right) - 1,$$

from which we find

$$D = d^{n-1}[N(u^{A+a}\ldots n)^{T+t}]\ldots \left(\begin{matrix} T+t \\ -t', -t'', -t''', \text{etc.} \end{matrix} \quad\vdots\quad \begin{matrix} A+a \\ -a', -a'', -a''', \text{etc.} \end{matrix} \quad\vdots\right.$$
$$\left.\begin{matrix} A+\underset{\prime}{a} \\ -\underset{\prime}{a}', -\underset{\prime}{a}'', -\underset{\prime}{a}''', \text{etc.} \end{matrix} \quad : \text{etc.}\right)$$
$$-d^{n-1}[N(u^{A}\ldots n)^{T}]\ldots \left(\begin{matrix} T \\ -t', -t'', -t''', \text{etc.} \end{matrix} \quad\vdots\quad \begin{matrix} A \\ -a', -a'', -a''', \text{etc.} \end{matrix} \quad\vdots\right.$$
$$\left.\begin{matrix} A \\ -\underset{\prime}{a}', -\underset{\prime}{a}'', -\underset{\prime}{a}''', \text{etc.} \end{matrix} \quad : \text{etc.}\right),$$

that is,

$$D = d^{n}[N(u^{A+a}\ldots n)^{T+t}]\ldots \left(\begin{matrix} T+t \\ -t, -t', -t'', -t''', \text{etc.} \end{matrix} \quad\vdots\right.$$
$$\begin{matrix} A+a \\ -a, -a', -a'', -a''', \text{etc.} \end{matrix} \quad\vdots\quad \begin{matrix} A+\underset{\prime}{a} \\ -\underset{\prime}{a}, -\underset{\prime}{a}', -\underset{\prime}{a}'', -\underset{\prime}{a}''', \text{etc.} \end{matrix} \quad : \text{etc.}\left.\right).$$

Thus we get the value of D by differentiating the quantity $N(u\ldots n)^{T+t}-$
$N(u\ldots n)^{T+t-A-a-1}-N(u\ldots n)^{T+t-\underset{\prime}{A}-\underset{\prime}{a}-1}-N(u\ldots n)^{T+t-\underset{\prime\prime}{A}-\underset{\prime\prime}{a}-1}$, etc.,
which from (59) is the value of $N(u^{A+a}\ldots n)^{T+t}$. We differentiate it n
times according to the rules given earlier (13) and (14) to finally get

$$D = tt't''t''',\text{etc.}\quad -(t-a).(t'-a').(t''-a'').(t'''-a'''),\text{etc.}$$
$$-(t-\underset{\prime}{a}).(t'-\underset{\prime}{a'}).(t''-\underset{\prime}{a''}).(t'''-\underset{\prime}{a'''}),\text{etc.}$$
$$-(t-\underset{\prime\prime}{a}).(t'-\underset{\prime\prime}{a'}).(t''-\underset{\prime\prime}{a''}).(t'''-\underset{\prime\prime}{a'''}),\text{etc.}$$
$$-(t-\underset{\prime\prime\prime}{a}).(t'-\underset{\prime\prime\prime}{a'}).(t''-\underset{\prime\prime\prime}{a''}).(t'''-\underset{\prime\prime\prime}{a'''}),\text{etc.}$$
$$-\text{etc.}$$

Each product contains as many factors as there are unknowns.

Thus if we have two unknowns only, the degree of the final equation is
$D = tt'-(t-a).(t'-a')-(t-\underset{\prime}{a})(t'-\underset{\prime}{a'})$.

(63.) Consider the case of two unknowns. If we assume moreover that
$a = t$, $a' = t'$, we obtain $D = tt'-(t-\underset{\prime}{a})(t'-\underset{\prime\prime}{a'})$; or if we assume $a = t$,
$\underset{\prime}{a'} = t'$, we obtain $D = tt'-(t-a).(t'-a')$, which means the same thing.

The last expression summarizes all that was known up to now concerning
incomplete equations involving two unknowns only.

(64.) The reader may remark that the general expression we found for D
contains the expression of D for complete equations as a specific case, a case
that occurs when $a = t$, $a' = t'$, $a'' = t''$, etc., $\underset{\prime}{a} = t$, $\underset{\prime}{a'} = t'$, $\underset{\prime}{a''} = t''$, etc.,
$\underset{\prime\prime}{a} = t$, $\underset{\prime\prime}{a'} = t'$, $\underset{\prime\prime}{a''} = t''$, etc.; it is clear that in this case we have $D = tt't''t'''$,
etc.

(65.) We will limit ourselves to one example to illustrate how much lower
the degree of the final equation can be in the case of the incomplete equations
of interest to this chapter. Let us assume that $n = 3$, $t = t' = t'' = 2$;
$a = a' = a'' = 1$, $\underset{\prime}{a} = \underset{\prime}{a'} = \underset{\prime}{a''} = 1$, $\underset{\prime\prime}{a} = \underset{\prime\prime}{a'} = \underset{\prime\prime}{a''} = 1$. We have $D =$
$8-1-1-1 = 5$. The degree of the final equation is therefore three degrees
lower than if the proposed equations were complete.

(66.) We remark that if only one of the proposed equations were complete,
the value of D would be the same as if they were all complete.

Remark

(67.) We must be careful that the value of D we just found holds under the
assumption that the equations $(u^a\ldots n)^t = 0$, $(u^{a'}\ldots n)^{t'} = 0$, etc. satisfy
the same conditions (60). Applying this value to other cases would be a
mistake.

For example, consider three equations such that $t = t' = t'' = 3$; $a = a' = a'' = 1$; $\underset{\prime}{a} = \underset{\prime}{a'} = \underset{\prime}{a''} = 1$; $\underset{\prime\prime}{a} = \underset{\prime\prime}{a'} = \underset{\prime\prime}{a''} = 1$. Using the present result yields
$D = 27-8-8-8 = 3$; this is not true, as we will see later. Indeed, these

equations do not meet the required conditions to use that value for D, since instead of having $a + a_{\prime} > t$; $a + a_{\prime\prime} > t$; $a_{\prime} + a_{\prime\prime} > t$; $a' + a_{\prime}' > t$, etc. we have $a + a_{\prime} < t$; $a + a_{\prime\prime} < t$, etc.

We will see later how we can determine the value of D for equations of the form $(u^a \ldots n)^t = 0$ when the conditions $a + a_{\prime} > t$, etc. are not satisfied. But for clarity's sake, this discussion must be postponed after we have studied various other polynomials and equations.

About the sum of some quantities necessary to determine the number of terms of various types of incomplete polynomials

(68.) From the preceding remarks, we can already see that determining the degree of the final equation essentially depends on the number of terms of the polynomials involved.

We have already seen that the latter is rather easy to compute for the first type of incomplete polynomials that we have just considered; but the other polynomials we propose to study require the computation of the sum of certain quantities; we believe we must present these computations before computing the number of terms.

The guidelines we have given in (18 ff) are still sufficient to achieve this goal. However, calculations get complicated as we make progress, and we can never pay enough attention to simplifying the results, to giving them the simplest, easiest way to handle, and most general, expression. Our effort is therefore less to find a new way to compute the sum of quantities (that we will present shortly), rather than finding easier expressions for the results towards which we will be led by scrupulously following the principles given in the introduction. Let us begin our investigation.

PROBLEM VIII

(69.) *Compute $\int N(u \ldots n - 1)^{P+s}$ from $s = Q$ to $s = R$; we assume $R > Q$ and s varies by increments of one.*

We know from (39) that

$$N(u \ldots n - 1)^{P+s} = \frac{(P + s + 1) \times (P + s + 2) \ldots (P + s + n - 1)}{1.2.3. \ldots n - 1}.$$

If we multiply the latter quantity, top and bottom, by $\frac{P+s+n-(P+s)}{P+s+n-(P+s)}$ or by $\frac{P+s+n-(P+s)}{n}$, we obtain

$$N(u \ldots n - 1)^{P+s} = \frac{(P+s+1).(P+s+2)\ldots(P+s+n-1).[P+s+n-(P+s)]}{1.2.3.\ldots(n-1).n}$$

$$= \frac{(P+s+1).(P+s+2)\ldots.(P+s+n)-(P+s).(P+s+1)\ldots(P+s+n-1)}{1.2.3\ldots n}$$

$$= N(u \ldots n)^{P+s} - N(u \ldots n)^{P+s-1} = dN(u \ldots n)^{P+s} \ldots \begin{pmatrix} s \\ -1 \end{pmatrix}.$$

We therefore have $N(u \ldots n - 1)^{P+s} = dN(u \ldots n)^{P+s} \begin{pmatrix} s \\ -1 \end{pmatrix}$.

Thus $\int N(u\ldots n-1)^{P+s} = N(u\ldots n)^{P+s} + C.$
Thus when $s = Q - 1$, we have

$$\int N(u\ldots n-1)^{P+s} = N(u\ldots n)^{P+Q-1} + C.$$

Hence from $s = Q$ included to $s = R$ included, we have

$$\int N(u\ldots n-1)^{P+s} = N(u\ldots n)^{P+R} - N(u\ldots n)^{P+Q-1}.$$

PROBLEM IX

(70.) *Compute the sum of* $N(u\ldots n-1)^{P-s}$, *from* $s = Q$ *to* $s = R$; *we assume* $R > Q$, *and* s *increases by increments of one*
From (39) we have

$$N(u\ldots n-1)^{P-s} = \frac{(P-s+1).(P-s+2).(P-s+3)\ldots(P-s+n-1)}{1.2.3.\ldots\ldots n-1}$$

$$= \frac{(P-s+1).(P-s+2)\ldots\ldots(P-s+n-1).[P-s+n-(P-s)]}{1.2.3.\ldots n}$$

$$= \frac{(P-s+1).(P-s+2)\ldots(P-s+n)-(P-s).(P-s+1)\ldots(P-s+n-1)}{1.2.3.\ldots n}$$

$$= N(u\ldots n)^{P-s} - N(u\ldots n)^{P-s-1} = -dN(u\ldots n)^{P-s-1}\ldots \begin{pmatrix} s \\ -1 \end{pmatrix}.$$

Thus $\int N(u\ldots n-1)^{P-s} = -N(u\ldots n)^{P-s-1} + C.$
Thus $\int N(u\ldots n-1)^{P-s} = -N(u\ldots n)^{P-Q} + C$ when $s = Q - 1$.
And when $s = R$, we have

$$\int N(u\ldots n-1)^{P-s} = -N(u\ldots n)^{P-R-1} + C.$$

Thus from $s = Q$ included to $s = R$ included, we have

$$\int N(u\ldots n-1)^{P-s} = -N(u\ldots n)^{P-R-1} + N(u\ldots n)^{P-Q}.$$

PROBLEM X

(71.) *Compute the sum of* $N(u)^{L+Ms} \times N(u\ldots n-2)^{P+s}$, *from* $s = Q$ *to* $s = R$.
I begin by recasting this quantity whose sum we have just computed in (69) as follows.
Since from (35) $N(u)^{L+Ms} = L + Ms + 1$, I have $N(u)^{L+Ms} \times N(u\ldots n-2)^{P+s} = (L + Ms + 1) \times N(u\ldots n-2)^{P+s}$.
I assume that the latter quantity is equal to $A.N(u\ldots n-2)^{P+s} + B.(P+s) \times N(u\ldots n-2)^{P+s}$; I therefore obtain $L + Ms + 1 = A + B.(P+s)$, from which I get $L + 1 = A + BP$ and $B = M$; thus $A = L + 1 - MP$. I therefore have $N(u)^{L+Ms} \times N(u\ldots n-2)^{P+s} = (L+1 - MP) \times N(u\ldots n-2)^{P+s} + M.(P+s) \times N(u\ldots n-2)^{P+s}$.
But $(P+s) \times N(u\ldots n-2)^{P+s} = (n-1) \times N(u\ldots n-1)^{P+s-1}$, which is easy to check. So finally $N(u)^{L+Ms} \times N(u\ldots n-2)^{P+s} = (L + 1 - $

$MP) \times N(u \ldots n-2)^{P+s} + M(n-1) \times N(u \ldots n-1)^{P+s-1} = N(u)^{L-MP} \times$
$N(u \ldots n-2)^{P+s} + M(n-1) \times N(u \ldots n-1)^{P+s-1}$.

Therefore, from (69) $\int N(u)^{L+Ms} \times N(u \ldots n-2)^{P+s} = N(u)^{L-MP} \times$
$[N(u \ldots n-1)^{P+R} - N(u \ldots n-1)^{P+Q-1}] + M(n-1) \times N(u \ldots n)^{P+R-1} -$
$M(n-1) \times N(u \ldots n)^{P+Q-2}$, where this sum is taken from $s = Q$ included
to $s = R$ included.

PROBLEM XI

(72.) *Compute the sum* $N(u)^{L+Ms} \times N(u \ldots n-2)^{P-s}$, *from* $s = Q$
included to $s = R$ *included.*

As before, we write $N(u)^{L+Ms} \times N(u \ldots n-2)^{P-s} = (L + Ms + 1) \times$
$N(u \ldots n-2)^{P-s} = A.N(u \ldots n-2)^{P-s} + B.(P-s) \times N(u \ldots n-2)^{P-s}$.
This leads to $A + BP = L+1$, $B = -M$. We therefore have $A = L+1+MP$,
and consequently $N(u)^{L+Ms} \times N(u \ldots n-2)^{P-s} = (L+1+MP) \times N(u \ldots n-$
$2)^{P-s} - M.(P-s) \times N(u \ldots n-2)^{P-s} = N(u)^{L+MP} \times N(u \ldots n-2)^{P-s} -$
$M.(n-1) \times N(u \ldots n-2)^{P-s-1}$.

Therefore, by (70) the sum from $s = Q$ included to $s = R$ included is

$$
\begin{aligned}
\int N(u)^{L+Ms} \times N(u \ldots n-2)^{P-s} &= N(u)^{L+MP} \\
\times \quad & [N(u \ldots n-1)^{P-Q} - N(u \ldots n-1)^{P-R-1}] \\
- \quad & M.(n-1) \times N(u \ldots n)^{P-Q-1} + M.(n-1) \times N(u \times n)^{P-R-2}.
\end{aligned}
$$

Remark

(73.) We will not examine any other forms at this point: We will pursue
them later. But we add a useful observation to shorten the calculations
involving these formulas.

When we must compute the sum of the same expression in different consecutive intervals, we can readily compute its sum for the complete interval
instead of summing it over each of these intervals.

For example, if I must compute the sum of $N(u \ldots n-1)^{P-s}$ from $s = 0$
to $s = A$, then from $s = A$ excluded to $s = B$, then from $s = B$ excluded to
$s = C$: the question reduces to computing the sum $N(u \ldots n-1)^{P-s}$ from
$s = 0$ to $s = C$, which by (70) yields

$$ N(u \ldots n)^P - N(u \ldots n)^{P-C-1}. $$

Thus if we had to compute the sum of

1. $N(u \ldots n-1)^{T-s} - N(u \ldots n-1)^{T-A-s-1} - N(u \ldots n-1)^{T-A-s-1}$,
   from $s = 0$ to $s = P$;

2. $N(u \ldots n-1)^{T-s} - N(u \ldots n-1)^{T-A-s-1} - N(u \ldots n-1)^{T-B}$, from
   $s = P$ excluded to $s = P'$;

3. $N(u \ldots n-1)^{T-s} - N(u \ldots n-1)^{T-B'-1} - N(u \ldots n-1)^{T-B}$, from
   $s = P'$ excluded to $s = P''$,

then we would sum

1. $N(u \ldots n-1)^{T-s}$ from $s=0$ to $s=P''$, yielding by (70)

$$N(u \ldots n)^T - N(u \ldots n)^{T-P''-1};$$

2. $-N(u \ldots n-1)^{T-A-s-1}$ from $s=0$ to $s=P'$, yielding by (70)
   $-N(u \ldots n)^{T-A-1} + N(u \ldots n)^{T-A-P'-2}$.

3. $-N(u \ldots n-1)^{T-B'-1}$, from $s=P'$ excluded to $s=P''$. This sum, by (35), is $-(P''-P') \times N(u \ldots n-1)^{T-B'-1}$ or $-N(u)^{P''-P'-1} \times N(u \ldots n-1)^{T-B'-1}$.

4. $-N(u \ldots n-1)^{T-A-s-1}_{\prime}$, from $s=0$, to $s=P$, which by (70) gives
   $-N(u \ldots n)^{T-A'-1} + N(u \ldots n)^{T-A-P-2}_{\prime}$.

5. $-N(u \ldots n-1)^{T-B}$, from $s=P$ excluded to $s=P''$, which by (35) gives $-(P''-P) \times N(u \ldots n-1)^{T-B}$ or $-N(u)^{P''-P-1} \times N(u \ldots n-1)^{T-B}$.

Thus the total sum is

$$N(u \ldots n)^T - N(u \ldots n)^{T-P''-1} - N(u \ldots n)^{T-A-1}$$
$$+ N(u \ldots n)^{T-A-P'-2} - N(u)^{P''-P'-1} \times N(u \ldots n-1)^{T-B'-1}$$
$$- N(u \ldots n)^{T-A-1}_{\prime} + N(u \ldots n)^{T-A-P-2}_{\prime} - N(u)^{P''-P-1}$$
$$\times N(u \ldots n-1)^{T-B}.$$

About incomplete polynomials and incomplete equations, in which two of the unknowns (the same in each polynomial or equation) share the following characteristics:

1. *The degree of each of these unknowns does not exceed a given number (different or the same for each unknown);*

2. *These two unknowns, taken together, do not exceed a given dimension;*

3. *The other unkowns do not exceed a given degree (different or the same for each), but, when combined in groups of two or three among themselves as well as with the first two, they reach all possible dimensions until that of the polynomial or the equation*

(74.) We represent such a polynomial as $[(u^A, x^{A}_{\prime})^B, y^{A}_{\prime\prime} \ldots n]^T$. In this expression, we understand that u does not exceed the degree A; x does not exceed the degree $A_{\prime}$; u and x, taken together, do not exceed a dimension higher than B; the other unknowns y, z, etc., of which there are $n-2$, cannot exceed the degrees $A_{\prime\prime}$, $A_{\prime\prime\prime}$, etc.. respectively; but when combined among each other as well as with the variables u and x, they reach all possible dimensions until T.

This polynomial is therefore characterized by

$$A < B; \underset{\prime\prime}{A} < B; B < T; \underset{\prime\prime}{A} < T; \underset{\prime\prime\prime}{A} < T, \text{ etc}$$

$$\underset{\prime}{A} + A > B; \underset{\prime\prime}{A} + B > T; \underset{\prime\prime\prime}{A} + B > T, \text{ etc.}$$

$$\underset{\prime\prime}{A} + \underset{\prime\prime\prime}{A} > T; \underset{\prime\prime\prime}{A} + \underset{IV}{A} > T; \underset{\prime\prime\prime}{A} + \underset{IV}{A} > T, \text{ etc.}$$

$$A + \underset{\prime\prime}{A} > T; A + \underset{\prime\prime\prime}{A} > T; \text{ etc.}$$

$$\underset{\prime}{A} + \underset{\prime\prime}{A} > T; \underset{\prime}{A} + \underset{\prime\prime\prime}{A} > T, \text{ etc.}$$

In order not to complicate our computations, we first determine the number of terms of this polynomial when $\underset{\prime\prime}{A} = \underset{\prime\prime\prime}{A} = \underset{IV}{A} = \text{etc.} = T$. It will then be easy to consider other cases.

PROBLEM XII

(75.) *Compute the value of* $N[(u^A, x^{\overset{A}{\prime}})^B, y \ldots n]^T$.

Let us assume this polynomial is ordered with respect to either u or x, with respect to x, for example. Denoting as s the exponent of x in any term of the polynomial, the form of all terms affected by x^s is given by the polynomial $(u^A, y \ldots n - 1)^{T-s}$; that is, any term will have the form $x^s(u^A, y \ldots n - 1)^{T-s}$ from $s = 0$ to $s + A = B$, since u and s cannot exceed the dimension B altogether.

Past $s + A = B$, that is, $s = B - A$, each term has the form $x^s(u^{B-s}, y \ldots n - 1)^{T-s}$ until $s = \underset{\prime}{A}$, since x must not exceed the degree $\underset{\prime}{A}$: And $\underset{\prime}{A}$ must satisfy $\underset{\prime}{A} > B - A$, which is true since our assumptions about the polynomial require $A + \underset{\prime}{A} > B$.

The question is to compute the sum of

1. $N(u^A, y \ldots n - 1)^{T-s}$ from $s = 0$ to $s = B - A$;

2. $N(u^{B-s}, y \ldots n - 1)^{T-s}$ from $s = B - A$ excluded to $s = \underset{\prime}{A}$.

However:

1. From (59), $N(u^A, y \ldots n - 1)^{T-s} = N(u \ldots n - 1)^{T-s} - N(u \ldots n - 1)^{T-A-s-1}$;

2. $N(u^{B-s}, y \ldots n - 1)^{T-s} = N(u \ldots n - 1)^{T-s} - N(u \ldots n - 1)^{T-B-1}$;

we therefore must compute the sum of

1. $N(u \ldots n - 1)^{T-s} - N(u \ldots n - 1)^{T-A-s-1}$ from $s = 0$ to $s = B - A$;

2. $N(u \ldots n - 1)^{T-s} - N(u \ldots n - 1)^{T-B-1}$ from $s = B - A$ excluded to $s = \underset{\prime}{A}$.

Thus from what was said in (70), we find

$$N[(u^A, x^{\overset{A}{\prime}})^B, y \ldots n]^T = N(u \ldots n)^T - N(u \ldots n)^{T-A-1} - N(u \ldots n)^{T-\overset{}{\underset{\prime}{A}}-1}$$
$$+N(u \ldots n)^{T-B-2} - N(u)^{A+\overset{}{\underset{\prime}{A}}-B-1} \times N(u \ldots n-1)^{T-B-1}.$$

COROLLARY

(76.) If $A + \underset{\prime}{A} < B$, then u and x would not reach the dimension B together; rather the highest dimension they can reach is $A + \underset{\prime}{A}$. The expression we just found would thus not be exact. But if we replace B by $A + \underset{\prime}{A}$, we obtain

$$N[(u^A, x^{\overset{A}{\prime}})^B, y \ldots n]^T = N(u \ldots n)^T - N(u \ldots n)^{T-A-1} - N(u \ldots n)^{T-\overset{}{\underset{\prime}{A}}-1}$$
$$+N(u \ldots n)^{T-A-\underset{\prime}{A}-2}$$

when $A + \underset{\prime}{A} < B$; this requires observing that $N(u)^{-1} = 0$, since in general

$$N(u)^{A+\underset{\prime}{A}-B-1} = A + \underset{\prime}{A} - B.$$

PROBLEM XIII

We ask for the value of $N((u^A, x^{\overset{A}{\prime}})^B, y^{\overset{A}{\prime\prime}} \ldots n)^T$, when this polynomial satisfies the conditions given in (74).

(77.) In this polynomial, all terms that may be divided by $y^{\overset{A+1}{\prime\prime}}$, $z^{\overset{A+1}{\prime\prime\prime}}$, etc., are missing. But under our assumptions, the absence of terms that may be divided by $y^{\overset{A+1}{\prime\prime}}$, for example, does not imply the absence of terms that may be divided by $z^{\overset{A+1}{\prime\prime\prime}}$; this argument is valid for all other unknowns; thus we conclude, as we have already done in (59), that the number of missing terms in y is $N(u \ldots n)^{T-\overset{}{\underset{\prime\prime}{A}}-1}$, that the number of missing terms in z, is $N(u \ldots n)^{T-\overset{}{\underset{\prime\prime\prime}{A}}-1}$, and so on. Therefore, and from what was said about the preceding problem, we conclude that

$$N[(u^A, x^{\overset{A}{\prime}})^B, y^{\overset{A}{\prime\prime}} \ldots n]^T = N(u \ldots n)^T - N(u \ldots n)^{T-A-1}$$
$$-N(u \ldots n)^{T-\overset{}{\underset{\prime}{A}}-1} - N(u \ldots n)^{T-\overset{}{\underset{\prime\prime}{A}}-1} - N(u \ldots n)^{T-\overset{}{\underset{\prime\prime\prime}{A}}-1}, \text{ etc.}$$
$$+N(u \ldots n)^{T-B-2} - N(u)^{A+\overset{}{\underset{\prime}{A}}-B-1} \times N(u \ldots n-1)^{T-B-1};$$

We abbreviate this expression by writing it as

$$N(u \ldots n)^T - N(u \ldots n)^{T-A-1}, \text{ etc.}$$
$$+N(u \ldots n)^{T-B-2} - N(u)^{A+\overset{}{\underset{\prime}{A}}-B-1} \times N(u \ldots n-1)^{T-B-1}.$$

Problem XIV

(78.) *Let*

$$\left[(u^{a'}, x^{\overset{d}{\prime}})^{b'}, y^{\overset{d'}{\prime\prime}} \ldots n \right]^{t'} = 0,$$

$$\left[(u^{a''}, x^{\overset{d''}{\prime}})^{b''}, y^{\overset{d''}{\prime\prime}} \ldots n \right]^{t''} = 0,$$

$$\left[(u^{a'''}, x^{\overset{d'''}{\prime}})^{b'''}, y^{\overset{d'''}{\prime\prime}} \ldots n \right]^{t'''} = 0, \ etc.,$$

an arbitrary number of n − 1 equations with n unknowns, satisfying the conditions indicated in (74). Let $[(u^A, x^{\overset{A}{\prime}})^B, y^{\overset{A}{\prime\prime}} \ldots n]^T$ *be a polynomial which not only satisfies these conditions, but is also such that the polynomial*

$$[(u^{A-a'-a''-a'''-etc.}, x^{\overset{A-d-d'-d''-etc.}{\prime}})^{B-b'-b''-b'''-etc.},$$
$$y^{\overset{A-d'-d''-d'''-etc.}{\prime\prime}} \ldots n]^{T-t'-t''-t'''-\ etc.}.$$

also satisfies these conditions: Using these equations, we ask how many terms can be cancelled in the polynomial $[(u^A, x^{\overset{A}{\prime}})^B, y^{\overset{A}{\prime\prime}} \ldots n]^T$, *without introducing new ones.*

Applying the previous argument (60) to this new type of polynomials and equations verbatim, the first equation can be used to cancel

$$N \left[(u^{A-a'}, x^{\overset{A-d}{\prime}})^{B-b'}, y^{\overset{A-d'}{\prime\prime}} \ldots n \right]^{T-t'}$$

terms; that using two equations, we can cancel

$$N \left[(u^{A-a'}, x^{\overset{A-d}{\prime}})^{B-b'}, y^{\overset{A-d'}{\prime\prime}} \ldots n \right]^{T-t'}$$

$$+N \left[(u^{A-a''}, x^{\overset{A-d''}{\prime}})^{B-b''}, y^{\overset{A-d''}{\prime\prime}} \ldots n \right]^{T-t''}$$

$$-N \left[(u^{A-a'-a''}, x^{\overset{A-d-d''}{\prime}})^{B-b'-b''}, y^{\overset{A-d'-d''}{\prime\prime}} \ldots n \right]^{T-t'-t''}$$

terms; that using three equations, we can cancel

$$N[(u^{A-a'}, x^{\overset{A-d}{\prime}})^{B-b'}, y^{\overset{A-d'}{\prime\prime}} \ldots n]^{T-t'}$$

$$+N[(u^{A-a''}, x^{\overset{A-d''}{\prime}})^{B-b''}, y^{\overset{A-d''}{\prime\prime}} \ldots n]^{T-t''}$$

$$-N[(u^{A-a'-a''}, x^{\overset{A-d-d''}{\prime}})^{B-b'-b''}, y^{\overset{A-d'-d''}{\prime\prime}} \ldots n]^{T-t'-t''}$$

$$+N \left[(u^{A-a'''}, x^{\overset{A-d'''}{\prime}})^{B-b'''}, y^{\overset{A-d'''}{\prime\prime}} \ldots n \right]^{T-t'''}$$

$$-N[(u^{A-a'-a''}, x^{\overset{A-d-d''}{\prime}})^{B-b'-b''}, y^{\overset{A-d'-d''}{\prime\prime}} \ldots n]^{T-t'-t'''}$$

$$-N[(u^{A-a'-a'''}, x^{\overset{A-d'-d'''}{\prime}})^{B-b''-b'''}, y^{\overset{A-d''-d'''}{\prime\prime}} \ldots n]^{T-t''-t'''}$$

$$+N[(u^{A-a'-a''-a'''}, x^{\overset{A-d-d'-d''}{\prime}})^{B-b'-b''-b'''},$$
$$y^{\overset{A-d'-d''-d'''}{\prime\prime}} \ldots n]^{T-t-t'-t'''}$$

terms, and so on.

PROBLEM XV

(79.) *We ask how many terms remain in the polynomial*
$[(u^A, x'^{\overset{A}{'}})^B, y''^{\overset{A}{'}} \ldots n]^T$ *after having cancelled all possible terms using* $n - 1$
equations of the same type as this polynomial.

From the previous problem, and keeping in mind that all polynomials
entering in the expressions we have found are of the same type, we easily see
that, if there is only one equation, the number of remaining terms is

$$d\left(N\left[(u^A, x'^{\overset{A}{'}})^B, y''^{\overset{A}{'}} \ldots n\right]^T\right) \ldots \left(\begin{array}{ccccc} T & : & B & : & A & : \\ t' & & -b' & & -a' & \\ & & A' & & A'' & \\ & & -\underset{'}{a'} & : & -\underset{''}{a'} & \text{etc.} \end{array}\right)$$

If there are two equations, the number of remaining terms is

$$dd(N\left((u^A, x'^{\overset{A}{'}})^B, y''^{\overset{A}{'}} \ldots n\right)^T \ldots \left(\begin{array}{ccccc} T & : & B & : & A & : \\ -t', -t'' & & -b', -b'' & & -a', -a'' & \\ & & A' & & A'' & \\ & & -\underset{'}{a'}, -\underset{'}{a''} & : & -\underset{''}{a'}, -\underset{''}{a''} & \text{etc.} \end{array}\right).$$

and that, in general, if there are $n - 1$ equations, the number of remaining
terms is

$$d^{n-1}\left(N\left[(u^A, x'^{\overset{A}{'}})^B, y''^{\overset{A}{'}} \ldots n\right]^T\right) \ldots \left(\begin{array}{c} T \\ -t', -t'', -t''', \text{ etc.} \end{array} :\right.$$

$$\begin{array}{ccccc} B & & A' & & A' \\ -b', -b'', -b''', \text{etc.} & : & -\underset{'}{a'}, -\underset{'}{a''}, -\underset{'}{a'''}, \text{etc.} & : & -\underset{'}{a'}, -\underset{'}{a''}, -\underset{'}{a'''}, \text{ etc.} & : \end{array}$$

$$\left.\begin{array}{c} A'' \\ -\underset{''}{a'}, -\underset{''}{a''}, -\underset{''}{a'''}, \text{etc.} \quad : \text{ etc.} \end{array}\right).$$

PROBLEM XVI

(80.) *We ask for the degree of the final equation resulting from an arbitrary
number of equations of the type* $([u^a, x'^{\overset{A}{'}}]^b, y''^{\overset{a}{'}} \ldots n)^t = 0$ *containing the same
number of unknowns.*

Assume we multiply one of these equations, say $[(u^a, x'^{\overset{A}{'}})^b, y''^{\overset{a}{'}} \ldots n]^t =$
0, by a polynomial $[(u^A, x'^{\overset{A}{'}})^B, y^A \ldots n]^T$ of the same type. The product
$[(u^{A+a}, x'^{\overset{A+A}{'}})^{B+b}, y''^{\overset{A+a}{'}} \ldots n]^{T+t}$ will be of the same type; thus using the

$n-1$ other equations, the number of terms to be cancelled is

$$d^{n-1}(N[(u^{A+a}, x'^{A+a}_{\prime})^{B+b}, y''^{A+a}_{\prime\prime}\ldots n]^{T+t})\ldots\left(\begin{array}{c} T+t \\ -t', -t'', -t''', \text{ etc.} \end{array}\right.:$$

$$\begin{array}{ccccc} B+b & & A+a & & A+A_{\prime}_{\prime} \\ -b', -b'', -b''', \text{etc.} & : & -a', -a'', -a''', \text{etc.} & : & -a'_{\prime}, -a''_{\prime}, -a'''_{\prime}, \text{etc.} \end{array} :$$

$$\begin{array}{cc} A_{\prime\prime}+a_{\prime\prime} & \\ -a'_{\prime\prime}, -a''_{\prime\prime}, -a'''_{\prime\prime}, \text{etc.} & : \text{etc.} \end{array}\left.\right).$$

Thus, denoting D as the degree of the final equation, the number of terms to be cancelled is

$$d^{n-1}(N[(u^{A+a}, x'^{A+a}_{\prime})^{B+b}, y''^{A+a}_{\prime\prime}\ldots n]^{T+t})\ldots\left(\begin{array}{c} T+t \\ -t', -t'', -t''', \text{ etc.} \end{array}\right.:$$

$$\begin{array}{ccccc} B+b & & A+a & & A+A_{\prime}_{\prime} \\ -b', -b'', -b''', \text{etc.} & : & -a', -a'', -a''', \text{etc.} & : & -a'_{\prime}, -a''_{\prime}, -a'''_{\prime}, \text{etc.} \end{array} :$$

$$\begin{array}{cc} A_{\prime\prime}+a_{\prime\prime} & \\ -a'_{\prime\prime}, -a''_{\prime\prime}, -a'''_{\prime\prime}, \text{etc.} & : \text{etc.} \end{array}\left.\right) - D - 1.$$

But from (79), there remain

$$d^{n-1}(N[(u^{A}, x'^{A}_{\prime})^{B}, y''^{A}_{\prime\prime}\ldots n]^{T})\ldots\left(\begin{array}{c} T \\ -t', -t'', -t''', \text{etc.} \end{array}\right.:$$

$$\begin{array}{ccccc} B & & A & & A_{\prime} \\ -b', -b'', -b''', \text{etc.} & : & -a', -a'', -a''', \text{etc.} & : & -a'_{\prime}, -a''_{\prime}, -a'''_{\prime}, \text{etc.} \end{array} :$$

$$\begin{array}{cc} A_{\prime\prime} & \\ -a'_{\prime\prime}, -a''_{\prime\prime}, -a'''_{\prime\prime}, \text{etc.} & : \text{etc.} \end{array}\left.\right)$$

terms in the polynomial multiplier, when all possible terms have been cancelled using the $n-1$ equations (without introducing new ones). We therefore obtain the following equation:

$$d^{n-1}(N[(u^A, x'^{\overset{A}{\prime}})^B, y''^{\overset{A}{\prime\prime}}\ldots n]^T)\ldots\left(\begin{array}{c} T \\ -t', -t'', -t''', \text{ etc.} \end{array} \right. :$$

$$\begin{array}{ccc} B & A & \overset{A}{\prime} \\ -b', -b'', -b''', \text{etc.} & \overset{:}{} \quad -a', -a'', -a''', \text{etc.} & \overset{:}{} \quad -\underset{\prime}{a}', -\underset{\prime}{a}'', -\underset{\prime}{a}''', \text{ etc.} \end{array} \quad :$$

$$\left. \begin{array}{cc} & \overset{A}{\prime\prime} \\ & -\underset{\prime\prime}{a}', -\underset{\prime\prime}{a}'', -\underset{\prime\prime}{a}''', \text{etc.} \quad \text{etc.} \end{array} \right) - 1$$

$$= d^{n-1}(N[(u^{A+a}, x'^{\overset{A+a}{\prime}})^{B+b}, y''^{\overset{A+a}{\prime\prime}}\ldots n]^{T+t})\ldots\left(\begin{array}{c} T+t \\ -t', -t'', -t''', \text{ etc.} \end{array} \right. :$$

$$\begin{array}{ccc} B+b & A+a & \overset{A+a}{\prime} \\ -b', -b'', -b''', \text{etc.} & \overset{:}{} \quad -a', -a'', -a''', \text{etc.} & \overset{:}{} \quad -\underset{\prime}{a}', -\underset{\prime}{a}'', -\underset{\prime}{a}''', \text{ etc.} \end{array} \quad :$$

$$\left. \begin{array}{cc} & \overset{A+a}{\underset{\prime\prime}{\prime\prime}} \\ & -\underset{\prime\prime}{a}', -\underset{\prime\prime}{a}'', -\underset{\prime\prime}{a}''', \text{etc.} \quad : \text{ etc.} \end{array} \right) - D - 1.$$

Thus

$$D = d^{n}(N[(u^{A+a}, x'^{\overset{A+A}{\prime}})^{B+b}, y''^{\overset{A+a}{\prime\prime}}\ldots n]^{T+t})\ldots\left(\begin{array}{c} T+t \\ -t', -t'', -t''', \text{ etc.} \end{array} \right. :$$

$$\begin{array}{cc} B+b & A+a \\ -b', -b'', -b''', \text{etc.} & \overset{:}{} \quad -a, -a', -a'', -a''', \text{etc.} \end{array} \quad :$$

$$\left. \begin{array}{cc} \overset{A+A}{\underset{\prime}{\prime}} & \overset{A+a}{\underset{\prime\prime}{\prime}} \\ -\underset{\prime}{A}, -\underset{\prime}{a}', -\underset{\prime}{a}'', -\underset{\prime}{a}''', \text{ etc.} \quad \overset{:}{} \quad -\underset{\prime\prime}{a}, -\underset{\prime\prime}{a}', -\underset{\prime\prime}{a}'', -\underset{\prime\prime}{a}''', \text{etc.} \quad : \text{ etc.} \end{array} \right).$$

Thus by differentiating the value found in (75) for
$N[(u^{A+a}, x'^{\overset{A+a}{\prime}})^{B+b}, y''^{\overset{A+a}{\prime\prime}}\ldots n]^{T+t}$ n times, we finally obtain

$$D = tt't''t''' \text{ etc.} - (t-a).(t'-a').(t''-a'').(t'''-a'''). \text{ etc.}$$
$$-(t-\underset{\prime}{A}).(t'-\underset{\prime}{a}').(t''-\underset{\prime}{a}'').(t'''-\underset{\prime}{a}'''). \text{ etc.}$$
$$-(t-\underset{\prime\prime}{a}).(t'-\underset{\prime\prime}{a}').(t''-\underset{\prime\prime}{a}'').(t'''-\underset{\prime\prime}{a}'''). \text{ etc.}$$
$$-(t-\underset{\prime\prime\prime}{a}).(t'-\underset{\prime\prime\prime}{a}').(t''-\underset{\prime\prime\prime}{a}'').(t'''-\underset{\prime\prime\prime}{a}'''). \text{ etc.}$$
$$- \text{ etc.}$$
$$+(t-b).(t'-b').(t''-b'').(t'''-b'''). \text{ etc.}$$
$$-(a+\underset{\prime}{a}-b).(t'-b').(t''-b'').(t'''-b'''). \text{ etc.}$$
$$-(a'+\underset{\prime}{a}'-b').(t-b).(t''-b'').(t'''-b'''). \text{ etc.}$$
$$-(a''+\underset{\prime}{a}''-b'').(t-b).(t'-b').(t'''-b'''). \text{ etc.}$$
$$-(a'''+\underset{\prime}{a}'''-b''').(t-b).(t'-b').(t''-b''). \text{ etc.}$$

In this expression, the number of factors in each product is always equal to the number of unknowns.

(81.) Assume $b = t$, $b' = t'$, $b'' = t''$, etc. We then obtain

$$D = tt't''t'''.\ \text{etc.} - (t - a).(t' - a').(t'' - a'').(t''' - a''').\ \text{etc.}$$
$$-(t - \underset{\prime}{a}).(t' - \underset{\prime}{a}').(t'' - \underset{\prime}{a}'').(t''' - \underset{\prime}{a}''').\ \text{etc.}$$
$$-(t - \underset{\prime\prime}{a}).(t' - \underset{\prime\prime}{a}').(t'' - \underset{\prime\prime}{a}'').(t''' - \underset{\prime\prime}{a}''').\ \text{etc.}$$
$$-(t - \underset{\prime\prime\prime}{a}).(t' - \underset{\prime\prime\prime}{a}').(t'' - \underset{\prime\prime\prime}{a}'').(t''' - \underset{\prime\prime\prime}{a}''').\ \text{etc.}$$

which agrees with what was found earlier in (62).

If there are two unknowns only, then we necessarily have $b = t$ and $b' = t'$, and consequently $D = tt' - (t - a).(t' - a') - (t - \underset{\prime}{a}).(t' - \underset{\prime}{a}')$, as was found earlier in (62).

If there are only three unknowns, we get

$$D = tt't'' - (t - a).(t' - a').(t'' - a'') - (t - \underset{\prime}{a}).(t' - \underset{\prime}{a}').(t'' - \underset{\prime}{a}'')$$
$$-(t - \underset{\prime\prime}{a}).(t' - \underset{\prime\prime}{a}').(t'' - \underset{\prime\prime}{a}'') + (t - b).(t' - b').(t'' - b'')$$
$$-(a + \underset{\prime}{a} - b).(t' - b').(t'' - b'') - (a' + \underset{\prime}{a}' - b').(t - b).(t'' - b'')$$
$$-(a'' + \underset{\prime}{a}'' - b'').(t - b).(t' - b').$$

This is the expression for the degree of the final equation resulting from three equations of the form $[(x^a, y^{\overset{a}{\prime}})^b, z^{\overset{a}{\prime\prime}}]^t = 0$.

About incomplete polynomials and equations, in which three of the unknowns share the following characteristics:

(1) The degree of each unknown does not exceed a given value, different or the same for each;

(2) The combination of two unknowns does not exceed a given dimension, different or the same for each combination of two of these three unknowns;

(3) The combination of the three unknowns does not exceed a given dimension.

We further assume that the degrees of the $n - 3$ other unknowns do not exceed given values; we also assume that the combination of two, three, four, etc. of these variables among themselves or with the first three reaches all possible dimensions, up to the dimension of the polynomial

(82.) We have encountered only one form of the polynomial multiplier until now, and therefore a unique expression for the degree of the final equation. This is not the case anymore, when considering more general problems. We will see that the polynomial multiplier can take more than one form, and that the expression of the degree of the final equation is not unique. But we do not know anything yet about the right way to deal with such different forms; thus we will operate as much as possible by analogy with what was done until now. The kind of polynomial under consideration is represented by the expression $([(u^A, x'^{A})^B, (u^A, y''^{A})^{B'}, (x'^{A}, y''^{A})^{B''}]^C, z'''^{A} \dots n)^T$ This expression means the three unknowns u, x, and y satisfy the following:

1. u does not exceed the degree A, x does not exceed the degree A', y does not exceed the degree A''.

2. The combination of u and x does not exceed the dimension B, u and y do not exceed the dimension B', x and y do not exceed the dimension B'';

3. The combination of u and x and y together cannot exceed a dimension larger than C.

Each of the $n - 3$ other unknowns does not exceed a given degree; for example, z does not exceed the degree A''', u' does not exceed the degree A_{IV}, x' does not exceed the degree A_V, but the combination of two, three, four, etc. of the variables z, u', x', etc. among themselves as well as with u, x, and y reaches all possible dimensions up to T, which is the degree of the polynomial.

PROBLEM XVII

(83.) *We ask for the expression of general conditions to guarantee the existence of the polynomial*

$$([(u^A, x'^{A})^B, (u^A, y''^{A})^{B'}, (x'^{A}, y''^{A})^{B''}]^C, z'''^{A} \dots n)^T.$$

There are three kinds of conditions; the first kind deals with the exponents taken separately; the second kind deals with combinations of two exponents; the third deals with combinations of three exponents.

Consider first exponents taken separately; the conditions are $A < B$; $A' < B$; $A'' < B$; $A' < B$; $A'' < B$; $A''' < B$; $A < T$; $A_{IV} < T$, etc.

Consider then combinations of two exponents; the conditions are $B < C$; $B' < C$; $B'' < C$; $A + A' > B$; $A + A'' > B$; $A' + A'' > B$; $A + A_{IV} > T$; $A + A_{IV} > T$; etc.; $A' + A''' > T$; $A' + A_{IV} > T$; etc.; $A'' + A''' > T$; $A'' + A_{IV} > T$; etc.; $A''' + A_{IV} > T$; $A''' + A_V > T$; etc.; $A_{IV} + A_V > T$; etc.

All these conditions are obviously necessary; indeed, assume we have $A + A' < B$. It is clear that u and x cannot not reach the dimension B together, and this is contrary to our assumptions.

Consider finally combinations of three exponents; the conditions are $C < T$; $A + B'' > C$; $A' + B' > C$; $A'' + B > C$; $A + B''' > T$; $A''' + B > T$; $A''' + B'' > T$;

$\underset{IV}{A} + B > T;\ \underset{IV}{A} + \underset{\prime}{B} > T;\ \underset{IV}{A} + \underset{\prime\prime}{B} > T$, etc.. $\underset{\prime}{B} + \underset{\prime\prime}{B} + B > 2C$. From these conditions, only $\underset{\prime}{B} + B + \underset{\prime\prime}{B} > 2C$ needs justifying: In order for u, x and y to reach the dimension C, the sum of the three lowest degrees of u, x and y when they find themselves in a term of dimension C must be less than the highest dimension where these three letters can find themselves together; this necessary condition is obvious. The lowest degree of u in a term of dimension C is $C - \underset{\prime\prime}{B}$; that of x is $C - \underset{\prime}{B}$; that of y is $C - B$; thus

$$C - \underset{\prime\prime}{B} + C - \underset{\prime}{B} + C - B < C, \text{ or } \underset{\prime}{B} + B + \underset{\prime\prime}{B} > 2C.$$

PROBLEM XVIII

(84.) *We ask for the value of*

$$N([(u^A, x^{\overset{A}{\prime}})^B, (u^A, y^{\overset{A}{\prime\prime}})^{\overset{B}{\prime}}, (x^{\overset{A}{\prime}}, y^{\overset{A}{\prime\prime}})^{\overset{B}{\prime\prime}}]^C, z^{\overset{A}{\prime\prime\prime}} \dots n)^T$$

when this polynomial satisfies the conditions mentioned in (83).
We first look for the value of

$$N([(u^A, x^{\overset{A}{\prime}})^B, (u^A, y^{\overset{A}{\prime\prime}})^{\overset{B}{\prime}}, (x^{\overset{A}{\prime}}, y^{\overset{A}{\prime\prime}})^{\overset{B}{\prime\prime}}]^C, z^{\overset{A}{\prime\prime\prime}} \dots n)^T;$$

that is, we assume $\underset{\prime\prime\prime}{A} = \underset{IV}{A} =$ etc. $= T$. It will then be easy, as we have seen in (77), to conclude about the value of these quantities when these other unknowns have the exponents $\underset{\prime\prime\prime}{A}$, $\underset{IV}{A}$, etc.

Let us assume the polynomial

$$([(u^A, x^{\overset{A}{\prime}})^B, (u^A, y^{\overset{A}{\prime\prime}})^{\overset{B}{\prime}}, (x^{\overset{A}{\prime}}, y^{\overset{A}{\prime\prime}})^{\overset{B}{\prime\prime}}]^C, z \dots n)^T$$

is ordered with respect to any one of the arbitrary unknowns u, x, y, with respect to y, for example; denoting s the exponent of y in an arbitrary term, we see each term can be written as $y^s \times [(u^A, x^{\overset{A}{\prime}})^B, z \dots n - 1]^{T-s}$ from $s = 0$ to s reaches the smallest of the values provided by the three equations $s + A = \underset{\prime}{B}$, $s + \underset{\prime}{A} = \underset{\prime\prime}{B}$, $s + B = C$, that is, until s is equal to the smallest of the three quantities $\underset{\prime}{B} - A$; $\underset{\prime\prime}{B} - \underset{\prime}{A}$; $C - B$. This is obvious, since u and y, for example, cannot exceed, together, the dimension $\underset{\prime}{B}$, as soon as $s + A$ becomes equal to B'; indeed s keeps increasing and thus x cannot bear the exponent $\underset{\prime}{A}$ anymore, and consequently the expression $y^s \times ([u^A, x^{\overset{A}{\prime}}]^B, z \dots n - 1)^{T-s}$ must change.

We must therefore consider the following six cases:

$$C - B \;<\; \underset{\prime}{B} - A \;<\; \underset{\prime\prime}{B} - \underset{\prime}{A},$$

$$C - B \;<\; \underset{\prime\prime}{B} - \underset{\prime}{A}\,b \;<\; \underset{\prime}{B} - A,$$

$$\underset{\prime}{B} - A \;<\; C - B \;<\; \underset{\prime\prime}{B} - \underset{\prime}{A},$$

$$\underset{\prime}{B} - A \;<\; \underset{\prime\prime}{B} - \underset{\prime}{A} \;<\; C - B,$$

$$\underset{\prime\prime}{B} - \underset{\prime}{A} \;<\; C - B \;<\; \underset{\prime}{B} - A,$$

$$\underset{\prime\prime}{B} - \underset{\prime}{A} \;<\; \underset{\prime}{B} - A \;<\; C - B.$$

First case
$$C - B < \underset{\prime}{B} - A < \underset{\prime\prime}{B} - \underset{\prime}{A}.$$

(85.) In this case, the form $y^s([u^A, x^{\overset{A}{\prime}}]^B, z \ldots n-1)^{T-s}$ occurs from $s = 0$ to $s = C - B$.

Past $s = C - B$, it is $y^s([u^A, x^{\overset{A}{\prime}}]^{C-s}, z \ldots n - 1)^{T-s}$, until $s = \underset{\prime}{B} - A$.

Past $s = \underset{\prime}{B} - A$, it is $y^s([u^{\overset{B-s}{\prime}}, x^{\overset{A}{\prime}}]^{C-s}, z \ldots n - 1)^{T-s}$, until $s = \underset{\prime\prime}{B} - \underset{\prime}{A}$.

Past $s = \underset{\prime\prime}{B} - \underset{\prime}{A}$, it is $y^s([u^{\overset{B-s}{\prime}}, x^{\overset{B-s}{\prime\prime}}]^{C-s}, z \ldots n - 1)^{T-s}$, until $s = \underset{\prime\prime}{A}$.

We therefore must compute the sum of

1. $N([u^A, x^{\overset{A}{\prime}}]^B, z \ldots n - 1)^{T-s}$ from $s = 0$ to $s = C - B$;

2. $N([u^A, x^{\overset{A}{\prime}}]^{C-s}, z \ldots n-1)^{T-s}$ from $s = C - B$ excluded to $s = \underset{\prime}{B} - A$;

3. $N([u^{\overset{B-s}{\prime}}, x^{\overset{A}{\prime}}]^{C-s}, z \ldots n - 1)^{T-s}$ from $s = \underset{\prime}{B} - A$ excluded to $s = \underset{\prime\prime}{B} - \underset{\prime}{A}$;

4. $N([u^{\overset{B-s}{\prime}}, x^{\overset{B-s}{\prime\prime}}]^{C-s}, z \ldots n-1)^{T-s}$ from $s = \underset{\prime\prime}{B}-\underset{\prime}{A}$ excluded, to $s = \underset{\prime\prime}{A}$.

But we have seen earlier in (75 and 76) that the value of $N([u^A, x^{\overset{A}{\prime}}]^B, z \ldots n - 1)^T$ (this expression contains all four we just mentioned) can take two different expressions, depending upon whether $A + \underset{\prime}{A} > B$ or $A + \underset{\prime}{A} < B$; thus we must determine first which one of these two cases occurs in each of the four expressions.

In the first expression, we have $A + \underset{\prime}{A} > B$ from the hypothesis itself; thus from $s = 0$ to $s = C - B$, we have from (75)

1. $N([u^A, x'^{\overset{A}{'}}]^B, z \ldots n - 1)^{T-s} = N(u \ldots n - 1)^{T-s}$
$- N(u \ldots n-1)^{T-A-s-1} - N(u \ldots n-1)^{T-\overset{A}{'}-s-1} + N(u \ldots n-1)^{T-B-s-2} -$
$N(u)^{A+\overset{A}{'}-B-1} \times N(u \ldots n - 2)^{2T-B-s-1}.$

In the second expression, we have $A + \underset{'}{A} > C - s$ from $s = C - B$ to

$s = \underset{'}{B} - A.$

Indeed, for this condition to occur, we must have $s > C - A - \underset{'}{A}$; however,

the smallest value of s is $C - B$ by hypothesis; we must therefore have
$C - B > C - A - \underset{'}{A}$ or $A + \underset{'}{A} > B$, which is true by hypothesis. Anyway,

it is easy to see a priori that $A + \underset{'}{A} > C - s$, since $A + \underset{'}{A} > B$, and $C - s$ is

smaller than B.

Thus, from $s = C - B$ to $s = B - A$, we have

2. $N([u^A, x'^{\overset{A}{'}}]^{C-s}, z \ldots n - 1)^{T-s} = N(u \ldots n - 1)^{T-s} - N(u \ldots n -$
$1)^{T-A-s-1} - N(u \ldots n - 1)^{T-\overset{A}{'}-s-1}$
$+ N(u \ldots n - 1)^{T-C-2} - N(u)^{A+\overset{A}{'}-C+s-1} \times N(u \ldots n - 2)^{T-C-1}.$

In the third expression, we have $\underset{'}{B} - s + \underset{'}{A} > C - s$ or $\underset{'}{B} + \underset{'}{A} > C$; this is

true from the general conditions justifying the existence of the polynomial.
Thus, from $s = \underset{'}{B} - A$ to $s = \underset{''}{B} - \underset{'}{A}$, we have

3. $N([u'^{\overset{B-s}{'}}, y'^{\overset{A}{'}}]^{C-s}, z \ldots n - 1)^{T-s} = N(u \ldots n - 1)^{T-s} - N(u \ldots n -$
$1)^{T-B-1}_{'} - N(u \ldots n - 1)^{T-\overset{A}{'}-s-1} + N(u \ldots n - 1)^{T-C-2}$
$- N(u)^{A+B-C-1}_{'\ '} \times N(u \ldots n - 2)^{T-C-1}.$

In the fourth expression, it may happen that $\underset{'}{B} - s + \underset{''}{B} - s > C - s$ or

that $\underset{'}{B} - s + \underset{''}{B} - s < C - s$, that is, $\underset{'}{B} + \underset{''}{B} - C > s$ or $\underset{'}{B} + \underset{''}{B} - C < s$; to

know the circumstances under which either one occurs, we remember that
the final value of s is $\underset{''}{A}$; the first case therefore occurs if $\underset{'}{B} + \underset{''}{B} - C > \underset{''}{A}$,

and the second case occurs if $\underset{'}{B} + \underset{''}{B} - C < \underset{''}{A}$.

We therefore face two cases, that is, $\underset{'}{B} + \underset{''}{B} - C > \underset{''}{A}$ and $\underset{'}{B} + \underset{''}{B} - C < \underset{''}{A}$,

or $\underset{''}{B} - \underset{''}{A} > C - \underset{'}{B}$ and $\underset{''}{B} - \underset{''}{A} < C - \underset{'}{B}$.

In the first case, we therefore have

4. $N[(u'^{\overset{B-s}{'}}, x''^{\overset{B-s}{''}})^{C-s}, z \ldots n - 1]^{T-s} = N(u \ldots n - 1)^{T-s} - N(u \ldots n -$
$1)^{T-B-1}_{'} - N(u \ldots n-1)^{T-B-1}_{''} + N(u \ldots n-1)^{T-C-2} - N(u)^{B+B-C-s-1}_{'\ ''} \times$
$N(u \ldots n - 2)^{T-C-1}$ from $s = \underset{'}{B} - \underset{'}{A}$ to $s = \underset{''}{A}$.

In the second case, this expression occurs from $s = \underset{''}{B} - \underset{'}{A}$ to $s = \underset{'}{B} + \underset{''}{B} - C$;

from $s = \underset{'}{B} + \underset{''}{B} - C$ excluded to $s = \underset{''}{A}$, we use, by (76), the expression

5. $N[(u_{'}^{B-s}, x_{''}^{B-s})^{C-s}, z \ldots n-1]^{T-s} = N(u \ldots n-1)^{T-s} - N(u \ldots n-1)_{'}^{T-B-1} - N(u \ldots n-1)_{''}^{T-B-1} + N(u \ldots n-1)_{'\ ''}^{T-B-B+s-2}$; summing these quantities within the appropriate intervals, and accounting for what was said in (73) we have that if

$$\underset{''}{B} - \underset{''}{A} > C - \underset{'}{B}$$

$$N([(u^A, x_{'}^A)^B, (u^A, y_{''}^A)_{'}^B, (x_{'}^A, y_{''}^A)_{''}^B]^C, z \ldots n)^T$$
$$= N(u \ldots n)^T - N(u \ldots n)^{T-A-1} - N(u \ldots n)_{'}^{T-A-1}$$
$$- N(u \ldots n)_{''}^{T-A-1} + N(u \ldots n)^{T-B-2} + N(u \ldots n)_{'}^{T-B-2}$$
$$+ N(u \ldots n)_{''}^{T-B-2} - N(u)_{'}^{A+A-B-1} \times N(u \ldots n-1)^{T-B-1}$$
$$- N(u)_{''\ '}^{A+A-B-1} \times N(u \ldots n-1)_{'}^{T-B-2}$$
$$- N(u)_{'\ ''\ ''}^{A+A-B-1} \times N(u \ldots n-1)_{''}^{T-B-1}$$
$$- N(u \ldots n)^{T-C-3} + N(u)_{'\ ''\ ''}^{A+A+A-C-1} \times N(u \ldots n-1)^{T-C-2}$$
$$+ (N(u \ldots 2)_{'}^{A+A-B-1} + N(u \ldots 2)_{'\ ''\ ''}^{B+B-A-C-2}$$
$$- N(u)_{'\ '}^{A+B-C-1} \times N(u)_{''}^{A+B-C-1}) \times N(u \ldots n-2)^{T-C-1},$$

and if $\underset{''}{B} - \underset{''}{A} < C - \underset{'}{B}$

$$N([(u^A, x_{'}^A)^B, (u^A, y_{''}^A)_{'}^B, (x_{'}^A, y_{''}^A)_{''}^B]^C, z \ldots n)^T$$
$$= N(u \ldots n)^T - N(u \ldots n)^{T-A-1} - N(u \ldots n)_{'}^{T-A-1}$$
$$- N(u \ldots n)_{''}^{T-A-1} + N(u \ldots n)^{T-B-2} + N(u \ldots n)_{'}^{T-B-2}$$
$$+ N(u \ldots n)_{''}^{T-B-2} - N(u)_{'}^{A+A-B-1} \times N(u \ldots n-1)^{T-B-1}$$
$$- N(u)_{''\ '}^{A+A-B-1} \times N(u \ldots n-1)_{'}^{T-B-1}$$
$$- N(u)_{'\ ''\ ''}^{A+A-B-1} \times N(u \ldots n-1)_{''}^{T-B-1}$$
$$+ N(u \ldots n)_{''\ ''\ ''}^{T+A-B-B-2} - N(u \ldots n)^{T-C-3} - N(u \ldots n)^{T-C-2}$$
$$+ N(u)_{'\ '\ ''}^{A+A+B+B-2C-1} \times N(u \ldots n-1)^{T-C-2}$$
$$+ (N(u \ldots 2)_{'}^{A+A-B-1} - N(u)_{'\ '}^{A+B-C-1} \times N(u)_{''}^{A+B-C-1})$$
$$\times N(u \ldots n-2)^{T-C-1}.$$

Second case

$$C - B < \underset{''}{B} - \underset{''}{A} < \underset{'}{B} - A.$$

(86.) This second case, like its predecessor, must be subdivided into the following two other cases: $\underset{''}{B} - \underset{''}{A} > C - \underset{'}{B}$ and $\underset{''}{B} - \underset{''}{A} < C - \underset{'}{B}$; but since it does not otherwise differ from the preceding case except by replacing $\underset{\prime\prime}{B}$ by $\underset{\prime}{B}$, $\underset{\prime}{A}$ by A, and vice versa, and that this permutation does not change the number of terms, it follows that the two values we have found for the preceding case are applicable to this one as well.

Third case

$$\underset{\prime}{B} - A < C - B < \underset{\prime\prime}{B} - \underset{\prime}{A}.$$

(87.) In this case the expression $y^s([u^A, x^{\overset{A}{\prime}}]^B, z \ldots n-1)^{T-s}$ occurs from $s = 0$ to $s = \underset{\prime}{B} - A$.

Past $s = \underset{\prime}{B} - A$, the expression is $y^s([u^{\overset{B-s}{\prime}}, x^{\overset{A}{\prime}}]^B, z \ldots n-1)^{T-s}$, until $s = C - B$.

Past $s = C - B$, it is $y^s([u^{\overset{B-s}{\prime}}, x^{\overset{A}{\prime}}]^{C-s}, z \ldots n-1)^{T-s}$, until $s = \underset{\prime\prime}{B} - \underset{\prime}{A}$.

Past $s = \underset{\prime\prime}{B} - \underset{\prime}{A}$, it is $y^s([u^{\overset{B-s}{\prime}}, x^{\overset{B-s}{\prime\prime}}]^{C-s}, z \ldots n-1)^{T-s}$, until $s = \underset{\prime\prime}{A}$.

Hence, since $A + \underset{\prime}{A} > B$, we have

$$
N([u^A, x^{\overset{A}{\prime}}]^B, z \ldots n-1)^{T-s} = N(u \ldots n-1)^{T-s} - N(u \ldots n-1)^{T-A-s-1}
$$
$$
-N(u \ldots n-1)^{\overset{T-A-s-1}{\prime}} + N(u \ldots n-1)^{T-B-s-2}
$$
$$
-N(u)^{\overset{A+A-B-1}{\prime}} \times N(u \ldots n-2)^{T-B-s-1}.
$$

From $s = \underset{\prime}{B} - A$ excluded, to $s = C - B$, we have $\underset{\prime}{B} - s + \underset{\prime}{A} > B$ or $s < \underset{\prime}{A} + \underset{\prime}{B} - B$. Indeed the largest value of s in that interval is $C - B$ and $C - B < \underset{\prime}{A} + \underset{\prime}{B} - B$ since we have $\underset{\prime}{A} + \underset{\prime}{B} > C$ from the conditions of existence of the polynomial (83).

Thus from $s = \underset{\prime}{B} - A$ excluded, to $s = C - B$, we have

$$
N([(u^{\overset{B-s}{\prime}}, x^{\overset{A}{\prime}})^B, z \ldots n-1]^{T-s}) = N(u \ldots n-1)^{T-s}
$$
$$
-N(u \ldots n-1)^{\overset{T-B-1}{\prime}} - N(u \ldots n-1)^{\overset{T-A-s-1}{\prime}} + N(u \ldots n-1)^{T-B-s-2}
$$
$$
-N(u)^{\overset{A+B-B-s-1}{\prime\prime}} \times N(u \ldots n-2)^{T-B-s-1}.
$$

From $s = C - B$ excluded to $s = \underset{\prime\prime}{B} - \underset{\prime}{A}$, we have $\underset{\prime}{B} - s + \underset{\prime}{A} > C - s$, that is, $\underset{\prime}{B} + \underset{\prime}{A} > C$. Thus

$$
N([(u^{\overset{B-s}{\prime}}, x^{\overset{A}{\prime}})^{C-s}, z \ldots n-1]^{T-s}) = N(u \ldots n-1)^{T-s}
$$
$$
-N(u \ldots n-1)^{\overset{T-B-1}{\prime}} - N(u \ldots n-1)^{\overset{T-A-s-1}{\prime}} + N(u \ldots n-1)^{T-C-2}
$$
$$
-N(u)^{\overset{A+B-C-1}{\prime\prime}} \times N(u \ldots n-2)^{T-C-1}.
$$

From $s = \underset{''}{B} - \underset{''}{A}$ excluded to $s = \underset{'}{A}$ we have $\underset{''}{B} - s + \underset{'}{B} - s > C - s$, when $\underset{''}{B} - \underset{''}{A} > C - \underset{'}{B}$; but if $\underset{''}{B} - \underset{''}{A} < C - \underset{'}{B}$, the inequality $\underset{'}{B} - s + \underset{''}{B} - s > C - s$ holds only from $s = \underset{''}{B} - \underset{'}{A}$ to $s = \underset{'}{B} + \underset{''}{B} - C$; past $s = \underset{'}{B} + \underset{''}{B} - C$, we have $\underset{'}{B} - s + \underset{''}{B} - s < C - s$.

Thus if $\underset{''}{B} - \underset{''}{A} > C - \underset{'}{B}$, we have

$$N([(u^{\overset{B-s}{'}}, x^{\overset{B-s}{''}})^{C-s}, z \dots n - 1]^{T-s}) = N(u \dots n - 1)^{T-s}$$
$$-N(u \dots n - 1)^{T-B-1} - N(u \dots n - 1)^{\overset{T-B-1}{''}} + N(u \dots n - 1)^{T-C-2}$$
$$-N(u)^{\overset{B+B-C-s-1}{'}{''}} \times N(u \dots n - 2)^{T-C-1},$$

until $s = \underset{''}{A}$.

But if $\underset{''}{B} - \underset{''}{A} < C - \underset{'}{B}$, this expression holds true only until $s = \underset{'}{B} + \underset{''}{B} - C$; past this term, we have

$$N([(u^{\overset{B-s}{'}}, x^{\overset{B-s}{''}})^{C-s}, z \dots n - 1]^{T-s}) = N(u \dots n - 1)^{T-s}$$
$$-N(u \dots n - 1)^{\overset{T-B-1}{'}} - N(u \dots n - 1)^{\overset{T-B-1}{''}}$$
$$+N(u \dots n - 1)^{\overset{T-B-B+s-2}{'}{''}}.$$

Adding these different quantities over their relevant intervals, we find that, if $\underset{''}{B} - \underset{''}{A} > C - \underset{'}{B}$, we have

$$N([(u^A, x^{\overset{A}{'}})^B, (u^A, y^{\overset{A}{''}})^{\overset{B}{'}}, (x^{\overset{A}{'}}, y^{\overset{A}{''}})^{\overset{B}{''}}]^C, z \dots n)^{T-s}$$
$$= N(u \dots n)^T - N(u \dots n)^{T-A-1} - N(u \dots n)^{\overset{T-A-1}{'}} - N(u \dots n)^{\overset{T-A-1}{''}}$$
$$+N(u \dots n)^{T-B-2} + N(u \dots n)^{\overset{T-B-2}{'}} + N(u \dots n)^{\overset{T-B-2}{''}}$$
$$-N(u)^{\overset{A+A-B-1}{'}} \times N(u \dots n - 1)^{T-B-1}$$
$$-N(u)^{\overset{A+A-B-1}{''}{'}} \times N(u \dots n - 1)^{\overset{T-B-1}{'}}$$
$$-N(u)^{\overset{A+A-B-1}{'}{''}{''}} \times N(u \dots n - 1)^{\overset{T-B-1}{''}} + N(u \dots n)^{\overset{T+A-B-B-2}{'}}$$
$$+N(u)^{\overset{A+A+B+B-2C-1}{'}{'}} \times N(u \dots n - 1)^{T-C-2}$$
$$-N(u \dots n)^{T-C-3} - N(u \dots n)^{T-C-2}$$
$$+[N(u \dots 2)^{\overset{B+B-A-C-2}{'}{''}{''}} - N(u \dots 2)^{\overset{B+A-C-2}{'}{'}}$$
$$-N(u)^{\overset{A+B-C-1}{'}{'}} \times N(u)^{\overset{B+B-A-C-1}{''}{''}}] \times N(u \dots n - 2)^{T-C-1}.$$

And if $\underset{''}{B} - \underset{''}{A} < C - \underset{'}{B}$, we have

$$N([(u^A, \overset{A}{x'})^B, (u^A, \overset{A}{y''})^{\overset{B}{'}}, (\overset{A}{x'}, \overset{A}{y''})^{\overset{B}{''}}]^C, z \ldots n)^T$$

$$= N(u \ldots n)^T - N(u \ldots n)^{T-A-1} - N(u \ldots n)^{\overset{T-A-1}{'}} - N(u \ldots n)^{\overset{T-A-1}{''}}$$

$$+ N(u \ldots n)^{T-B-2} + N(u \ldots n)^{\overset{T-B-2}{'}} + N(u \ldots n)^{\overset{T-B-2}{''}}$$

$$- N(u)^{\overset{A+A-B-1}{'}} \times N(u \ldots n-1)^{T-B-1}$$

$$- N(u)^{\overset{A+A-B-1}{'}} \times N(u \ldots n-1)^{\overset{T-B-1}{'}}$$

$$- N(u)^{\overset{A+A-B-1}{'}} \times N(u \ldots n-1)^{\overset{T-B-1}{''}} + N(u \ldots n)^{\overset{T+A-B-B-2}{'}}$$

$$+ N(u \ldots n)^{\overset{T+A-B-B-2}{''}} - N(u \ldots n)^{T-C-3} - 2N(u \ldots n)^{T-C-2}$$

$$+ N(u)^{\overset{A+2B+B+B-3C-1}{'}} \times N(u \ldots n-2)^{T-C-2}$$

$$- [N(u \ldots 2)^{\overset{A+B-C-2}{'}} + N(u)^{\overset{B+B-A-C-1}{'}} \times N(u)^{\overset{A+B-C-1}{'}}]$$

$$\times N(u \ldots n-2)^{T-C-1}.$$

We observe here that these two results do not directly arise from what was said in (73); rather, they have been simplified according to the following ideas:

Applying immediately what was said in (73), we find results such as $(n-1) \times N(u \ldots n)^{T-C-3} + N(u)^{\overset{A+B-T}{'}} \times N(u \ldots n-1)^{T-C-2}$.

The latter term is nothing but

$$-(T - \underset{'}{A} - \underset{'}{B} - 1) \times N(u \ldots n-1)^{T-C-2}$$

$$= -(T - C - 2) \times N(u \ldots n-1)^{T-C-2}$$
$$+ (\underset{'}{A} + \underset{'}{B} - C - 1) \times N(u \ldots n-1)^{T-C-2}.$$

But $-(T - C - 2) \times N(u \ldots n-1)^{T-C-2} = -nN(u \ldots n)^{T-C-3}$; the quantity $(n-1) \times N(u \ldots n)^{T-C-3} + N(u)^{\overset{A+B-T}{'}} \times N(u \ldots n-1)^{T-C-2}$ thus reduces to $(n-1) \times N(u \ldots n)^{T-C-3} - nN(u \ldots n)^{T-C-3} + (\underset{'}{A} + \underset{'}{B} - C-1) \times N(u \ldots n-1)^{T-C-2}$; that is, it reduces to $-N(u \ldots n)^{T-C-3} + (\underset{'}{A} + \underset{'}{B} - C) \times N(u \ldots n-1)^{T-C-2} - N(u \ldots n-1)^{T-C-2}$. Moreover, we note that $N(u \ldots n)^{T-C-3} + N(u \ldots n-1)^{T-C-2} = N(u \ldots n)^{T-C-2}$, in such a way that we finally have

$$(n-1) \times N(u \ldots n)^{T-C-3} + N(u)^{\overset{A+B-T}{'}} \times N(u \ldots n-1)^{T-C-2}$$
$$= -N(u \ldots n)^{T-C-2} + (\underset{'}{A} + \underset{'}{B} - C) \times N(u \ldots n-1)^{T-C-2}$$

$$= -N(u \ldots n)^{T-C-2} + N(u)^{\overset{A+B-C-1}{'}} \times N(u \ldots n-1)^{T-C-2}.$$

This example is sufficient to elucidate all the reductions that occur within all the results of the different cases under consideration.

Fourth case

$$B_{\prime} - A < B_{\prime\prime} - A_{\prime} < C - B.$$

(88.) For this case the expression $y^s([u^A, x^{A_{\prime}}]^B, z \ldots n-1)^{T-s}$ holds from $s = 0$ to $s = B_{\prime} - A$.

Past $s = B_{\prime} - A$, it is $y^s([u^{B_{\prime}-s}, x^{A_{\prime}}]^B, z \ldots n-1)^{T-s}$ until $s = B_{\prime\prime} - A_{\prime}$.

Past $s = B_{\prime\prime} - A_{\prime}$, it is $y^s([u^{B_{\prime}-s}, x^{B_{\prime\prime}-s}]^B, z \ldots n-1)^{T-s}$ until $s = C - B$.

Past $s = C - B$, it is $y^s([u^{B_{\prime}-s}, x^{B_{\prime\prime}-s}]^{C-s}, z \ldots n-1)^{T-s}$ until $s = A_{\prime\prime}$.

Since $A + A_{\prime} > B$, we have $N([u^A, x^{A_{\prime}}]^B, z \ldots n-1)^{T-s} = N_{(}u \ldots n-1)^{T-s} - N(u \ldots n-1)^{T-A-s-1} - N(u \ldots n-1)^{T-A_{\prime}-s-1} + N(u \ldots n-1)^{T-B-s-2} - N(u)^{A+A_{\prime}-B-1} \times N(u \ldots n-1)^{T-B-s-1}$ from $s = 0$ to $s = B_{\prime} - A$. From $s = B_{\prime} - A$ to $s = B_{\prime\prime} - A_{\prime}$, we have $B_{\prime} - s + A_{\prime} > B$ or $s < A_{\prime} + B_{\prime} - B$; indeed, the largest value of s within this interval is $B_{\prime\prime} - A_{\prime}$; however, $B_{\prime\prime} - A_{\prime} < A_{\prime} + B_{\prime} - B$, because $A_{\prime} + B_{\prime} > C$ from (83); thus $A_{\prime} + B_{\prime} - B > C - B$. But by hypothesis, $C - B > B_{\prime\prime} - A_{\prime}$; therefore $A_{\prime} + B_{\prime} - B > B_{\prime\prime} - A_{\prime}$.

Thus from $s = B_{\prime} - A$ to $s = B_{\prime\prime} - A_{\prime}$, we have

$$N([u^{B_{\prime}-s}, x^{A_{\prime}}]^B, z \ldots n-1)^{T-s} = N(u \ldots n-1)^{T-s} - N(u \ldots n-1)^{T-B-1}$$
$$-N(u \ldots n-1)^{T-A_{\prime}-s-1} + N(u \ldots n-1)^{T-B-s-2}$$
$$-N(u)^{A_{\prime}+B_{\prime}-s-1} \times N(u \ldots n-2)^{T-B-s-1}.$$

From $s = B_{\prime\prime} - A_{\prime}$ to $s = C - B$, we have $B_{\prime} - s + B_{\prime\prime} - s > B$ or $s < \dfrac{B_{\prime} + B_{\prime\prime} - B}{2}$; indeed the highest value of s is $C - B$. However, we have $C - B < \dfrac{B_{\prime} + B_{\prime\prime} - B}{2}$, that is, $2C < B_{\prime} + B_{\prime\prime} + B$, from (83). Therefore, from $s = B_{\prime\prime} - A_{\prime}$ to $s = C - B$, we have

$$N[(u^{B_{\prime}-s}, x^{B_{\prime\prime}-s})^B, z \ldots n-1]^{T-s} = N(u \ldots n-1)^{T-s}$$
$$-N(u \ldots n-1)^{T-B_{\prime}-1} - N(u \ldots n-1)^{T-B_{\prime\prime}-1} + N(u \ldots n-1)^{T-B-s-2}$$
$$-N(u)^{B_{\prime}+B_{\prime\prime}-B-2s-1} \times N(u \ldots n-1)^{T-B-s-1}.$$

From $s = C - \underset{''}{B}$ to $s = \underset{''}{A}$, we can use the same arguments to show that, if $\underset{''}{B} - \underset{''}{A} > C - \underset{'}{B}$,

$$
\begin{aligned}
N[(u'^{\underset{}{B}-s}, x''^{\underset{}{B}-s})^{C-s}, z \ldots n-1]^{T-s} &= N(u \ldots n-1)^{T-s} \\
-N(u \ldots n-1)^{T-\underset{'}{B}-1} - N(u \ldots n-1)^{T-\underset{''}{B}-1} &+ N(u \ldots n-1)^{T-C-2} \\
-N(u)'^{\underset{}{+}B''-C-s-1} \times N(u \ldots n-2)^{T-C-1}.&
\end{aligned}
$$

But if $\underset{''}{B} - \underset{''}{A} < C - \underset{'}{B}$, this expression is valid only until $s = \underset{'}{B} + \underset{''}{B} - C$; from $s = \underset{'}{B} + \underset{''}{B} - C$ to $s = \underset{''}{A}$, this expression becomes

$$
\begin{aligned}
N[(u'^{\underset{}{B}-s}, x''^{\underset{}{B}-s})^{C-s}, z \ldots n-1]^{T-s} &= N(u \ldots n-1)^{T-s} \\
-N(u \ldots n-1)^{T-\underset{'}{B}-1} - N(u \ldots n-1)^{T-\underset{''}{B}-1} & \\
+N(u \ldots n-1)^{T-\underset{'}{B}-\underset{''}{B}+s-2}.&
\end{aligned}
$$

Summing these various quantities over the intervals within which they are valid, we find that if $\underset{''}{B} - \underset{''}{A} > C - \underset{'}{B}$, we have

$$
\begin{aligned}
N([(u^A, x'^A)^B, (u^A, y''^A)^B, (x'^A, y'^A)^B]^C, z \ldots n)^T & \\
= N(u \ldots n)^T - N(u \ldots n)^{T-A-1} - N(u \ldots n)'^{T-\underset{'}{A}-1} - N(u \ldots n)''^{T-\underset{''}{A}-1} & \\
+N(u \ldots n)^{T-B-2} + N(u \ldots n)'^{T-\underset{'}{B}-2} + N(u \ldots n)''^{T-\underset{''}{B}-2} & \\
-N(u)'^{A+\underset{}{A}-B-1} \times N(u \ldots n-1)^{T-\underset{'}{B}-1} & \\
-N(u)''^{A+\underset{}{A}-B-1} \times N(u \ldots n-1)^{T-\underset{''}{B}-1} & \\
-N(u)'^{\underset{}{A}+''\underset{}{A}-''B-1} \times N(u \ldots n-1)^{T-\underset{''}{B}-1} + N(u \ldots n)^{T+A-\underset{'}{B}-\underset{''}{B}-2} & \\
+N(u \ldots n)'^{T+\underset{}{A}-\underset{}{B}-''B-2} & \\
+N(u)''^{A+2B+\underset{'}{B}+\underset{''}{B}-3C-1} \times N(u \ldots n-1)^{T-C-2} & \\
-N(u \ldots n)^{T-C-3} - 2N(u \ldots n)^{T-C-2} & \\
+[N(u \ldots 2)'^{\underset{}{B}+''B-\underset{}{A}-C-2} - N(u \ldots 2)'^{B+\underset{''}{B}+\underset{}{B}-2C-2}] & \\
\times N(u \ldots n-2)^{T-C-1}.&
\end{aligned}
$$

If $\underset{''}{B} - \underset{''}{A} < C - \underset{'}{B}$, we have

$$N([(u^A, \overset{A}{\underset{'}{x}})^B, (u^A, \overset{A}{\underset{''}{y}})^{\underset{'}{B}}, (\overset{A}{\underset{'}{x}}, \overset{A}{\underset{''}{y}})^{\underset{''}{B}}]^C, z \dots n)^T$$

$$= N(u \dots n)^T - N(u \dots n)^{T-A-1} - N(u \dots n)^{T-\underset{'}{A}-1} - N(u \dots n)^{T-\underset{''}{A}-1}$$

$$+ N(u \dots n)^{T-B-2} + N(u \dots n)^{T-\underset{'}{B}-2} + N(u \dots n)^{T-\underset{''}{B}-2}$$

$$- N(u)^{A+\underset{'}{A}-\underset{'}{B}-1} \times N(u \dots n-1)^{T-\underset{'}{B}-1}$$

$$- N(u)^{A+\underset{''}{A}-\underset{''}{B}-1} \times N(u \dots n-1)^{T-\underset{''}{B}-1}$$

$$- N(u)^{\underset{'}{A}+\underset{''}{A}-\underset{''}{B}-1} \times N(u \dots n-1)^{T-\underset{''}{B}-1} + N(u \dots n)^{T+A-B-\underset{'}{B}-2}$$

$$+ N(u \dots n)^{T+\underset{'}{A}-B-\underset{''}{B}-2} + N(u \dots n)^{T+\underset{''}{A}-\underset{'}{B}-\underset{''}{B}-2}$$

$$+ 2N(u)^{B+\underset{'}{B}+\underset{''}{B}-2C-1} \times N(u \dots n-1)^{T-C-2}$$

$$- N(u \dots 2)^{B+\underset{'}{B}+\underset{''}{B}-2C-2} \times N(u \dots n-2)^{T-C-1} - N(u \dots n)^{T-C-3}$$

$$- 3N(u \dots n)^{T-C-2}.$$

Fifth case

$$\underset{''}{B} - \underset{'}{A} < C - B < \underset{'}{B} - A.$$

(89.) The fifth case can be subdivided into the following two: $\underset{''}{B} - \underset{''}{A} > C - \underset{'}{B}$ and $\underset{''}{B} - \underset{''}{A} < C - \underset{'}{B}$. However, since this case does not differ from the third case, except for replacing B by $\underset{'}{B}$, A by $\underset{''}{A}$, and vice versa, it follows that we only need to perform these substitutions in the expressions obtained for the third case to obtain the expression of the number of terms for the fifth.

Sixth case

$$\underset{''}{B} - \underset{'}{A} < \underset{'}{B} - A < C - B.$$

(90.) The sixth case can also be subdivided into the following two: $\underset{''}{B} - \underset{'}{A} > C - \underset{'}{B}$ and $\underset{''}{B} - \underset{''}{A} < C - \underset{'}{B}$. However, it is no different from the fourth case except for changing $\underset{'}{B}$ into $\underset{''}{B}$, A into $\underset{''}{A}$, and vice versa; since it is quite easy to see that these substitutions do not impact their value, it follows that these expressions are also valid for the sixth case.

Summary and table of the different values of the number of terms sought in the preceding polynomial and in related quantities

(91.) When we compare the conditions where each expression we have found is valid, we see that, in general, the question can be subdivided into twelve cases. But if, for brevity's sake, we define

$$P = N([(u^A, \overset{A}{\underset{'}{x}})^B, (u^A, \overset{A}{\underset{''}{y}})^{\underset{'}{B}}, (\overset{A}{\underset{'}{x}}, \overset{A}{\underset{''}{y}})^{\underset{''}{B}}]^C, z \dots n)^T,$$

we easily see that P can have only eight different values. Thus we only need to distinguish among the eight following cases. Since these different cases determine, so to speak, as many different forms, we name them *forms*, because their expression reflects the conditions that determine the number of terms of the polynomial.

First form

$$C - B < \underset{\prime}{B} - A; \; C - B < \underset{\prime\prime}{B} - \underset{\prime}{A}; \; C - \underset{\prime}{B} < \underset{\prime\prime}{B} - \underset{\prime\prime}{A}.$$

(92.)

$$P = N(u \ldots n)^T - N(u \ldots n)^{T-A-1} - N(u \ldots n)^{\overset{T-A-1}{\prime}}$$
$$-N(u \ldots n)^{\overset{T-A-1}{\prime\prime}} + N(u \ldots n)^{T-B-2} + N(u \ldots n)^{\overset{T-B-2}{\prime}}$$
$$+N(u \ldots n)^{\overset{T-B-2}{\prime\prime}} - N(u)^{\overset{A+A-B-1}{\prime}} \times N(u \ldots n-1)^{T-B-1}$$
$$-N(u)^{\overset{A+A-B-1}{\prime\prime}{}_{\prime}} \times N(u \ldots n-1)^{\overset{T-B-1}{\prime}}$$
$$-N(u)^{\overset{A+A-B-1}{\prime}{}_{\prime\prime}{}_{\prime\prime}} \times N(u \ldots n-1)^{\overset{T-B-1}{\prime\prime}}$$
$$-N(u \ldots n)^{T-C-3} + N(u)^{\overset{A+A+A-C-1}{\prime}{}_{\prime}{}_{\prime\prime}} \times N(u \ldots n-1)^{T-C-2}$$
$$+[N(u \ldots 2)^{\overset{A+A-B-1}{\prime}} + N(u \ldots 2)^{\overset{B+B-A-C-2}{\prime}{}_{\prime\prime}{}_{\prime\prime}}$$
$$-N(u)^{\overset{A+B-C-1}{\prime}{}_{\prime}} \times N(u)^{\overset{A+B-C-1}{\prime}{}_{\prime\prime}}] \times N(u \ldots n-2)^{T-C-1}.$$

Second form

$$C - B < \underset{\prime}{B} - A; \; C - B < \underset{\prime\prime}{B} - \underset{\prime}{A}; \; C - \underset{\prime}{B} < \underset{\prime\prime}{B} - \underset{\prime\prime}{A}.$$

(93.)

$$P = N(u \ldots n)^T - N(u \ldots n)^{T-A-1} - N(u \ldots n)^{\overset{T-A-1}{\prime}}$$
$$-N(u \ldots n)^{\overset{T-A-1}{\prime\prime}} + N(u \ldots n)^{T-B-2} + N(u \ldots n)^{\overset{T-B-2}{\prime}}$$
$$+N(u \ldots n)^{\overset{T-B-2}{\prime\prime}} - N(u)^{\overset{A+A-B-1}{\prime}} \times N(u \ldots n-1)^{T-B-1}$$
$$-N(u)^{\overset{A+A-B-1}{\prime\prime}{}_{\prime}} \times N(u \ldots n-1)^{\overset{T-B-1}{\prime}}$$
$$-N(u)^{\overset{A+A-B-1}{\prime}{}_{\prime\prime}{}_{\prime\prime}} \times N(u \ldots n-1)^{\overset{T-B-1}{\prime\prime}}$$
$$+N(u \ldots n)^{\overset{T+A-B-B-2}{\prime\prime}{}_{\prime}{}_{\prime\prime}} - N(u \ldots n)^{T-C-3}$$
$$-N(u \ldots n)^{T-C-2} + N(u)^{\overset{A+A+B+B-2C-1}{\prime}{}_{\prime}{}_{\prime\prime}} \times N(u \ldots n-1)^{T-C-2}$$
$$+[N(u \ldots 2)^{\overset{A+A-B-1}{\prime}} - N(u)^{\overset{A+B-C-1}{\prime}{}_{\prime}} \times N(u)^{\overset{A+B-C-1}{\prime}{}_{\prime\prime}}]$$
$$\times N(u \ldots n-2)^{T-C-1}.$$

Third form

$$C - B > B_{/} - A; \quad C - B < B_{//} - A_{/}; \quad C - B_{/} < B_{//} - A_{//}.$$

(94.)

$$P = N(u\ldots n)^T - N(u\ldots n)^{T-A-1} - N(u\ldots n)^{T-A_{/}-1}$$
$$-N(u\ldots n)^{T-A_{//}-1} + N(u\ldots n)^{T-B-2} + N(u\ldots n)^{T-B_{/}-2}$$
$$+N(u\ldots n)^{T-B_{//}-2} - N(u)^{A+A_{/}-B-1} \times N(u\ldots n-1)^{T-B-1}$$
$$-N(u)^{A+A_{//}-B-1} \times N(u\ldots n-1)^{T-B_{/}-1}$$
$$-N(u)^{A_{/}+A_{//}-B_{//}-1} \times N(u\ldots n-1)^{T-B_{/}-1} + N(u\ldots n)^{T+A-B-B_{/}-2}$$
$$-N(u\ldots n)^{T-C-3} - N(u\ldots n)^{T-C-2}$$
$$+N(u)^{A+A_{/}+B+B_{/}-2C-1} \times N(u\ldots n-1)^{T-C-2}$$
$$+[N(u\ldots 2)^{B_{/}+B_{//}-A_{//}-C-2} - N(u\ldots 2)^{B_{/}+A_{/}-C-2}$$
$$-N(u)^{A_{/}+B-C-1} \times N(u)^{B_{//}+B_{/}-A_{/}-C-1}] \times N(u\ldots n-2)^{T-C-1}.$$

Fourth form

$$C - B > B_{/} - A; \quad C - B < B_{//} - A_{/}; \quad C - B_{/} > B_{//} - A_{//}.$$

(95.)

$$P = N(u\ldots n)^T - N(u\ldots n)^{T-A-1} - N(u\ldots n)^{T-A_{/}-1}$$
$$-N(u\ldots n)^{T-A_{//}-1} + N(u\ldots n)^{T-B-2} + N(u\ldots n)^{T-B_{/}-2}$$
$$+N(u\ldots n)^{T-B_{//}-2} - N(u)^{A+A_{/}-B-1} \times N(u\ldots n-1)^{T-B-1}$$
$$-N(u)^{A+A_{//}-B-1} \times N(u\ldots n-1)^{T-B_{/}-1}$$
$$-N(u)^{A_{/}+A_{//}-B_{//}-1} \times N(u\ldots n-1)^{T-B_{//}-1} + N(u\ldots n)^{T+A-B-B_{/}-2}$$
$$+N(u\ldots n)^{T+A_{//}-B_{/}-B-2} - N(u\ldots n)^{T-C-3} - 2N(u\ldots n)^{T-C-2}$$
$$+N(u)^{A+B+2B_{/}+B-3C-1} \times N(u\ldots n-1)^{T-C-2}$$
$$-[N(u\ldots 2)^{A+B-C-2} + N(u)^{B_{//}+B_{/}-A_{/}-C-1} \times N(u)^{A+B_{/}-C-1}]$$
$$\times N(u\ldots n-2)^{T-C-1}.$$

Fifth form

$$C - B > B_{\prime} - A;\ C - B > B_{\prime\prime} - A_{\prime};\ C_{\prime} - B < B_{\prime\prime} - A_{\prime\prime}.$$

(96.)

$$
P = N(u\ldots n)^T - N(u\ldots n)^{T-A-1} - N(u\ldots n)_{\prime}^{T-A-1}
$$
$$
- N(u\ldots n)_{\prime\prime}^{T-A-1} + N(u\ldots n)^{T-B-2} + N(u\ldots n)_{\prime}^{T-B-2}
$$
$$
+ N(u\ldots n)_{\prime\prime}^{T-B-2} - N(u)_{\prime}^{A+A-B-1} \times N(u\ldots n-1)^{T-B-1}
$$
$$
+ N(u)_{\prime\prime\ \prime}^{A+A-B-1} \times N(u\ldots n-1)_{\prime}^{T-B-1}
$$
$$
- N(u)_{\prime\ \prime\prime}^{A+A-B-1} \times N(u\ldots n-1)_{\prime\prime}^{T-B-1} + N(u\ldots n)^{T+A-B-B-2}
$$
$$
+ N(u\ldots n)_{\prime\ \prime\prime}^{T+A-B-B-2} - N(u\ldots n)^{T-C-3} - 2N(u\ldots n)^{T-C-2}
$$
$$
+ N(u)_{\prime\ \prime\prime}^{A+2B+B+B-3C-1} \times N(u\ldots n-1)^{T-C-2}
$$
$$
+ [N(u\ldots 2)_{\prime\ \prime\prime\ \prime\prime}^{B+B-A-C-2} - N(u\ldots 2)_{\prime\ \prime\ \prime\prime}^{B+B+B-2C-2}]
$$
$$
\times N(u\ldots n-2)^{T-C-1}.
$$

Sixth form

$$C - B > B_{\prime} - A;\ C - B > B_{\prime\prime} - A_{\prime};\ C_{\prime} - B > B_{\prime\prime} - A_{\prime\prime}.$$

(97.)

$$
P = N(u\ldots n)^T - N(u\ldots n)^{T-A-1} - N(u\ldots n)_{\prime}^{T-A-1}
$$
$$
- N(u\ldots n)_{\prime\prime}^{T-A-1} + N(u\ldots n)^{T-B-2} + N(u\ldots n)_{\prime}^{T-B-2}
$$
$$
+ N(u\ldots n)_{\prime\prime}^{T-B-2} - N(u)_{\prime}^{A+A-B-1} \times N(u\ldots n-1)^{T-B-1}
$$
$$
- N(u)_{\prime\prime\ \prime}^{A+A-B-1} \times N(u\ldots n-1)_{\prime}^{T-B-1}
$$
$$
- N(u)_{\prime\ \prime\prime\ \prime\prime}^{A+A-B-1} \times N(u\ldots n-1)_{\prime\prime}^{T-B-1} + N(u\ldots n)^{T+A-B-B-2}
$$
$$
+ N(u\ldots n)_{\prime}^{T+A-B-B-2} + N(u\ldots n)_{\prime\prime\ \prime\ \prime\prime}^{T+A-B-B-2}
$$
$$
+ 2N(u)_{\prime\ \prime\prime}^{B+B+B-2C-1} \times N(u\ldots n-1)^{T-C-2}
$$
$$
- N(u\ldots 2)_{\prime\ \prime\prime}^{B+B+B-2C-2} \times N(u\ldots n-2)^{T-C-1}
$$
$$
- N(u\ldots n)^{T-C-3} - 3N(u\ldots n)^{T-C-2}.
$$

Seventh form

$$C - B < B' - A;\ C - B > B'' - A';\ C - B' < B'' - A''.$$

(98.)

$$P = N(u\ldots n)^T - N(u\ldots n)^{T-A-1} - N(u\ldots n)^{T-A-1}_{'}$$

$$-N(u\ldots n)^{T-A-1}_{''} + N(u\ldots n)^{T-B-2} + N(u\ldots n)^{T-B-2}_{'}$$

$$+N(u\ldots n)^{T-B-2}_{''} - N(u)^{A+A'-B-1}_{'} \times N(u\ldots n-1)^{T-B-1}$$

$$-N(u)^{A+A'-B-1}_{''} \times N(u\ldots n)^{T-B-1}_{'}$$

$$-N(u)^{A+A''-B''-1}_{'\,''} \times N(u\ldots n-1)^{T-B-1}_{''} + N(u\ldots n)^{T+A-B'-B''-2}$$

$$-N(u\ldots n)^{T-C-3} - N(u\ldots n)^{T-C-2}$$

$$+N(u)^{A+A'+B+B''-2C-1}_{''} \times N(u\ldots n-1)^{T-C-2}$$

$$+[N(u\ldots 2)^{B+B''-A'-C-2}_{'\,''\,''} - N(u\ldots 2)^{B+A'-C-2}_{''}$$

$$-N(u)^{A+B-C-1}_{''} \times N(u)^{B+B'-A'-C-1}_{'}\,]\times (u\ldots n-2)^{T-C-1}.$$

Eigth form

$$C - B < B' - A;\ C - B > B'' - A';\ C - B' > B'' - A''.$$

(99.)

$$P = N(u\ldots n)^T - N(u\ldots n)^{T-A-1} - N(u\ldots n)^{T-A-1}_{'}$$

$$-N(u\ldots n)^{T-A-1}_{''} + N(u\ldots n)^{T-B-2} + N(u\ldots n)^{T-B-2}_{'}$$

$$+N(u\ldots n)^{T-B-2}_{''} - N(u)^{A+A'-B-1}_{'} \times N(u\ldots n-1)^{T-B-1}$$

$$-N(u)^{A+A'-B-1}_{'\,''} \times N(u\ldots n-1)^{T-B-1}_{'}$$

$$-N(u)^{A+A'-B-1}_{'\,''} \times N(u\ldots n-1)^{T-B-1}_{''} + N(u\ldots n)^{T+A-B'-B''-2}_{'}$$

$$+N(u\ldots n)^{T+A-B'-B''-2}_{''} - N(u\ldots n)^{T-C-3} - 2N(u\ldots n)^{T-C-2}$$

$$+N(u)^{A+B+2B'+B''-3C-1}_{''\,'} \times N(u\ldots n-1)^{T-C-2}$$

$$-[N(u\ldots 2)^{A+B-C-2}_{''} + N(u)^{B+B'-A-C-1}_{'} \times N(u)^{A+B-C-1}_{''}\,]$$

$$\times N(u\ldots n-2)^{T-C-1}.$$

COROLLARY

(100.) By reasoning as we did in (77), we therefore see that the value of

$$N([u^A, x'^A)^B, (u^A, y''^A)^B)^{'}, (x'^A, y''^A)^B)^{''}]^C, z'''^A \ldots n)^T$$

may be computed based on the previous conditions (82) and using the previously computed quantities, adding $-N(u\ldots n)^{T-A-1}_{'''}$, $-N(u\ldots n)^{T-A-1}_{iv}$, etc., to them.

Problem XIX

(101.) *We ask for a method to compute the value of*

$$N([[(u^A, x^{\overset{A}{\prime}})^B, (u^A, y^{\overset{A}{\prime\prime}})^{\overset{B}{\prime}}, (x^{\overset{A}{\prime}}, y^{\overset{A}{\prime\prime}})^{\overset{B}{\prime\prime}}]^C, z^{\overset{A}{\prime\prime\prime}} \ldots n)^T,$$

when some of the necessary conditions for the existence of the polynomial
$([[(u^A, x^{\overset{A}{\prime}})^B, (u^A, y^{\overset{A}{\prime\prime}})^{\overset{B}{\prime}}, (x^{\overset{A}{\prime}}, y^{\overset{A}{\prime\prime}})^{\overset{B}{\prime\prime}}]^C, z^{\overset{A}{\prime\prime\prime}} \ldots n)^T$ *do not hold.*

This question is of little interest for the kind of equations we are currently interested in, but it is necessary for subsequent classes of incomplete equations.

We will consider only a few representative cases, to illustrate how things should be done in the other cases.

Let us assume, for example, that $B + \underset{\prime}{B} + \underset{\prime\prime}{B} < 2C$, while all other conditions remain true. It is then clear that u, x and y cannot reach the dimension C together; thus we must reduce the value of C until $B + \underset{\prime}{B} + \underset{\prime\prime}{B}$ become larger than twice this value; that is, we must assume $C = \dfrac{B + \underset{\prime}{B} + \underset{\prime\prime}{B} - r}{2}$, where r equals 0 or 1 depending upon whether $B + \underset{\prime}{B} + \underset{\prime\prime}{B}$ is even or odd.

If P represents the number of terms of the above polynomial, we determine the value of P by computing the relative values of the quantities $\underset{\prime}{B} - A$, $\underset{\prime\prime}{B} - A$, $\underset{\prime}{B} - \underset{\prime}{A}$, $C - B$ and $C - \underset{\prime}{B}$; that is, the quantities $\underset{\prime}{B} - A$, $\underset{\prime\prime}{B} - \underset{\prime}{A}$, $\underset{\prime\prime}{B} - \underset{\prime\prime}{A}$, $\dfrac{B + \underset{\prime}{B} + \underset{\prime\prime}{B} - r}{2} - B$, or $\dfrac{B + \underset{\prime\prime}{B} - \underset{\prime}{B} - r}{2}$ and $\dfrac{B + \underset{\prime\prime}{B} - \underset{\prime}{B} - r}{2}$. Likewise, we must replace C by its value $\dfrac{B + \underset{\prime}{B} + \underset{\prime\prime}{B} - r}{2}$ in all expressions found in the preceding problem.

Assume we have $B + \underset{\prime}{B} + \underset{\prime\prime}{B} < 2C$ and $A + \underset{\prime\prime}{B} < C$ at the same time; then we first see that we must reduce the value of C until $C = \dfrac{B + \underset{\prime}{B} + \underset{\prime\prime}{B} - r}{2}$. But since $A + \underset{\prime\prime}{B} < C$, we must also reduce the value of C until $C = A + \underset{\prime\prime}{B}$; therefore, C must be equal to the smaller of the two quantities $\dfrac{B + \underset{\prime}{B} + \underset{\prime\prime}{B} - r}{2}$ and $A + \underset{\prime\prime}{B}$. If we had $B + \underset{\prime}{B} + \underset{\prime\prime}{B} < 2C$, $\underset{\prime}{A} + \underset{\prime}{B} < C$, and $A + \underset{\prime\prime}{B} < C$, we would make C equal to the smaller of the three quantities $\dfrac{B + \underset{\prime}{B} + \underset{\prime\prime}{B} - r}{2}$, $\underset{\prime}{A} + \underset{\prime}{B}$, and $A + \underset{\prime\prime}{B}$.

If we had $B + \underset{\prime}{B} + \underset{\prime\prime}{B} < 2C$ and $A + \underset{\prime}{A} < B$, we would first set $B = A + \underset{\prime}{A}$; if $A + \underset{\prime}{A} + \underset{\prime}{B} + \underset{\prime\prime}{B} > 2C$, there would be no other change to be made. But if $A + \underset{\prime}{A} + \underset{\prime}{B} + \underset{\prime\prime}{B} < 2C$, we should, in addition, set $C = \dfrac{A + \underset{\prime}{A} + \underset{\prime}{B} + \underset{\prime\prime}{B} - r}{2}$, with r equal to 0 or 1 depending on whether $A + \underset{\prime}{A} + \underset{\prime}{B} + \underset{\prime\prime}{B}$ is even or odd.

These examples are sufficient to see how to deal with all cases.

Problem XX

(102.) *Consider $n - 1$ equations of the form*

$$([(u^{\overset{a}{\prime}}, x^{\overset{a'}{\prime}})^{\overset{b'}{\prime}}, (u^{a'}, y^{\overset{a'}{\prime\prime}})^{\overset{b'}{\prime}}, (x^{\overset{a'}{\prime}}, y^{\overset{a'}{\prime\prime}})^{\overset{b'}{\prime\prime}}]^{c'}, z^{\overset{a'}{\prime\prime\prime}} \ldots n)^{t'} = 0$$

satisfying the general conditions mentioned in (83), and containing n unknowns. In addition, let

$$([(u^A, x^{\overset{A}{\prime}})^B, (u^A, y^{\overset{A}{\prime\prime}})^{\overset{B}{\prime}}, (x^{\overset{A}{\prime}}, y^{\overset{A}{\prime\prime}})^{\overset{B}{\prime\prime}}]^C, z^{\overset{A}{\prime\prime\prime}} \ldots n)^T$$

be a polynomial satisfying the same general conditions, along with the specific conditions that determine one of the eight forms presented in (91 ff). Let us assume moreover that the polynomial satisfies the same conditions when replacing A by $A - a'$, $\underset{\prime}{A}$ by $\underset{\prime}{A} - \underset{\prime}{a'}$, etc.; B by $B - b'$, $\underset{\prime}{B}$ by $\underset{\prime}{B} - \underset{\prime}{b'}$, etc.; C by $C - c'$, T by $T - t'$. Let us also assume that the same property holds when replacing A by $A - a' - a''$, $\underset{\prime}{A}$ by $\underset{\prime}{A} - \underset{\prime}{a'} - \underset{\prime}{a''}$, etc.; $B - b' - b''$, $\underset{\prime}{B}$ by $\underset{\prime}{B} - \underset{\prime}{b'} - \underset{\prime}{b''}$, etc., etc., and so on. We ask for the number of terms that can be cancelled in the first of these polynomials without introducing new ones using these $n - 1$ equations.

From what was said in (60) and by using the same argument, the number of terms that can be cancelled using a single equation is

$$N([(u^{A-a'}, x^{\overset{A-a'}{\prime}})^{B-b'}, (u^{A-a'}, y^{\overset{A-a'}{\prime\prime}})^{\overset{B-b'}{\prime}},$$
$$(x^{\overset{A-a'}{\prime}}, y^{\overset{A-a'}{\prime\prime}})^{\overset{B-b'}{\prime\prime}}]^{C-c'}, z^{\overset{A-a'}{\prime\prime\prime}} \ldots n)^{T-t'}.$$

If there are two equations, the number of terms that can be cancelled, without introducing new ones, is

$$N([(u^{A-a'}, x^{\overset{A-a'}{\prime}})^{B-b'}, (u^{A-a'}, y^{\overset{A-a'}{\prime\prime}})^{\overset{B-b'}{\prime}},$$
$$(x^{\overset{A-a'}{\prime}}, y^{\overset{A-a'}{\prime\prime}})^{\overset{B-b'}{\prime\prime}}]^{C-c'}, z^{\overset{A-a'}{\prime\prime\prime}} \ldots n)^{T-t'}$$

$$+ N([(u^{A-a''}, x^{\overset{A-a''}{\prime}})^{B-b''}, (u^{A-a''}, y^{\overset{A-a''}{\prime\prime}})^{\overset{B-b''}{\prime}},$$
$$(x^{\overset{A-a''}{\prime}}, y^{\overset{A-a''}{\prime\prime}})^{\overset{B-b''}{\prime\prime}}]^{C-c''}, z^{\overset{A-a''}{\prime\prime\prime}} \ldots n)^{T-t''}$$

$$- N([(u^{A-a'-a''}, x^{\overset{A-a'-a''}{\prime}})^{B-b'-b''},$$
$$(u^{A-a'-a''}, y^{\overset{A-a'-a''}{\prime\prime}})^{\overset{B-b'-b''}{\prime}},$$
$$(x^{\overset{A-a'-a''}{\prime}}, y^{\overset{A-a'-a''}{\prime\prime}})^{\overset{B-b'-b''}{\prime\prime}}]^{C-c'-c''},$$
$$z^{\overset{A-a'-a''}{\prime\prime\prime}} \ldots n)^{T-t'-t''}.$$

If there are three equations, the number of terms is expressed by a function which is lengthy to write down, but easy to imagine from what was said in (60). The same is true for four, five, etc. equations.

Problem XXI

(103.) *Given the same assumptions as in the preceding problem, we ask, what is the number of terms remaining in the polynomial, when all possible terms are cancelled using the $n-1$ equations, without introducing new ones?*

From what was said in (60) and (102), and from the stated conditions for the preceding problem (which make all polynomials entering in the expression of the number of terms that can be cancelled similar), the number of remaining terms that may be cancelled without introducing new ones is

$$d^{n-1}(N([u^A, x'^{\overset{A}{'}})^B, (u^A, y''^{\overset{A}{''}})^{\overset{B}{'}}, (x'^{\overset{A}{'}}, y''^{\overset{A}{''}})^{\overset{B}{'}}]^C, z'''^{\overset{A}{'''}} \ldots n)^T \ldots$$

$$\left(\begin{array}{cccccc} T & & C & & B & & \overset{B}{'} \\ -t', -t'', \text{ etc.} & : & -c', -c'', \text{ etc.} & : & -b', -b'', \text{ etc.} & : & -\underset{'}{b}', -\underset{'}{b}'', \text{ etc.} & : \end{array}\right.$$

$$\begin{array}{ccccc} \overset{B}{''} & & A & & \overset{A}{'} \\ -\underset{''}{b}', -\underset{''}{b}'', \text{ etc.} & : & -a', -a'', \text{ etc.} & : & -\underset{'}{a}', -\underset{'}{a}'', \text{ etc.} & : \end{array}$$

$$\left.\begin{array}{cccc} \overset{A}{''} & & \overset{A}{'''} & \\ -\underset{''}{a}', -\underset{''}{a}'', \text{ etc.} & : & -\underset{'''}{a}', -\underset{'''}{a}'', \text{ etc.} & \text{etc.} \end{array}\right).$$

Problem XXII

(104.) *Consider n equations of the form*

$$([(u^a, x'^{\overset{a}{'}})^b, (u^a, y''^{\overset{a}{''}})^{\overset{b}{'}}, (x'^{\overset{a}{'}}, y''^{\overset{a}{''}})^{\overset{b}{'}}]^c, z'''^{\overset{a}{'''}} \ldots n)^t = 0,$$

satisfying the general conditions (83), and containing n unknowns: We ask for the degree of the final equation resulting from the elimination of $n-1$ of these unknowns.

Assume we multiply any of these equations by the polynomial

$$([(u^A, x'^{\overset{A}{'}})^B, (u^A, y''^{\overset{A}{''}})^{\overset{B}{'}}, (x'^{\overset{A}{'}}, y''^{\overset{A}{''}})^{\overset{B}{'}}]^C, z'''^{\overset{A}{'''}} \ldots n)^T.$$

In addition to the conditions expressed in (102), we assume this polynomial also satisfies the same conditions relative to the product equation

$$([(u^{A+a}, x'^{\overset{A+a}{'}})^{B+b}, (u^{A+a}, y''^{\overset{A+a}{''}}{}'')^{\overset{B+b}{'}}, (x'^{\overset{A+a}{'}}, y''^{\overset{A+a}{''}}{}'')^{\overset{B+b}{'}}]^{C+c},$$
$$z'''^{\overset{A+a}{'''}} \ldots n)^{T+t}$$

and relative to the polynomials that can express the number of terms that may be cancelled using the $n-1$ other equations.

Then all polynomials entering in the expression of the number of remaining terms both in the polynomial multiplier and in the product equation share the same characteristics; thus it is enough to apply what was said in (62) *verbatim*. (Note that it is not necessary for the equations to be of the same nature, that is, to all fall within one of the forms previously presented in (91 ff)). We therefore easily see that the degree of the final equation is

$$D = d^n N([(u^{A+a}, x'^{A+a}_{\,'})^{B+b}, (u^{A+a}, y''^{A+a}_{\,\,''})^{B+b}_{\,\,''},$$
$$(x'^{A+a}_{\,'}, y''^{A+a}_{\,\,''})^{B+b}_{\,\,''}]^{C+c}, z'''^{A+a}_{\,\,\,'''} \ldots n)^{T+t} \ldots$$
$$\ldots \left(\begin{array}{c} T+t \\ -t, -t', \text{ etc.} \end{array} : \begin{array}{c} C+c \\ -c, -c', \text{ etc.} \end{array} : \begin{array}{c} B+b \\ -b, -b', \text{ etc.} \end{array} : \text{ etc} : \right.$$
$$\left. \begin{array}{c} A+a \\ -a, -a', \text{ etc.} \end{array} : \text{ etc.} \right).$$

(105.) But several matters need to be discussed at this point.

First, we must justify why we have constrained the polynomial multiplier to satisfy the conditions (102).

Asking for the degree of the final equation is asking for a rational expression of the quantities a $a_{\,'}$ $a_{\,''}$ $a_{\,'''}$, etc. b $b_{\,'}$ $b_{\,''}$; c; t; a' $a'_{\,'}$ $a'_{\,''}$, etc. b' $b'_{\,'}$ $b'_{\,''}$; c'; t'; a'' $a''_{\,'}$ $a''_{\,''}$ $a''_{\,'''}$, etc. b'' $b''_{\,'}$ $b''_{\,''}$; c''; t'';, etc., etc., which is independent of the quantities A $A_{\,'}$ $A_{\,''}$ $A_{\,'''}$, etc. B $B_{\,'}$ $B_{\,''}$; C; T, and which is the lowest possible degree of the final equation without assuming any specific relation among the coefficients of the given equations.

The function that gives the expression for D must therefore be such that the quantities A, $A_{\,'}$, etc. B, $B_{\,'}$, etc. vanish by themselves. But it is obvious that all the various polynomials that express the degree D of the final equation via the number of their terms must be of the same nature for this to happen; if this were not the case, the expression of the number of terms of one of them would be of one of the forms expressed in (91 ff). And the expression of the number of terms of another would have another form: But then these two expressions could become one another by simply swapping the quantities a $a_{\,'}$ $a_{\,''}$, etc. of one of the equations with those belonging to the other equation. This property is indeed absolutely necessary for the result of these various number of terms to be an exact differential whose order is equal to the number of equations, and to become a function of the quantities a, $a_{\,'}$, $a_{\,''}$ etc. that is independent of the quantities A, $A_{\,'}$, $A_{\,''}$, etc.

Any expression for D where the quantities A, $A_{\,'}$, $A_{\,''}$, etc. appear would indicate that the form of the polynomial multiplier, or that of the polynomials contributing to the expression of D is inappropriate.

(106.) Second, we have found eight different expressions for the value of the number of terms for the kind of polynomial we are currently dealing with in (91 ff); thus it follows that we will also have eight different expressions for the value of the degree of the final equation. But are these eight expressions for D equally acceptable, or do they belong to different cases for the equations, and, if so, how can we distinguish these cases?

Undoubtedly, these eight expressions for the value of D belong to different cases for the given equations, while satisfying the general conditions (83). But to distinguish these cases, we must deal with a question we did not raise until now to spare the reader's attention.

About the largest number of terms that can be cancelled in a given polynomial by using a given number of equations, without introducing new terms

(107.) In order not to burden our text with cumbersome calculations, we base our argument on a very simple polynomial, and we also assume that the given equations are of the form $(u^A \ldots n)^T$, where the exponents A, $A_{\prime}$, $A_{\prime\prime}$, etc. are not subject to any condition. It will be easy for the reader to see that the following can be applied to more general polynomial forms.

When there is one equation only, such as $(u^{a'} \ldots n)^{t'} = 0$, the largest number of terms that can be cancelled using this equation and without introducing new terms is $N(u^{A-a'} \ldots n)^{T-t'}$. This is easy to compute from (60).

When there are two equations only, can we say that the largest number of terms that may be cancelled in the polynomial $(u^A \ldots n)^T$, using these two equations, without introducing new terms, is always expressed by

$$N(u^{A-a'} \ldots n)^{T-t'} + N(u^{A-a''} \ldots n)^{T-t''} - N(u^{A-a'-a''} \ldots n)^{T-t'-t''}$$

as we seem to have assumed up until now?

We have indeed assumed this statement to be true for the polynomials used in the current theory. But if we generalize the question, then the expression above does not necessarily provide the largest number of terms that may be cancelled without introducing new ones.

For example, consider the complete polynomial $(x, y, z)^3$ and two incomplete equations such as $[x, (y, z)^1]^2 = 0$; that is, two equations whose incompleteness is relative to y and z, since they cannot reach the dimension 1 either together or separately.

From what was said until now, the largest number of terms that may be cancelled without introducing new ones appears to be

$$N[x, (y, z)^{3-1}]^{3-2} + N[x, (y, z)^{3-1}]^{3-2} - N[x, (y, z)^{3-2}]^{3-4}$$

or $2N[x, (y, z)^2]^1 - N[x, (y, z)^1]^{-1}$; but $[x, (y, z)^2]^1$ has no other term than $[x, (y, z)^1]^1$ or, equivalently, $(x, y, z)^1$, and $N[x, (y, z)^1]^{-1} = N(x, y, z)^{-1} = 0$; thus we have $2N(x, y, z)^1$, or eight as the largest number of remaining terms that may be cancelled in the polynomial $(x, y, z)^3$ using two equations of the form $[x, (y, z)^1]^2 = 0$. Therefore, multiplying the first equation by $Ax + By + Cz + E$ and the second by $A'x + B'y + C'z + E'$, we should be able to cancel eight terms in the polynomial $(x, y, z)^3$ with arbitrary coefficients. However, we can cancel nine such terms, without introducing new ones. We just have to multiply the first equation by the polynomial $(x, y, z)^2$, and the second by a similar polynomial; the first equation only provides $N(x, y, z)^2 - N(x, y, z)^0$ terms because of the terms that may be cancelled using the second equation. The second equation provides $N(x, y, z)^2$ terms; therefore, using the two equations, we can cancel $2N(x, y, z)^2 - N(x, y, z)^0$

terms. But we have introduced ten terms in the fourth dimension and we will need ten coefficients to cancel them. Thus the number of terms that may be cancelled without introducing new ones is $2N(x, y, z)^2 - N(x, y, z)^0 - 10 = 20 - 1 - 10 = 9$.

(108.) Thus there are two ways to not intoduce new terms: The first way is to introduce neither apparent nor real additional terms: This is done by using polynomial multipliers such that the product equation does not include terms whose order is higher than that of the proposed polynomial. The second is to introduce new apparent terms; that is, we use polynomial multipliers whose product with the equations does indeed introduce terms whose degree is higher than that of the proposed polynomial, but we will be able to cancel these terms later on.

(109.) Assume now that we ask for the largest number of terms that may be cancelled in the polynomial $(u^A \dots n)^T$ using the two equations $(u^{a'} \dots n)^{t'} = 0$, $(u^{a''} \dots n)^{t''} = 0$, without introducing new terms; assume we have multiplied the first equation by the polynomial $(u^{A'} \dots n)^{T'}$, and the second equation by the polynomial $(u^{A''} \dots n)^{T''}$; 'assume also that we have added the two products $(u^{A'+a'} \dots n)^{T'+t'}$, $(u^{A''+a''} \dots n)^{T''+t''}$ to the proposed polynomial $(u^A \dots n)^T$.

Then, assuming that $T' + t' > T$, $T'' + t'' > T$, $A' + a' > A$, etc., we need to use one of the polynomials to cancel terms introduced by the other. For this to happen, we must assume $T'' + t'' = T' + t'$, $A'' + a'' = A' + a'$, etc., which leads to $T'' = T' + t' - t''$, $A'' = A' + a' - a''$, etc. This results in $N(u^{A'} \dots n)^{T'} + N(u^{A'+a'-a''} \dots n)^{T'+t'-t''}$ coefficients.

But we can cancel $N(u^{A'-a''} \dots n)^{T'-t''}$ terms in the first polynomial without introducing new ones using the second equation; thus our two polynomial multipliers really provide only $N(u^{A'} \dots n)^{T'} + N(u^{A'+a'-a''} \dots n)^{T+t'-t''} - N(u^{A'-a''} \dots n)^{T'-t''}$ coefficients. However, we need $N(u^{A'+a'} \dots n)^{T'+t'} - N(u^A \dots n)^T$ coefficients to cancel the newly introduced terms; thus the true number of terms that we are able to cancel without introducing new ones is

$$N(u^A \dots n)^T - [N(u^{A'+a'} \dots n)^{T'+t'} - N(u^{A'} \dots n)^{T'} \\ -N(u^{A'+a'-a''} \dots n)^{T'+t'-t''} + N(u^{A'-a''} \dots n)^{T'-t''}].$$

(110.) From now on, we will say that the terms we introduce only to cancel them later are *fictitious*.

(111.) In order to cancel more terms in the polynomial using fictitious terms than without using them, the number of terms remaining in the polynomial in the first case must be smaller than in the second case. We must therefore have

$$N(u^{A'+a'} \dots n)^{T'+t'} - N(u^{A'} \dots n)^{T'} - N(u^{A'+a'-a''} \dots n)^{T'+t'-t''} \\ +N(u^{A'-a''} \dots n)^{T'-t''} < N(u^A \dots n)^T \\ -N(u^{A-a'} \dots n)^{T-t'} - N(u^{A-a''} \dots n)^{T-t''} + N(u^{A-a'-a''} \dots n)^{T-t'-t''}$$

with $T' > T - t'$; $A' > A - a'$, etc.

Thus we are able to cancel more terms by introducing fictitious terms than by not introducing them, if this condition is satisfied.

The quantities T', A', etc. must be chosen so that

$$N(u^{A'+a'}\ldots n)^{T'+t'} - N(u^{A'}\ldots n)^{T'} - N(u^{A'+a'-a''}\ldots n)^{T'+t'-t''}$$
$$+N(u^{A'-a''}\ldots n)^{T'-t''}$$

reaches a *minimum* to cancel the largest possible number of terms.

(112.) No matter what, we note that the latter expression is precisely that of the number of remaining terms in the polynomial $(u^{A'+a'}\ldots n)^{T'+t'}$, when we use the two equations to cancel all terms that can be cancelled without introducing new ones and without introducing fictitious terms.

Thus it is always possible to find a polynomial $(u^A\ldots n)^T$ such that the number of remaining terms is as small as possible, by cancelling, without using fictitious terms, as many terms as possible without introducing new ones; that is, the number of terms cannot be made lower by introducing fictitious terms.

Let us now consider the case of three equations.

(113.) Let us assume that we multiply the first equation by the polynomial $(u^{A'}\ldots n)^{T'}$, the second by the polynomial $(u^{A''}\ldots n)^{T''}$, and the third by the polynomial $(u^{A'''}\ldots n)^{T'''}$, and that we add the three products to the proposed polynomial $(u^A\ldots n)^T$; altogether, the number of coefficients is $N(u^{A'}\ldots n)^{T'} + N(u^{A''}\ldots n)^{T''} + N(u^{A'''}\ldots n)^{T'''}$, but all these coefficients are equally appropriate to cancel terms in the proposed polynomial.

For the sake of generality, we assume $T'+t' > T$; $A'+a' > A$; etc. $T''+t'' > T$; $A''+a'' > A$; etc. $T'''+t''' > T$; $A'''+a''' > A$; etc.

We first remark that one of the essential conditions to cancel fictitious terms is that at least two of the quantities $T'+t'$, $T''+t''$, $T'''+t'''$, be equal. The same must hold regarding the quantities

$$A'+a', A''+a'', A'''+a''', \text{ etc.}$$

In addition, these three quantities must be equal to each other to maximize the number of terms that can be cancelled. Indeed, it is clear that if one of them were smaller than the other two, we would obviously have fewer coefficients available than if we assumed all three to be equal.

We therefore have $T'' = T'+t'-t''$, $T''' = T'+t'-t'''$, $A'' = A'+a'-a''$, $A''' = A'+a'-a'''$, etc.

Let us suppose for now (and this is always possible) that the polynomial $(u^{A'}\ldots n)^{T'}$ is such that no advantage may be gained by introducing fictitious terms; then the number of useful coefficients of the polynomial $(u^{A'}\ldots n)^{T'}$ is

$$N(u^{A'}\ldots n)^{T'} - N(u^{A'-a''}\ldots n)^{T'-t''} - N(u^{A'-a'''}\ldots n)^{T'-t'''}$$
$$+N(u^{A'-a''-a'''}\ldots n)^{T'-t''-t'''},$$

considering the number of terms that may be cancelled using the last two equations.

The number of useful coefficients of the polynomial $(u^{A''}\ldots n)^{T''}$, that is, of the polynomial $(u^{A'+a'-a''}\ldots n)^{T'+t'-t''}$, is

$$N(u^{A'+a'-a''}\ldots n)^{T'+t'-t''} - N(u^{A'+a'-a''-a'''}\ldots n)^{T'+t'-t''-t'''}$$

because of the terms that may be cancelled using the last equation.

Finally, the number of useful coefficients of the polynomial $(u^{A'''} \ldots n)^{T'''}$ is $N(u^{A'+a'-a'''} \ldots n)^{T'+t'-t'''}$.

We must use some of these useful coefficients to cancel fictitious terms, whose number equals $N(u^{A'+a'} \ldots n)^{T'+t'} - N(u^A \ldots n)^T$; removing the rest from $N(u^A \ldots n)^T$, the total number of remaining terms without introducing new ones is

$$
\begin{aligned}
&N(u^{A'+a'} \ldots n)^{T'+t'} - N(u^{A'} \ldots n)^{T'} - N(u^{A'+a'-a''} \ldots n)^{T'+t'-t''} \\
&+N(u^{A'-a''} \ldots n)^{T'-t''} - N(u^{A'+a'-a'''} \ldots n)^{T'+t'-t'''} \\
&+N(u^{A'-a'''} \ldots n)^{T'-t'''} + N(u^{A'+a'-a''-a'''} \ldots n)^{T'+t'-t''-t'''} \\
&-N(u^{A'-a'-a'''} \ldots n)^{T'-t''-t'''}.
\end{aligned}
$$

This quantity will therefore have to be *minimized*.

We remark that this expression is precisely the expression of the number of terms that would remain in the polynomial $(u^{A'+a'} \ldots n)^{T'+t'}$, after having cancelled all possible terms, using the three equations without introducing fictitious terms.

(114.) Therefore it is always possible to find a polynomial $(u^A \ldots T)$ such that the remaining number of terms is as small as possible, after having cancelled all possible terms without introducing fictitious terms; that is, the number of remaining terms cannot be made lower by using fictitious terms. We now see how to generalize our approach to larger sets of equations.

But although we have shown it is always possible to find such a polynomial, it may happen that the partial polynomials entering in the expression of the number of remaining terms not be of the same nature. But since finding the degree of the final equation (105) necessarily requires this condition, it follows that we will be able to recognize the appropriate value of D (106) based on the possibility or impossibility for these polynomials to be of the same nature.

(115.) We have established that the smallest number of remaining terms in the polynomial multiplier must necessarily be expressed as $d^{n-1} N(u^A \ldots n)^T$; thus this quantity must be at a *minimum*.

Assuming it is indeed a minimum, introducing fictitious terms using either similar polynomials or polynomials of a different kind must result in fewer cancelled terms, or in more remaining terms. Let us therefore assume a polynomial of a similar nature represented by $(u^{A'} \ldots n)^{T'}$ such that $T' > T$, and $A' > A$, etc.

We must have

$$
d^{n-1} N(u^{A'} \ldots n)^{T'} > d^{n-1} N(u^A \ldots n)^T
$$

or

$$
\begin{aligned}
&d^{n-1} N(u^{A'} \ldots n)^{T} \ldots \left(\begin{matrix} T' \\ -t', -t'', \text{ etc.} \end{matrix} : \begin{matrix} A' \\ -a', -a'', \text{ etc.} \end{matrix} \text{ etc.} \right) \\
&> d^{n-1} N(u^A \ldots n)^{T} \ldots \left(\begin{matrix} T \\ -t', -t'', \text{ etc.} \end{matrix} : \begin{matrix} A \\ -a', -a'', \text{ etc.} \end{matrix} \text{ etc.} \right).
\end{aligned}
$$

Therefore

$$d^n N(u^{A'} \dots n)^{T'} \dots$$

$$\dots \left(\begin{matrix} T' \\ -(T'-T), -t', -t'', \text{ etc.} \end{matrix} : \begin{matrix} A' \\ -(A'-A), -a', -a'', \text{ etc.} \end{matrix} \text{ etc.} \right) > 0$$

or

$$d^n N(u^{A'} \dots n)^{T'} \dots$$

$$\dots \left(\begin{matrix} T' \\ (T'-T), t', t'', \text{ etc.} \end{matrix} : \begin{matrix} A' \\ (A'-A), a', a'', \text{ etc.} \end{matrix} \text{ etc.} \right) > 0.$$

(116.) Assume we differentiate the quantity $N(u^{A'} \dots n)^{T'}$ n times, where $(u^{A'} \dots n)^{T'}$ is an arbitrary polynomial; assume T' varies by the amounts t', t'', etc. and the amount $T' - T$, successively; assume A' varies by the amounts a', a'', etc. and the amount $A' - A$, etc., successively. The result of these differentiations must be larger than 0, for any $T' - T$, $A' - A$, etc.

Therefore, if we bring together all the terms containing $T' - T$, their sum must be positive or greater than 0. The same property must hold for the sum of the terms containing $A' - A$, and so on.

Determination of the symptoms indicating which value of the degree of the final equation must be chosen or rejected, among the different available expressions

(117.) There are as many conditions to be satisfied as the number of exponents contained in the polynomial multiplier. If all these conditions are satisfied, the value of D is acceptable; if any one of them is not, the value for D must be rejected.

But we must observe that, since nothing is available to help choose one of the proposed equations rather than another, we must inspect these conditions as many times as there are equations or unknowns. We know the value of D is acceptable only if it passes all these tests.

We need not be concerned about not finding any satisfactory expression, because we know that the value of D always exists a priori. But it may happen that several sets of conditions are satisfied; then all the corresponding values for D are equally acceptable.

In that regard we must observe:

1. that all the values of D resulting from a new combination of the equations, that is, from the tacit change of the sum equation, are the same. This is what we will soon see by expanding the general value of D found in (104); this expansion will easily reveal that swapping the exponents of an equation with those of another does not change the value of D.

2. that the values of D are equal again when the equations belong to one form or another.

3. that if we find several different values for D, they must reduce to a single value, by examining how the conditions hold, when swapping

the exponents. Intuition suggests this must be true, since there can only be one final equation; but we could also think that it is possible for one of these forms to introduce a superfluous factor in the final equation, leading in effect to different values for D. We must therefore prove this is impossible; that is, if several values of D are possible, they must all be the same.

Indeed, if two diferent values for D could co-exist, then the two corresponding equations would be such that one is a factor of the other; that one would then have at least one superfluous root. Therefore, it would be possible to cancel one more term than what was done in the corresponding product equation; but the conditions under which the value of D is valid are such that the largest possible number of terms has been cancelled; thus there is no superfluous root, and therefore there cannot be different possible values for D. Thus, if testing the conditions leads to several values of D, these values must be identical, and the proposed equations simultaneously belong to many forms.

Expansion of the various values of the degree of the final equation, resulting from the general expression found in (104), and expansion of the set of conditions that justify these values

(118.) We therefore see

1. that to obtain the different values of D that may occur for the incomplete equations of interest in (82 ff), we only need to substitute the exponents of the product equation in the value of P for any of the forms, from (91 ff); to differentiate this value n times, while varying each exponent of the product equation by all corresponding exponents in the given equations.

2. that to obtain the conditions justifying this value of D, we must from (116) differentiate n times the value of P belonging to the same form, while varying successively each of its exponents by an amount corresponding to each exponent of all equations, other than that we implicitly take for the multiplicand equation, and of an arbitrary quantity; then we must assume that the sum of terms that multiply each arbitrary quantity is greater than zero.

But it is easy to see that the results of the first computation immediately yield those of the second, and that the computations to determine the appropriate conditions reduce to choosing the sum of terms which multiply one of the exponents of the multiplicand equation in the value of D, and to require this sum to be greater than zero.

In order to limit the proliferation of computations, we limit our developments to the case when we have three equations in three unknowns. The computations are exactly the same for a larger number of equations and unknowns, and there is no difference, except for the proliferation of expressions in the results. We therefore lose no generality by taking this approach; it contributes to making things clearer.

Application of the preceding theory to equations in three unknowns

(119.) The eight expressions we have found for P in (91 ff) do simplify, when there are three unknowns only. This is because $C = T$, and terms containing $T - C$ vanish: Indeed, $N(u\ldots n)^{T-C-1}$, $N(u\ldots n)^{T-C-2}$ and $N(u\ldots n)^{T-C-3}$ become $N(u\ldots n)^{-1}$, $N(u\ldots n)^{-2}$, $N(u\ldots n)^{-3}$, and these quantities are all 0 by (39).

We can use this observation to simplify the expression of P; differentiating as we just said in (118), we find the various values of D and the corresponding conditions for each of the eight forms.

First form

(120.) We assume the polynomial multiplier satisfies

$$C - B < \underset{\prime}{B} - A;\quad C - B < \underset{\prime\prime}{B} - A';\quad C - \underset{\prime}{B} < \underset{\prime\prime}{B} - \underset{\prime\prime}{A}.$$

We then have

$$D = tt't'' - (t - a).(t' - a').(t'' - a'') - (t - \underset{\prime}{a}).(t' - \underset{\prime}{a'}).(t'' - \underset{\prime}{a''})$$
$$-(t - \underset{\prime\prime}{a}).(t' - \underset{\prime\prime}{a'}).(t'' - \underset{\prime\prime}{a''}) + (t - b).(t' - b').(t'' - b'')$$
$$+(t - \underset{\prime}{b}).(t' - \underset{\prime}{b'}).(t'' - \underset{\prime}{b''}) + (t - \underset{\prime\prime}{b}).(t' - \underset{\prime\prime}{b'}).(t'' - \underset{\prime\prime}{b''})$$
$$-(a + \underset{\prime}{a} - b).(t' - b').(t'' - b'') - (a' + \underset{\prime}{a'} - b').(t - b).(t'' - b'')$$
$$-(a'' + \underset{\prime}{a''} - b'').(t - b).(t' - b') - (a + \underset{\prime\prime}{a} - b).(t' - \underset{\prime}{b'}).(t'' - \underset{\prime}{b''})$$
$$-(a' + \underset{\prime\prime}{a'} - b').(t - \underset{\prime}{b}).(t'' - \underset{\prime}{b''}) - (a'' + \underset{\prime\prime}{a''} - b'').(t - \underset{\prime}{b}).(t' - \underset{\prime}{b'})$$
$$-(\underset{\prime}{a} + \underset{\prime\prime}{a} - \underset{\prime\prime}{b}).(t' - \underset{\prime\prime}{b'}).(t'' - \underset{\prime\prime}{b''})$$
$$-(\underset{\prime}{a'} + \underset{\prime\prime}{a'} - \underset{\prime\prime}{b'}).(t - \underset{\prime\prime}{b}).(t'' - \underset{\prime\prime}{b''})$$
$$-(\underset{\prime}{a''} + \underset{\prime\prime}{a''} - \underset{\prime\prime}{b''}).(t - \underset{\prime\prime}{b}).(t' - \underset{\prime\prime}{b'}).$$

Conditions for this value of D to be valid

From what was said in (118), these conditions can be found by isolating all factors of t and requiring the sum of these terms to be positive; likewise, by isolating all factors of b and requiring their sum to be positive, and so on.

$$\left.\begin{array}{r}
t't'' - (t'-a').(t''-a'') - (t'-\underset{,}{a}').(t''-\underset{,}{a}'') \\[4pt]
-(t'-\underset{,}{a}').(t''-\underset{,}{a}'') + (t'-b').(t''-b'') + (t'-\underset{,}{b}').(t''-\underset{,}{b}'') \\[4pt]
+(t'-\underset{,}{b}').(t''-\underset{,}{b}'') - (a'+\underset{,}{a}'-b').(t''-b'') \\[4pt]
-(a''+\underset{,}{a}''-b'').(t'-b') - (a'+\underset{,}{a}'-\underset{,}{b}').(t''-\underset{,}{b}'') \\[4pt]
-(a''+\underset{,}{a}''-\underset{,}{b}'').(t'-\underset{,}{b}') - (\underset{,}{a}'+\underset{,}{a}'-\underset{,}{b}').(t''-\underset{,}{b}'') \\[4pt]
-(\underset{,}{a}''+\underset{,}{a}''-\underset{,}{b}'').(t'-\underset{,}{b}')
\end{array}\right\} > 0,$$

$$(a'+\underset{,}{a}'-b').(t''-b'') - (a''+\underset{,}{a}''-b'').(t'-b') > 0,$$

$$(a'+\underset{,}{a}'-\underset{,}{b}').(t''-b'') + (a''+\underset{,}{a}''-\underset{,}{b}'').(t'-b') > 0,$$

$$(\underset{,}{a}'+\underset{,}{a}'-\underset{,}{b}').(t''-\underset{,}{b}'') + (\underset{,}{a}''+\underset{,}{a}''-\underset{,}{b}'').(t'-\underset{,}{b}') > 0,$$

$$(t'-a').(t''-a'') - (t'-b').(t''-b'') - (t'-\underset{,}{b}').(t''-\underset{,}{b}'') > 0,$$

$$(t'-\underset{,}{a}').(t''-\underset{,}{a}'') - (t'-b').(t''-b'') - (t'-\underset{,}{b}').(t''-\underset{,}{b}'') > 0,$$

$$(t'-\underset{,}{a}').(t''-\underset{,}{a}'') - (t'-\underset{,}{b}').(t''-\underset{,}{b}'') - (t'-\underset{,}{b}').(t''-\underset{,}{b}'') > 0.$$

The second, third and fourth conditions always hold since, from (83), the general conditions for the existence of the equations under study require $a'+\underset{,}{a}' > b'$; $a''+\underset{,}{a}'' > b''$; $a'+\underset{,}{a}' > \underset{,}{b}'$; $a''+\underset{,}{a}'' > \underset{,}{b}''$; $\underset{,}{a}'+\underset{,}{a}' > \underset{,}{b}'$; $a''+\underset{,}{a}'' > \underset{,}{b}''$.

Second form

$$C - B < \underset{,}{B} - A; \quad C - B < \underset{,,}{B} - \underset{,}{A}; \quad C - \underset{,}{B} > \underset{,,}{B} - \underset{,,}{A}.$$

(121.)

$$D = tt't'' - (t-a).(t'-a').(t''-a'') - (t-a).(t'-\underset{,}{a}').(t''-\underset{,}{a}'')$$
$$-(t-\underset{,}{a}).(t'-\underset{,}{a}').(t''-\underset{,}{a}'') + (t-b).(t'-b').(t''-b'')$$
$$+(t-\underset{,}{b}).(t'-\underset{,}{b}').(t''-\underset{,}{b}'') + (t-\underset{,}{b}).(t'-\underset{,}{b}').(t''-\underset{,}{b}'')$$
$$-(a+\underset{,}{a}-b).(t'-b').(t''-b'') - (a'+\underset{,}{a}'-b').(t-b).(t''-b'')$$
$$-(a''+\underset{,}{a}''-b'').(t-b).(t'-b') - (a+\underset{,}{a}-b).(t'-\underset{,}{b}').(t''-\underset{,}{b}'')$$
$$-(a'+\underset{,}{a}'-\underset{,}{b}').(t-\underset{,}{b}).(t''-\underset{,}{b}'') - (a''+\underset{,}{a}''-\underset{,}{b}'').(t-\underset{,}{b}).(t'-\underset{,}{b}')$$
$$-(\underset{,}{a}+\underset{,}{a}-\underset{,}{b}).(t'-\underset{,}{b}').(t''-\underset{,}{b}'')$$
$$-(\underset{,}{a}'+\underset{,}{a}'-\underset{,}{b}').(t-\underset{,}{b}).(t''-\underset{,}{b}'')$$
$$-(\underset{,}{a}''+\underset{,}{a}''-\underset{,}{b}'').(t-\underset{,}{b}).(t'-\underset{,}{b}')$$
$$+(t+\underset{,}{a}-b-\underset{,}{b}).(t'+\underset{,}{a}'-b-\underset{,}{b}').(t''+\underset{,}{a}''-b''-\underset{,}{b}'').$$

Conditions

$$
\left.
\begin{aligned}
& t't'' - (t'-a').(t''-a'') - (t'-\underset{\prime}{a}').(t''-\underset{\prime}{a}'') \\
-(t'-\underset{\prime\prime}{a}').(t''-\underset{\prime\prime}{a}'') & + (t'-b').(t''-b'') + (t'-\underset{\prime}{b}').(t''-\underset{\prime}{b}'') \\
& +(t'-\underset{\prime\prime}{b}').(t''-\underset{\prime\prime}{b}'') - (a'+\underset{\prime}{a}'-b').(t''-b'') \\
& -(a''+\underset{\prime}{a}''-b'').(t'-b') - (a'+\underset{\prime\prime}{a}'-\underset{\prime}{b}').(t''-\underset{\prime}{b}'') \\
& -(a''+\underset{\prime\prime}{a}''-\underset{\prime}{b}'').(t'-\underset{\prime}{b}') - (\underset{\prime}{a}'+\underset{\prime\prime}{a}'-\underset{\prime\prime}{b}').(t''-\underset{\prime\prime}{b}'') \\
& -(\underset{\prime}{a}''+\underset{\prime\prime}{a}''-\underset{\prime\prime}{b}'').(t'-\underset{\prime\prime}{b}') \\
& +(t'+\underset{\prime\prime}{a}'-b'-\underset{\prime}{b}').(t''+\underset{\prime\prime}{a}''-b''-\underset{\prime}{b}'')
\end{aligned}
\right\} > 0,
$$

$$
(a'+\underset{\prime}{a}'-b').(t''-b'') - (a''+\underset{\prime}{a}''-b'').(t'-b') > 0,
$$

$$
\left.
\begin{aligned}
(a'+\underset{\prime\prime}{a}'-b').(t''-b'') + (a''+\underset{\prime}{a}''-b'').(t'-b') & \\
-(t'+\underset{\prime\prime}{a}'-b'-\underset{\prime}{b}').(t''+\underset{\prime\prime}{a}''-b''-\underset{\prime}{b}'') &
\end{aligned}
\right\} > 0,
$$

$$
\left.
\begin{aligned}
(\underset{\prime}{a}'+\underset{\prime\prime}{a}'-\underset{\prime}{b}').(t''-\underset{\prime}{b}'') + (\underset{\prime}{a}''+\underset{\prime\prime}{a}''-\underset{\prime}{b}'').(t'-\underset{\prime}{b}') & \\
-(t'+\underset{\prime}{a}'-b'-\underset{\prime}{b}').(t''+\underset{\prime}{a}''-b''-\underset{\prime}{b}'') &
\end{aligned}
\right\} > 0,
$$

$$
(t'-a').(t''-a'') - (t'-b').(t''-b'') - (t'-\underset{\prime}{b}').(t''-\underset{\prime}{b}'') > 0,
$$

$$
(t'-\underset{\prime}{a}').(t''-\underset{\prime}{a}'') - (t'-b').(t''-b'') - (t'-\underset{\prime\prime}{b}').(t''-\underset{\prime\prime}{b}'') > 0,
$$

$$
\left.
\begin{aligned}
(t'-\underset{\prime\prime}{a}').(t''-\underset{\prime\prime}{a}'') - (t'-\underset{\prime}{b}').(t''-\underset{\prime}{b}'') & \\
-(t'-\underset{\prime\prime}{b}').(t''-\underset{\prime\prime}{b}'') & \\
+(t'+\underset{\prime\prime}{a}'-\underset{\prime}{b}'-\underset{\prime\prime}{b}').(t''+\underset{\prime\prime}{a}''-\underset{\prime}{b}''-\underset{\prime\prime}{b}'') &
\end{aligned}
\right\} > 0.
$$

Third form

$$C - B > \underset{\prime}{B} - A;\quad C - B < \underset{\prime\prime}{B} - \underset{\prime}{A};\quad C - \underset{\prime}{B} < \underset{\prime\prime}{B} - \underset{\prime\prime}{A}.$$

(122.)

$$D = tt't'' - (t-a).(t'-a').(t''-a'') - (t-\underset{\prime}{a}).(t'-\underset{\prime}{a}').(t''-\underset{\prime}{a}'')$$
$$-(t-\underset{\prime\prime}{a}).(t'-\underset{\prime\prime}{a}').(t''-\underset{\prime\prime}{a}'') + (t-b).(t'-b').(t''-b'')$$
$$+(t-\underset{\prime}{b}).(t'-\underset{\prime}{b}').(t''-\underset{\prime}{b}'') + (t-\underset{\prime\prime}{b}).(t'-\underset{\prime\prime}{b}').(t''-\underset{\prime\prime}{b}'')$$
$$-(a+\underset{\prime}{a}-b).(t'-b').(t''-b'') - (a'+\underset{\prime}{a}'-b').(t-b).(t''-b'')$$
$$-(a''+\underset{\prime}{a}''-b'').(t-b).(t'-b') - (a+\underset{\prime\prime}{a}-b).(t'-\underset{\prime}{b}').(t''-\underset{\prime}{b}'')$$
$$-(a'+\underset{\prime\prime}{a}'-\underset{\prime}{b}').(t-\underset{\prime}{b}).(t''-\underset{\prime}{b}'')$$
$$-(\underset{\prime}{a}''+\underset{\prime\prime}{a}''-\underset{\prime}{b}'').(t-\underset{\prime}{b}).(t'-\underset{\prime}{b}')$$
$$-(\underset{\prime}{a}+\underset{\prime\prime}{a}-\underset{\prime}{b}).(t'-\underset{\prime\prime}{b}').(t''-\underset{\prime\prime}{b}'')$$
$$-(\underset{\prime}{a}'+\underset{\prime\prime}{a}'-\underset{\prime}{b}').(t-\underset{\prime\prime}{b}).(t''-\underset{\prime\prime}{b}'')$$
$$-(\underset{\prime}{a}''+\underset{\prime\prime}{a}''-\underset{\prime\prime}{b}'').(t-\underset{\prime\prime}{b}).(t'-\underset{\prime\prime}{b}')$$
$$+(t+a-b-\underset{\prime}{b}).(t'+a'-b'-\underset{\prime}{b}').(t''+a''-b''-\underset{\prime}{b}'').$$

Conditions

$$\left.\begin{array}{r}
t't'' - (t'-a').(t''-a'') - (t'-\underset{\prime}{a}').(t''-\underset{\prime}{a}'') \\
-(t'-\underset{\prime\prime}{a}').(t''-\underset{\prime\prime}{a}'') + (t'-b').(t''-b'') \\
+(t'-\underset{\prime}{b}').(t''-\underset{\prime}{b}'') + (t'-\underset{\prime\prime}{b}').(t''-\underset{\prime\prime}{b}'') \\
-(a'+\underset{\prime}{a}'-b').(t''-b'') - (a''+\underset{\prime}{a}''-b'').(t'-b') \\
-(a'+\underset{\prime\prime}{a}'-\underset{\prime}{b}').(t''-\underset{\prime}{b}'') - (a''+\underset{\prime\prime}{a}''-\underset{\prime}{b}'').(t'-\underset{\prime}{b}') \\
-(\underset{\prime}{a}'+\underset{\prime\prime}{a}'-\underset{\prime}{b}').(t''-\underset{\prime\prime}{b}'') - (\underset{\prime}{a}''+\underset{\prime\prime}{a}''-\underset{\prime\prime}{b}'').(t'-\underset{\prime\prime}{b}') \\
+(t'+a'-b'-\underset{\prime}{b}').(t''+a''-b''-\underset{\prime}{b}'')
\end{array}\right\} > 0,$$

$$\left.\begin{array}{r}
(a'+\underset{\prime}{a}'-b').(t''-b'') - (a''+\underset{\prime}{a}''-b'').(t'-b') \\
-(t'+a'-b'-\underset{\prime}{b}').(t''+a''-b''-\underset{\prime}{b}'')
\end{array}\right\} > 0,$$

$$\left.\begin{array}{r}
(a'+\underset{\prime\prime}{a}'-\underset{\prime}{b}').(t''-\underset{\prime}{b}'') + (a''+\underset{\prime}{a}''-\underset{\prime}{b}'').(t'-\underset{\prime}{b}') \\
-(t'+a'-b'-\underset{\prime}{b}').(t''+a''-b''-\underset{\prime}{b}'')
\end{array}\right\} > 0,$$

$$(\underset{\prime}{a}'+\underset{\prime\prime}{a}'-\underset{\prime}{b}').(t''-\underset{\prime\prime}{b}'') + (\underset{\prime}{a}''+\underset{\prime\prime}{a}''-\underset{\prime}{b}'').(t'-\underset{\prime\prime}{b}') > 0,$$

$$\left.\begin{array}{r}
(t'-a').(t''-a'') - (t'-b').(t''-b'') - (t'-\underset{\prime}{b}').(t''-\underset{\prime}{b}'') \\
+(t'+a'-b'-\underset{\prime}{b}').(t''+a''-b''-\underset{\prime}{b}'')
\end{array}\right\} > 0,$$

$$(t'-\underset{\prime}{a}').(t''-\underset{\prime}{a}'') - (t'-b').(t''-b'') - (t'-\underset{\prime}{b}').(t''-\underset{\prime}{b}'') > 0,$$

$$(t'-\underset{\prime\prime}{a}').(t''-\underset{\prime\prime}{a}'') - (t'-b').(t''-b'') - (t'-\underset{\prime\prime}{b}').(t''-\underset{\prime\prime}{b}'') > 0.$$

Fourth form

(123.)
$$C - B > \underset{,}{B} - A; \ C - B < \underset{,,}{B} - \underset{,}{A}; \ C - \underset{,}{B} > \underset{,,}{B} - \underset{,,}{A}.$$

$$D = tt't'' - (t-a).(t'-a').(t''-a'') - (t-\underset{,}{a}).(t'-\underset{,}{a}').(t''-\underset{,}{a}'')$$
$$-(t-\underset{,,}{a}).(t'-\underset{,,}{a}').(t''-\underset{,,}{a}'') + (t-b).(t'-b').(t''-b'')$$
$$+(t-\underset{,}{b}).(t'-\underset{,}{b}').(t''-\underset{,}{b}'') + (t-\underset{,,}{b}).(t'-\underset{,,}{b}').(t''-\underset{,,}{b}'')$$
$$-(a+\underset{,}{a}-b).(t'-b').(t''-b'') - (a'+\underset{,}{a}'-b').(t-b).(t''-b'')$$
$$-(a''+\underset{,}{a}''-b'').(t-b).(t'-b') - (a+\underset{,,}{a}-b).(t'-\underset{,}{b}').(t''-\underset{,}{b}'')$$
$$-(a'+\underset{,,}{a}'-\underset{,}{b}').(t-\underset{,}{b}).(t''-\underset{,}{b}'') - (a''+\underset{,,}{a}''-\underset{,}{b}'').(t-\underset{,}{b}).(t'-\underset{,}{b}')$$
$$-(\underset{,}{a}+\underset{,,}{a}-\underset{,,}{b}).(t'-\underset{,,}{b}').(t''-\underset{,,}{b}'') - (\underset{,}{a}'+\underset{,,}{a}'-\underset{,,}{b}').(t-\underset{,,}{b}).(t'-\underset{,,}{b}')$$
$$-(\underset{,}{a}''+\underset{,,}{a}''-\underset{,,}{b}'').(t-\underset{,,}{b}).(t'-\underset{,,}{b}')$$
$$+(t+a-b-\underset{,}{b}).(t'+a'-b'-\underset{,}{b}').(t''+a''-b''-\underset{,}{b}'')$$
$$+(t+\underset{,,}{a}-\underset{,}{b}-\underset{,,}{b}).(t'+\underset{,,}{a}'-\underset{,}{b}'-\underset{,,}{b}').(t''+\underset{,,}{a}''-\underset{,}{b}''-\underset{,,}{b}'').$$

Conditions

$$\left. \begin{array}{r}
t't'' - (t'-a').(t''-a'') - (t'-\underset{,}{a}').(t''-\underset{,}{a}'') - (t'-\underset{,,}{a}').(t''-\underset{,,}{a}'') \\
+(t'-b').(t''-b'') + (t'-\underset{,}{b}').(t''-\underset{,}{b}'') + (t'-\underset{,,}{b}').(t''-\underset{,,}{b}'') \\
-(a'+\underset{,}{a}'-b').(t''-b'') - (a''+\underset{,}{a}''-b'').(t'-b') \\
-(a'+\underset{,,}{a}'-\underset{,}{b}').(t''-\underset{,}{b}'') - (a''+\underset{,,}{a}''-\underset{,}{b}'').(t'-\underset{,}{b}') \\
-(\underset{,}{a}'+\underset{,,}{a}'-\underset{,,}{b}').(t''-\underset{,,}{b}'') - (\underset{,}{a}''+\underset{,,}{a}''-\underset{,,}{b}'').(t'-\underset{,,}{b}') \\
+(t'+a'-b'-\underset{,}{b}').(t''+a''-b''-\underset{,}{b}'') \\
+(t'+\underset{,,}{a}'-\underset{,}{b}'-\underset{,,}{b}').(t''+\underset{,,}{a}''-\underset{,}{b}''-\underset{,,}{b}'')
\end{array} \right\} > 0,$$

$$\left. \begin{array}{r}
(a'+\underset{,}{a}'-b').(t''-b'') + (a''+\underset{,}{a}''-b'').(t'-b') \\
-(t'+a'-b'-\underset{,}{b}').(t''+a''-b''-\underset{,}{b}'')
\end{array} \right\} > 0,$$

$$\left. \begin{array}{r}
(a'+\underset{,,}{a}'-\underset{,}{b}').(t''-\underset{,}{b}'') + (a''+\underset{,,}{a}''-\underset{,}{b}'').(t'-\underset{,}{b}') \\
-(t'+a'-b'-\underset{,}{b}').(t''+a''-b''-\underset{,}{b}'') \\
-(t'+\underset{,,}{a}'-\underset{,}{b}'-\underset{,,}{b}').(t''+\underset{,,}{a}''-\underset{,}{b}''-\underset{,,}{b}'')
\end{array} \right\} > 0,$$

$$\left. \begin{array}{r}
(\underset{,}{a}'+\underset{,,}{a}'-\underset{,,}{b}').(t''-\underset{,,}{b}'') + (\underset{,}{a}''+\underset{,,}{a}''-\underset{,,}{b}'').(t'-\underset{,,}{b}') \\
-(t'+\underset{,,}{a}'-\underset{,}{b}'-\underset{,,}{b}').(t''+\underset{,,}{a}''-\underset{,}{b}''-\underset{,,}{b}'')
\end{array} \right\} > 0,$$

$$\left. \begin{array}{r}
(t'-a').(t''-a'') - (t'-b').(t''-b'') - (t'-\underset{,}{b}').(t''-\underset{,}{b}'') \\
+(t'+a'-b'-\underset{,}{b}').(t''+a''-b''-\underset{,}{b}'')
\end{array} \right\} > 0,$$

$$(t'-\underset{,}{a}').(t''-\underset{,}{a}'') - (t'-b').(t''-b'') - (t'-\underset{,}{b}').(t''-\underset{,}{b}'') > 0,$$

$$\left. \begin{array}{r}
(t'-\underset{,,}{a}').(t''-\underset{,,}{a}'') - (t'-\underset{,}{b}').(t''-\underset{,}{b}'') - (t'-\underset{,,}{b}').(t''-\underset{,,}{b}'') \\
+(t'+\underset{,,}{a}'-\underset{,}{b}'-\underset{,,}{b}').(t''+\underset{,,}{a}''-\underset{,}{b}''-\underset{,,}{b}'')
\end{array} \right\} > 0.$$

Fifth form

(124.)

$$C - B > \underset{,}{B} - A; \quad C - B > \underset{,,}{B} - A; \quad C - B < \underset{,}{B} - \underset{,,}{A}.$$

$$D = tt't'' - (t-a).(t'-a').(t''-a'') - (t-\underset{,}{a}).(t'-\underset{,}{a}').(t''-\underset{,}{a}'')$$
$$-(t-\underset{,,}{a}).(t'-\underset{,,}{a}').(t''-\underset{,,}{a}'') + (t-b).(t'-b').(t''-b'')$$
$$+(t-\underset{,}{b}).(t'-\underset{,}{b}').(t''-\underset{,}{b}'') + (t-\underset{,,}{b}).(t'-\underset{,,}{b}').(t''-\underset{,,}{b}'')$$
$$-(a+\underset{,}{a}-b).(t'-b').(t''-b'') - (a'+\underset{,}{a}'-b').(t-b).(t''-b'')$$
$$-(a''+\underset{,}{a}''-b'').(t-b).(t'-b') - (a+\underset{,,}{a}-b).(t'-\underset{,}{b}').(t''-\underset{,}{b}'')$$
$$-(a'+\underset{,,}{a}'-\underset{,}{b}').(t-\underset{,}{b}).(t''-\underset{,}{b}'') - (a''+\underset{,,}{a}''-\underset{,}{b}'').(t-\underset{,}{b}).(t'-\underset{,}{b}')$$
$$-(\underset{,}{a}+\underset{,,}{a}-\underset{,}{b}).(t'-\underset{,,}{b}').(t''-\underset{,,}{b}'') - (\underset{,}{a}'+\underset{,,}{a}'-\underset{,}{b}').(t-\underset{,,}{b}).(t''-\underset{,,}{b}'')$$
$$-(\underset{,}{a}''+\underset{,,}{a}''-\underset{,}{b}'').(t-\underset{,,}{b}).(t'-\underset{,,}{b}')$$
$$+(t+a-b-\underset{,}{b}).(t'+a'-b'-\underset{,}{b}').(t''+a''-b''-\underset{,}{b}'')$$
$$+(t+\underset{,}{a}-b-\underset{,,}{b}).(t'+\underset{,}{a}'-b'-\underset{,,}{b}').(t''+\underset{,}{a}''-b''-\underset{,,}{b}'').$$

Conditions

$$\left.\begin{aligned}
&t't'' - (t'-a').(t''-a'') - (t'-\underset{,}{a}').(t''-\underset{,}{a}'') - (t'-\underset{,,}{a}').(t''-\underset{,,}{a}'')\\
&\quad +(t'-b').(t''-b'') + (t'-\underset{,}{b}').(t''-\underset{,}{b}'') + (t'-\underset{,,}{b}').(t''-\underset{,,}{b}'')\\
&\qquad -(a'+\underset{,}{a}'-b').(t''-b'') - (a''+\underset{,}{a}''-b'').(t'-b')\\
&\qquad -(a'+\underset{,,}{a}'-\underset{,}{b}').(t''-\underset{,}{b}'') - (a''+\underset{,,}{a}''-\underset{,}{b}'').(t'-\underset{,}{b}')\\
&\qquad -(\underset{,}{a}'+\underset{,,}{a}'-\underset{,}{b}').(t''-\underset{,,}{b}'') - (\underset{,}{a}''+\underset{,,}{a}''-\underset{,}{b}'').(t'-\underset{,,}{b}')\\
&\qquad\quad +(t'+a'-b'-\underset{,}{b}').(t''-a''-b''-\underset{,}{b}'')\\
&\qquad\quad +(t'+\underset{,}{a}'-b'-\underset{,,}{b}').(t''+\underset{,}{a}''-b''-\underset{,,}{b}'')
\end{aligned}\right\} > 0,$$

$$\left.\begin{aligned}
&(a'+\underset{,}{a}'-b').(t''-b'') + (a''+\underset{,}{a}''-b'').(t'-b')\\
&\quad -(t'+a'-b'-\underset{,}{b}').(t''+a''-b''-\underset{,}{b}'')\\
&\quad -(t'+\underset{,}{a}'-b'-\underset{,,}{b}').(t''+\underset{,}{a}''-b''-\underset{,,}{b}'')
\end{aligned}\right\} > 0,$$

$$\left.\begin{aligned}
&(a'+\underset{,}{a}'-b').(t''-\underset{,}{b}'') + (a''+\underset{,}{a}''-b'').(t'-\underset{,}{b}')\\
&\quad -(t'+a'-b'-\underset{,}{b}').(t''+a''-b''-\underset{,}{b}'')
\end{aligned}\right\} > 0,$$

$$\left.\begin{aligned}
&(\underset{,}{a}'+\underset{,,}{a}'-\underset{,}{b}').(t''-\underset{,,}{b}'') + (\underset{,}{a}''+\underset{,,}{a}''-\underset{,}{b}'').(t'-\underset{,,}{b}')\\
&\quad -(t'+\underset{,}{a}'-b'-\underset{,,}{b}').(t''+\underset{,}{a}''-b''-\underset{,,}{b}'')
\end{aligned}\right\} > 0,$$

$$\left.\begin{aligned}
&(t'-a').(t''-a'') - (t'-b').(t''-b'') - (t'-\underset{,}{b}').(t''-\underset{,}{b}'')\\
&\quad +(t'+a'-b'-\underset{,}{b}').(t''+a''-b''-\underset{,}{b}'')
\end{aligned}\right\} > 0,$$

$$\left.\begin{aligned}
&(t'-\underset{,}{a}').(t''-\underset{,}{a}'') - (t'-b').(t''-b'') - (t'-\underset{,,}{b}').(t''-\underset{,,}{b}'')\\
&\quad +(t'+\underset{,}{a}'-b'-\underset{,,}{b}').(t''+\underset{,}{a}''-b''-\underset{,,}{b}'')
\end{aligned}\right\} > 0,$$

$$(t'-\underset{,,}{a}').(t''-\underset{,,}{a}'') - (t'-\underset{,}{b}').(t''-\underset{,}{b}'') - (t'-\underset{,,}{b}').(t''-\underset{,,}{b}'') > 0.$$

$$\textit{Sixth form}$$

$$C - B > \underset{\prime}{B} - A;\ C - B > \underset{\prime\prime}{B} - \underset{\prime}{A};\ C - \underset{\prime}{B} > \underset{\prime\prime}{B} - \underset{\prime}{A}.$$

(125.)

$$D = tt't'' - (t-a).(t'-a').(t''-a'') - (t-\underset{\prime}{a}).(t'-\underset{\prime}{a}').(t''-\underset{\prime}{a}'')$$
$$-(t-\underset{\prime\prime}{a}).(t'-\underset{\prime\prime}{a}').(t''-\underset{\prime\prime}{a}'') + (t-b).(t'-b').(t''-b'')$$
$$+(t-\underset{\prime}{b}).(t'-\underset{\prime}{b}').(t''-\underset{\prime}{b}'') + (t-\underset{\prime\prime}{b}).(t'-\underset{\prime\prime}{b}').(t''-\underset{\prime\prime}{b}'')$$
$$-(a+\underset{\prime}{a}-b).(t'-b').(t''-b'') - (a'+\underset{\prime}{a}'-b').(t-b).(t''-b'')$$
$$-(a''+\underset{\prime}{a}''-b'').(t-b).(t'-b') - (a+\underset{\prime\prime}{a}-b).(t'-\underset{\prime}{b}').(t''-\underset{\prime}{b}'')$$
$$-(a'+\underset{\prime\prime}{a}'-\underset{\prime}{b}').(t-\underset{\prime}{b}).(t''-\underset{\prime}{b}'') - (a''+\underset{\prime\prime}{a}''-\underset{\prime}{b}'').(t-\underset{\prime}{b}).(t'-\underset{\prime}{b}')$$
$$-(\underset{\prime}{a}+\underset{\prime\prime}{a}-\underset{\prime}{b}).(t'-\underset{\prime\prime}{b}').(t''-\underset{\prime\prime}{b}'') - (\underset{\prime}{a}'+\underset{\prime\prime}{a}'-\underset{\prime}{b}').(t-\underset{\prime\prime}{b}).(t''-\underset{\prime\prime}{b}'')$$
$$-(\underset{\prime}{a}''+\underset{\prime\prime}{a}''-\underset{\prime}{b}'').(t-\underset{\prime\prime}{b}).(t'-\underset{\prime\prime}{b}')$$
$$+(t+a-b-\underset{\prime}{b}).(t'+a'-b'-\underset{\prime}{b}').(t''+a''-b''-\underset{\prime}{b}'')$$
$$+(t+\underset{\prime}{a}-b-\underset{\prime\prime}{b}).(t'+\underset{\prime}{a}'-b'-\underset{\prime\prime}{b}').(t''+\underset{\prime}{a}''-b''-\underset{\prime\prime}{b}'')$$
$$+(t+\underset{\prime\prime}{a}-\underset{\prime}{b}-\underset{\prime\prime}{b}).(t'+\underset{\prime\prime}{a}'-\underset{\prime}{b}'-\underset{\prime\prime}{b}').(t''+\underset{\prime\prime}{a}''-\underset{\prime}{b}''-\underset{\prime\prime}{b}'').$$

$$\textit{Conditions}$$

$$\left.\begin{array}{c}
t't'' - (t'-a').(t''-a'') - (t'-\underset{\prime}{a}').(t''-\underset{\prime}{a}'') - (t'-\underset{\prime\prime}{a}').(t''-\underset{\prime\prime}{a}'') \\[4pt]
+(t'-b').(t''-b'') + (t'-\underset{\prime}{b}').(t''-\underset{\prime}{b}'') + (t'-\underset{\prime\prime}{b}').(t''-\underset{\prime\prime}{b}'') \\[4pt]
-(a'+\underset{\prime}{a}'-b').(t''-b'') - (a''+\underset{\prime}{a}''-b'').(t'-b') \\[4pt]
-(a'+\underset{\prime\prime}{a}'-\underset{\prime}{b}').(t''-\underset{\prime}{b}'') - (a''+\underset{\prime\prime}{a}''-\underset{\prime}{b}'').(t'-\underset{\prime}{b}') \\[4pt]
-(\underset{\prime}{a}'+\underset{\prime\prime}{a}'-\underset{\prime}{b}').(t''-\underset{\prime\prime}{b}'') - (a''+\underset{\prime\prime}{a}''-\underset{\prime\prime}{b}'').(t'-\underset{\prime}{b}') \\[4pt]
+(t'+a'-b'-\underset{\prime}{b}').(t''-a''-b''-\underset{\prime}{b}'') \\[4pt]
+(t'+\underset{\prime}{a}'-b'-\underset{\prime\prime}{b}').(t''+\underset{\prime}{a}''-b''-\underset{\prime\prime}{b}'') \\[4pt]
+(t'+\underset{\prime\prime}{a}'-\underset{\prime}{b}''-\underset{\prime\prime}{b}').(t''+\underset{\prime\prime}{a}''-\underset{\prime}{b}''-\underset{\prime\prime}{b}'')
\end{array}\right\} > 0,$$

$$\left.\begin{array}{c}
(a'+\underset{\prime}{a}'-b').(t''-b'') + (a''+\underset{\prime}{a}''-b'').(t'-b') \\[4pt]
-(t'+a'-b'-\underset{\prime}{b}').(t''+a''-b''-\underset{\prime}{b}'') \\[4pt]
-(t'+\underset{\prime}{a}'-b'-\underset{\prime\prime}{b}').(t''+\underset{\prime}{a}''-b''-\underset{\prime\prime}{b}'')
\end{array}\right\} > 0,$$

$$\left.\begin{array}{c}
(a'+\underset{\prime\prime}{a}'-\underset{\prime}{b}').(t''-\underset{\prime}{b}'') + (a''+\underset{\prime\prime}{a}''-\underset{\prime}{b}'').(t'-\underset{\prime}{b}') \\[4pt]
-(t'+a'-b'-\underset{\prime}{b}').(t''+a''-b''-\underset{\prime}{b}'') \\[4pt]
-(t'+\underset{\prime}{a}'-b'-\underset{\prime\prime}{b}').(t''+\underset{\prime}{a}''-b''-\underset{\prime\prime}{b}'')
\end{array}\right\} > 0,$$

$$\left.\begin{array}{r} (a'_{\prime} + a'_{\prime\prime} - b'_{\prime}).(t'' - b''_{\prime\prime}) + (a''_{\prime} + a''_{\prime\prime} - b''_{\prime}).(t' - b'_{\prime\prime}) \\[4pt] -(t' + a'_{\prime} - b' - b'_{\prime}).(t'' + a''_{\prime} - b'' - b''_{\prime\prime}) \\[4pt] -(t' + a'_{\prime\prime} - b' - b'_{\prime\prime}).(t'' + a''_{\prime\prime} - b'' - b''_{\prime\prime}) \end{array}\right\} > 0,$$

$$\left.\begin{array}{r} (t' - a'_{\prime}).(t'' - a''_{\prime}) - (t' - b').(t'' - b'') - (t' - b'_{\prime}).(t'' - b''_{\prime}) \\[4pt] +(t' + a'_{\prime} - b' - b'_{\prime}).(t'' + a''_{\prime} - b'' - b''_{\prime}) \end{array}\right\} > 0,$$

$$\left.\begin{array}{r} (t' - a'_{\prime}).(t'' - a''_{\prime}) - (t' - b').(t'' - b'') - (t' - b'_{\prime\prime}).(t'' - b''_{\prime\prime}) \\[4pt] +(t' + a'_{\prime} - b' - b'_{\prime}).(t'' + a''_{\prime} - b'' - b''_{\prime}) \end{array}\right\} > 0,$$

$$\left.\begin{array}{r} (t' - a'_{\prime\prime}).(t'' - a''_{\prime\prime}) - (t' - b'_{\prime}).(t'' - b''_{\prime}) - (t' - b'_{\prime}).(t'' - b''_{\prime\prime}) \\[4pt] +(t' + a'_{\prime\prime} - b' - b'_{\prime\prime}).(t'' + a''_{\prime\prime} - b'' - b''_{\prime\prime}) \end{array}\right\} > 0.$$

Seventh form

$$C - B_{\prime} < B - A; \quad C - B > B_{\prime\prime} - A; \quad C - B_{\prime} < B_{\prime\prime} - A_{\prime\prime}.$$

(126.)

$$D = tt't'' - (t - a).(t' - a').(t'' - a'') - (t - a_{\prime}).(t' - a'_{\prime}).(t'' - a''_{\prime})$$

$$-(t - a_{\prime\prime}).(t' - a'_{\prime\prime}).(t'' - a''_{\prime\prime}) + (t - b).(t' - b').(t'' - b'')$$

$$+(t - b_{\prime}).(t' - b'_{\prime}).(t'' - b''_{\prime}) + (t - b_{\prime\prime}).(t' - b'_{\prime\prime}).(t'' - b''_{\prime\prime})$$

$$-(a + a_{\prime} - b).(t' - b').(t'' - b'') - (a' + a'_{\prime} - b').(t - b).(t'' - b'')$$

$$-(a'' + a''_{\prime} - b'').(t - b).(t' - b') - (a + a_{\prime\prime} - b).(t' - b'_{\prime}).(t'' - b''_{\prime})$$

$$-(a' + a'_{\prime} - b').(t - b_{\prime}).(t'' - b''_{\prime}) - (a'' + a''_{\prime\prime} - b'').(t - b_{\prime}).(t' - b'_{\prime})$$

$$-(a_{\prime} + a_{\prime\prime} - b_{\prime}).(t' - b'_{\prime}).(t'' - b''_{\prime\prime})$$

$$-(a' + a'_{\prime\prime} - b'_{\prime}).(t - b_{\prime}).(t'' - b''_{\prime\prime})$$

$$-(a''_{\prime} + a''_{\prime\prime} - b''_{\prime}).(t - b_{\prime\prime}).(t' - b'_{\prime\prime})$$

$$+(t + a_{\prime} - b - b_{\prime}).(t' + a'_{\prime\prime} - b' - b'_{\prime\prime}).(t'' + a''_{\prime} - b'' - b''_{\prime\prime}).$$

Conditions

$$
\left.
\begin{aligned}
& t't'' - (t'-a').(t''-a'') - (t'-\underset{\prime}{a}').(t''-\underset{\prime}{a}'') \\
& -(t'-\underset{\prime\prime}{a}').(t''-\underset{\prime\prime}{a}'') + (t'-b').(t''-b'') + (t'-\underset{\prime}{b}').(t''-\underset{\prime}{b}'') \\
& \quad +(t'-\underset{\prime\prime}{b}').(t''-\underset{\prime\prime}{b}'') - (a'+\underset{\prime}{a}'-b').(t''-b'') \\
& \quad\quad -(a''+\underset{\prime}{a}''-b'').(t'-b') - (a'+\underset{\prime\prime}{a}'-\underset{\prime}{b}').(t''-\underset{\prime}{b}'') \\
& \quad\quad\quad -(a''+\underset{\prime\prime}{a}''-\underset{\prime}{b}'').(t'-\underset{\prime}{b}') - (\underset{\prime}{a}'+\underset{\prime\prime}{a}'-\underset{\prime\prime}{b}').(t''-\underset{\prime\prime}{b}'') \\
& \quad\quad\quad\quad -(\underset{\prime}{a}''+\underset{\prime\prime}{a}''-\underset{\prime\prime}{b}'').(t'-\underset{\prime\prime}{b}') \\
& \quad\quad +(t'+\underset{\prime}{a}'-b'-\underset{\prime\prime}{b}').(t''+\underset{\prime}{a}''-b''-\underset{\prime\prime}{b}'')
\end{aligned}
\right\} > 0,
$$

$$
\left.
\begin{aligned}
& (a'+\underset{\prime}{a}'-b').(t''-b'') + (a''+\underset{\prime}{a}''-b'').(t'-b') \\
& \quad -(t'+\underset{\prime}{a}'-b'-\underset{\prime\prime}{b}').(t''+\underset{\prime}{a}''-b''-\underset{\prime\prime}{b}'')
\end{aligned}
\right\} > 0,
$$

$$
(a'+\underset{\prime}{a}'-\underset{\prime}{b}').(t''-\underset{\prime}{b}'') + (a''+\underset{\prime}{a}''-\underset{\prime}{b}'').(t'-\underset{\prime}{b}') > 0,
$$

$$
\left.
\begin{aligned}
& (\underset{\prime}{a}'+\underset{\prime\prime}{a}'-\underset{\prime\prime}{b}').(t''-\underset{\prime\prime}{b}'') + (\underset{\prime}{a}''+\underset{\prime\prime}{a}''-\underset{\prime\prime}{b}'').(t'-\underset{\prime\prime}{b}') \\
& \quad -(t'+\underset{\prime}{a}'-b'-\underset{\prime\prime}{b}').(t''+\underset{\prime}{a}''-b''-\underset{\prime\prime}{b}'')
\end{aligned}
\right\} > 0,
$$

$$
(t'-a').(t''-a'') - (t'-b').(t''-b'') - (t'-\underset{\prime}{b}').(t''-\underset{\prime}{b}'') > 0,
$$

$$
\left.
\begin{aligned}
& (t'-\underset{\prime}{a}').(t''-\underset{\prime}{a}'') - (t'-b').(t''-b'') - (t'-\underset{\prime\prime}{b}').(t''-\underset{\prime\prime}{b}'') \\
& \quad +(t'+\underset{\prime}{a}'-b'-\underset{\prime\prime}{b}').(t''+\underset{\prime}{a}''-b''-\underset{\prime\prime}{b}'')
\end{aligned}
\right\} > 0,
$$

$$
(t'-\underset{\prime\prime}{a}').(t''-\underset{\prime\prime}{a}'') - (t'-\underset{\prime}{b}').(t''-\underset{\prime}{b}'') - (t'-\underset{\prime\prime}{b}').(t''-\underset{\prime\prime}{b}'') > 0.
$$

Eighth form

$$
C - B < \underset{\prime}{B} - A; \quad C - B > \underset{\prime\prime}{B} - \underset{\prime}{A}; \quad C - \underset{\prime}{B} > \underset{\prime\prime}{B} - \underset{\prime\prime}{A}.
$$

(127.)

$$
\begin{aligned}
D = {}& tt't'' - (t-a).(t'-a').(t''-a'') - (t-\underset{\prime}{a}).(t'-\underset{\prime}{a}').(t''-\underset{\prime}{a}'') \\
& -(t-\underset{\prime\prime}{a}).(t'-\underset{\prime\prime}{a}').(t''-\underset{\prime\prime}{a}'') + (t-b).(t'-b').(t''-b'') \\
& +(t-\underset{\prime}{b}).(t'-\underset{\prime}{b}').(t''-\underset{\prime}{b}'') + (t-\underset{\prime\prime}{b}).(t'-\underset{\prime\prime}{b}').(t''-\underset{\prime\prime}{b}'') \\
& -(a+\underset{\prime}{a}-b).(t'-b').(t''-b'') - (a'+\underset{\prime}{a}'-b').(t-b).(t''-b'') \\
& -(a''+\underset{\prime}{a}''-b'').(t-b).(t'-b') - (a+\underset{\prime\prime}{a}-b).(t'-\underset{\prime}{b}').(t''-\underset{\prime}{b}'') \\
& -(a'+\underset{\prime\prime}{a}'-\underset{\prime}{b}').(t-b).(t''-\underset{\prime}{b}'') - (a''+\underset{\prime\prime}{a}''-\underset{\prime}{b}'').(t-b).(t'-\underset{\prime}{b}') \\
& -(\underset{\prime}{a}+\underset{\prime\prime}{a}-\underset{\prime\prime}{b}).(t'-\underset{\prime}{b}').(t''-\underset{\prime\prime}{b}'') - (\underset{\prime}{a}'+\underset{\prime\prime}{a}'-\underset{\prime\prime}{b}').(t-\underset{\prime}{b}).(t''-\underset{\prime\prime}{b}'') \\
& -(\underset{\prime}{a}''+\underset{\prime\prime}{a}''-\underset{\prime\prime}{b}'').(t-\underset{\prime}{b}).(t'-\underset{\prime}{b}') \\
& +(t+\underset{\prime}{a}-b-\underset{\prime\prime}{b}).(t'+\underset{\prime}{a}'-b'-\underset{\prime\prime}{b}').(t''+\underset{\prime}{a}''-b''-\underset{\prime\prime}{b}'') \\
& +(t+\underset{\prime\prime}{a}-\underset{\prime}{b}-\underset{\prime\prime}{b}).(t'+\underset{\prime\prime}{a}'-\underset{\prime}{b}'-\underset{\prime\prime}{b}').(t''+\underset{\prime\prime}{a}''-\underset{\prime}{b}''-\underset{\prime\prime}{b}'').
\end{aligned}
$$

Conditions

$$
\left.
\begin{aligned}
t't'' - (t' - a').(t'' - a'') - (t' - a'_{,}).(t'' - a''_{,}) - (t' - a'_{,,}).(t'' - a''_{,,}) \\
+ (t' - b').(t'' - b'') + (t' - b'_{,}).(t'' - b''_{,}) + (t' - b'_{,,}).(t'' - b''_{,,}) \\
- (a' + a'_{,} - b').(t'' - b'') - (a'' + a''_{,} - b'').(t' - b') \\
- (a' + a'_{,,} - b').(t'' - b'') - (a'' + a''_{,,} - b'').(t' - b') \\
- (a'_{,} + a'_{,,} - b'_{,}).(t'' - b''_{,}) - (a''_{,} + a''_{,,} - b''_{,}).(t' - b'_{,}) \\
+ (t' + a'_{,} - b' - b'_{,,}).(t'' + a''_{,} - b'' - b''_{,,}) \\
+ (t' + a'_{,,} - b'_{,} - b'_{,,}).(t'' + a''_{,,} - b''_{,} - b''_{,,})
\end{aligned}
\right\} > 0,
$$

$$
\left.
\begin{aligned}
(a' + a'_{,} - b').(t'' - b'') + (a'' + a''_{,} - b'').(t' - b') \\
- (t' + a'_{,} - b' - b'_{,,}).(t'' + a''_{,} - b'' - b''_{,,})
\end{aligned}
\right\} > 0,
$$

$$
\left.
\begin{aligned}
(a' + a'_{,,} - b').(t'' - b'') + (a'' + a''_{,,} - b'').(t' - b') \\
- (t' + a'_{,} - b' - b'_{,,}).(t'' + a''_{,} - b'' - b''_{,,})
\end{aligned}
\right\} > 0,
$$

$$
\left.
\begin{aligned}
(a'_{,} + a'_{,,} - b'_{,}).(t'' - b''_{,}) + (a''_{,} + a''_{,,} - b''_{,}).(t' - b'_{,}) \\
- (t' + a'_{,} - b' - b'_{,,}).(t'' + a''_{,} - b'' - b''_{,,}) \\
- (t' + a'_{,,} - b'_{,} - b'_{,,}).(t'' + a''_{,,} - b''_{,} - b''_{,,})
\end{aligned}
\right\} > 0,
$$

$$
(t' - a').(t'' - a'') - (t' - b').(t'' - b'') - (t' - b'_{,}).(t'' - b''_{,}) > 0,
$$

$$
\left.
\begin{aligned}
(t' - a'_{,}).(t'' - a''_{,}) - (t' - b').(t'' - b'') - (t' - b'_{,,}).(t'' - b''_{,,}) \\
+ (t' + a'_{,} - b' - b'_{,,}).(t'' + a''_{,} - b'' - b''_{,,})
\end{aligned}
\right\} > 0,
$$

$$
\left.
\begin{aligned}
(t' - a').(t'' - a''_{,}) - (t' - b'_{,}).(t'' - b''_{,}) - (t' - b'_{,,}).(t'' - b''_{,,}) \\
+ (t' + a'_{,,} - b'_{,} - b'_{,,}).(t'' + a''_{,,} - b''_{,} - b''_{,,})
\end{aligned}
\right\} > 0.
$$

General Remark

(128.) The method we use to find the degree of the final equation consists of enumerating the number of terms of the product equation and of the largest number of terms that can be cancelled from that equation using the $n - 1$ other equations. The resulting value of D, augmented by 1, expresses the smallest number of terms in the product equation. There is no indication as to whether these terms should be in x only, y only, or in x and y, or in x and z, etc. or in x, y, z, etc.

We can therefore conclude that, in general: *The degree of the final equation resulting from an arbitrary number of equations with the same number of unknowns is the same for each of these unknowns.* We assume here that the polynomials entering the problem are completely arbitrary; that is, we do not assume any specific relation within the coefficients of the proposed

equations. We will see in the Second Book some relations among the coefficients that may lead to lowering the degree of the final equation for some of
the unknowns, without lowering the degree of some others.

Applications to various examples

(129.) Let us first assume that one out of the three proposed equations
has degree 1. Let us assume, for example,

$$a'' = a''_{,} = a''_{,,} = b'' = b''_{,} = b''_{,,} = t'' = 1.$$

Then all the sums computed in (120 ff) give

$$D = tt' - (t-b).(t'-b') - (t-b_{,}).(t'-b'_{,}) - (t-b_{,,}).(t'-b'_{,,});$$

the corresponding conditions all reduce to the single condition $b' + b'_{,} + b'_{,,} >$
$2t'$, which is always true from (83).

Let us now compare this result with the one that could be expected from
the Method of Successive Eliminations.

The three equations are

$$[(x^a, y^{\overset{a}{\prime}})^b, (x^a, z''^{\overset{a}{\prime\prime}})^{\overset{b}{\prime}}, (y^{\overset{a}{\prime}}, z''^{\overset{a}{\prime\prime}})^{\overset{b}{\prime\prime}}]^t = 0,$$
$$[(x^{a'}, y^{\overset{a'}{\prime}})^{b'}, (x^{a'}, z''^{\overset{a'}{\prime\prime}})^{\overset{b'}{\prime}}, (y^{\overset{a'}{\prime}}, z''^{\overset{a'}{\prime\prime}})^{\overset{b'}{\prime\prime}}]^{t'} = 0,$$
$$(x, y, z)^1 = 0.$$

Assume that z is replaced by the value given by the third equation in the
first two equations. The first two equations become

$$(x^{\overset{b}{\prime}}, y''^{\overset{b}{\prime\prime}})^t = 0,$$
$$(x^{\overset{b'}{\prime}}, y''^{\overset{b'}{\prime\prime}})^{t'} = 0.$$

However, the degree of the final equation resulting from these two equations must by (62) be $tt' - (t-b_{,}).(t'-b') - (t-b_{,}).(t'-b'_{,,})$; it therefore
exceeds the true degree by the amount $(t-b)(t'-b')$.

If, instead of assuming $a'' = a''_{,} = a''_{,,} = b'' = b''_{,} = t'' = 1$, we had assumed
$a' = a'_{,} = a'_{,,} = b' = b'_{,} = b'_{,,} = t' = 1$, all forms would have led to the same
value of D; but the conditions would in general not be the same. This shows
that the polynomial multiplier would not be of the same kind for each of
the eight forms in that case; however, by (117) there is always one condition
that is satisfied.

(130.) Assume $b = b_{,} = b_{,,} = t$; $b' = b'_{,} = b'_{,,} = t'$; $b'' = b''_{,} = b''_{,,} = t''$; this
generates equations of the form $(x^a \ldots 3)^t = 0$ subject to the conditions (58);
that is, the combinations of two and three of the unknowns x, y, z reach all
possible dimensions up to the dimension t of the equation; but taken alone,
they can only reach the degrees a, $a_{,}$, $a_{,,}$

In the present case, we find that, among the eight forms presented in (120 ff), only the first can occur; in each of the other seven, only some of the conditions can be satisfied. This first form gives

$$D = tt't'' - (t - a).(t' - a').(t'' - a'')$$
$$-(t - a).(t' - a').(t'' - a'') - (t - a).(t' - a').(t'' - a'').$$

This is in agreement with what was found in (62), and the conditions for the existence of this value of D reduce to the single condition

$$t't'' - (t' - a'), (t'' - a'') - (t' - a').(t'' - a'') - (t' - a').(t'' - a'') > 0.$$

This condition must be satisfied since we have assumed $a' + a' > t'$, $a'' + a'' > t''$, $a' + a' > t'$, $a'' + a'' > t''$, etc.

Indeed the minimum value of

$$t't'' - (t' - a').(t'' - a'') - (t' - a').(t'' - a'') - (t' - a').(t'' - a'')$$

is reached when $t' - a'$, $t'' - a''$, $t' - a'$, etc. are as large as possible, that is, when $t' - a' = a'$, $t'' - a'' = a''$, $t' - a' = a'$, etc. But in that case the condition reduces to $a'a'' > 0$.

Things would be different if any one of the conditions $a' + a' > t'$, etc. was not satisfied; we would then find that none of the eight forms would be adequate. This is easy to understand, since we would then have erroneously assumed $b' = t'$: Indeed, we have $a' + a' < t'$ by hypothesis; thus we cannot have $b' = t'$ since it is less than $a' + a'$ from (83).

Let us ask, for example, for the degree of the final equation resulting from three equations of the form

$$axy + bxz + cyz + dx + ey + fz + g = 0,$$

that is, three equations such that

$$[(x^1, y^1)^2, (x^1, z^1)^2, (y^1, z^1)^2]^2 = 0;$$

we have $D = 8 - 1 - 1 - 1 = 5$ and the sole condition above reduces to $4 - 1 - 1 - 1 > 0$, or $1 > 0$, which is true.

But we would be wrong in using the same formula to compute D for three equations of the form

$$[(x^1, y^1)^2, (x^1, z^1)^2, (y^1, z^1)^2]^3 = 0,$$

that is, for three equations of the form

$$axyz + bxy + cxz + dyz + ex + fy + gz + h = 0;$$

indeed the combinations of two or three unknowns do not reach the dimension of the equation.

To obtain the value of D for these equations, we must use the general expressions for the value of D found in (120 ff); defining

$$b = a + a_{\prime}, \qquad b_{\prime} = a + a_{\prime\prime}, \qquad b_{\prime\prime} = a_{\prime} + a_{\prime\prime};$$
$$b' = a' + a'_{\prime}, \qquad b'_{\prime} = a' + a'_{\prime\prime}, \qquad b'_{\prime\prime} = a'_{\prime} + a'_{\prime\prime};$$
$$b'' = a'' + a''_{\prime}, \qquad b''_{\prime} = a'' + a''_{\prime\prime}, \qquad b''_{\prime\prime} = a''_{\prime} + a''_{\prime\prime};$$

we find $D = 6$.

Assume we wanted to use the Method of Successive Eliminations for three equations such as those described in this example; we would replace z in two of these three equations by the value found from the third equation (for example); we would then obtain two equations in x and y of the form $(x^2, y^2)^4 = 0$. Then, eliminating y, we would be led by (62) to an equation of degree $16 - 4 - 4$, that is, of degree 8.

(131.) Assume $b = b_{\prime} = t$; $b' = b'_{\prime} = t'$; $b'' = b''_{\prime} = t''$. Only the first form

(120) is acceptable; it yields

$$D = tt't'' - (t - a).(t' - a').(t'' - a'') - (t - a_{\prime}).(t' - a'_{\prime}).(t'' - a''_{\prime})$$
$$- (t - a_{\prime\prime}).(t' - a'_{\prime\prime}).(t'' - a''_{\prime\prime}) + (t - b).(t' - b').(t'' - b'')$$
$$- (a_{\prime} + a_{\prime\prime} - b_{\prime\prime}).(t' - b'_{\prime\prime}).(t'' - b''_{\prime\prime})$$
$$- (a'_{\prime} + a'_{\prime\prime} - b'_{\prime\prime}).(t - b_{\prime\prime}).(t'' - b''_{\prime\prime})$$
$$- (a''_{\prime} + a''_{\prime\prime} - b''_{\prime\prime}).(t - b_{\prime\prime}).(t' - b'_{\prime\prime})$$

and the conditions reduce to the single inequality

$$t't'' - (t' - a').(t'' - a'') - (t' - a'_{\prime}).(t'' - a''_{\prime}) - (t' - a'_{\prime\prime}).(t'' - a''_{\prime\prime})$$
$$+ (t' - b'_{\prime}).(t'' - b''_{\prime}) - (a'_{\prime} + a'_{\prime\prime} - b'_{\prime\prime}).(t'' - b''_{\prime\prime})$$
$$- (a''_{\prime} + a''_{\prime\prime} - b''_{\prime\prime}).(t' - b'_{\prime\prime}) > 0$$

since all other conditions are trivially satisfied. This inequality must also be automatically satisfied, following the discussion (117).

In general, this condition may not be satisfied; but that occurs only when the necessary conditions of existence of the proposed equation (83) do not hold. For example, assume we have $a_{\prime} + a_{\prime\prime} < b_{\prime\prime}$, $a'_{\prime} + a'_{\prime\prime} < b'_{\prime\prime}$, and so on; it then becomes obvious that the expression of the form of the equations would become wrong and reducible to another (See 101). Thus the value of D we just provided is the general and only expression for the degree of the final equation in three equations of the form $[x^a, (y^{\overset{a}{\prime}}, z^{\overset{a}{\prime\prime}})^{\overset{b}{\prime\prime}}]^t = 0$.

Let us assume moreover that $a_{\prime} = a_{\prime\prime} = b_{\prime\prime} = 1$; $a'_{\prime} = a'_{\prime\prime} = b'_{\prime\prime} = 1$; $a''_{\prime} = a''_{\prime\prime} = b''_{\prime\prime} = 1$. Then a can take only one of the two values $a = t$ and $a = t - 1$; likewise $a' = t'$ or $a' = t' - 1$, and $a'' = t''$ or $a'' = t'' - 1$. In the first case, the value of D is $D = t + t' + t'' - 2$; and in the second case $D = t + t' + t'' - 3$.

Indeed, in the first case, the three equations can be represented by

$$(x\ldots1)^t + (x\ldots1)^{t-1}.y + (x\ldots1)^{t-1}.z = 0,$$
$$(x\ldots1)^{t'} + (x\ldots1)^{t'-1}.y + (x\ldots1)^{t'-1}.z = 0,$$
$$(x\ldots1)^{t''} + (x\ldots1)^{t''-1}.y + (x\ldots1)^{t''-1}.z = 0.$$

But it is easy to see that we get an equation in x whose degree is $t+t'+t''-2$, if we substitute the values of y and z from the last two equations into the first equation. We do not see as easily that the same must be true of the equation in y and the equation in z. Instead, the method by which we reach a general value for D shows that the degree of the final equation is always the same for each of the three unknowns, assuming no specific relation among the coefficients.

(132.) Assume only x and y appear in the first equation, among the three unknowns x, y and z; that only x and z appear in the second; and that only y and z appear in the third.

We have $\underset{''}{a} = 0,\ \underset{'}{a'} = 0,\ a'' = 0.$

It follows that $\underset{''}{b} = a,\ \underset{'}{b} = a,\ b = t,\ b' = a',\ \underset{'}{b'} = t',\ \underset{''}{b'} = \underset{''}{a'},\ b'' = \underset{''}{a''},$
$\underset{'}{b''} = \underset{''}{a''},\ \underset{''}{b''} = t''.$

Subsituting these values in each of the forms (120 ff) given for D, we get

$$D = tt't'' - t.(t' - \underset{''}{a'}).(t'' - \underset{''}{a''}) - t'.(t - \underset{'}{a}).(t'' - \underset{'}{a''})$$
$$-t''.(t - a).(t' - a') - (a + \underset{''}{a} - t).(t' - a').(t'' - \underset{'}{a''})$$
$$-(a' + \underset{''}{a'} - t').(t - a).(t'' - \underset{''}{a''}) - (\underset{'}{a''} + \underset{''}{a''} - t'').(t - \underset{'}{a}).(t' - \underset{''}{a'})$$

for all forms. The conditions then determine the form of the polynomial multiplier.

As a conclusion to our study of the equations similar to those considered until now, we give an idea of how to determine the number of terms in polynomials of this kind, since this is how we determine the degree of the final equation.

General considerations about the degree of the final equation, when
considering the other incomplete equations similar to those considered up
until now

(133.) Following our discussion, the reader can see what we mean by
"equations that are similar to those considered up until now": These are
equations such that n' out of the n unknowns satisfy the following properties:

1. Each of the n' unknowns does not exceed a given degree; this degree
   may be different or the same for each unknown.

2. The combinations of two of these unknowns cannot exceed given dimensions.

3. The combinations of three of these unknowns cannot exceed given dimensions.

4. The combinations of four of these unknowns cannot exceed given dimensions; and so on until the combinations of n' unknowns.

5. Finally, the other $(n - n')$ unknowns can reach all possible dimensions
   up to the highest dimension of the equation, including their combinations with the n' first unknowns.

(134.) By polynomials or equations of the *same form*, we understand those
whose contents are similar, as described above; by polynomials or equations
of the *same nature*, we understand those whose expression for the number
of their terms has the same form, that is, whose contents are similar.

For example, considering the equations (82), we saw that the expression
for the number of terms of the polynomial multiplier can take eight different
forms, while the polynomial always has the form

$$([(u^A, x'^{\overset{A}{'}})^B, (u^A, y''^{\overset{A}{''}})^{\overset{B}{'}}, (x'^{\overset{A}{'}}, y''^{\overset{A}{''}})^{\overset{B}{''}}]^C, z'''^{\overset{A}{'''}} \dots n)^T.$$

Consider the other polynomial

$$([(u^{A-a}, x'^{\overset{A-a}{'}})^{B-b}, (u^{A-a}, y''^{\overset{A-a}{''}})^{\overset{B-b}{'}}, (x'^{\overset{A-a}{'}}, y''^{\overset{A-a}{''}})^{\overset{B-b}{''}}]^{C-c},$$
$$z'''^{\overset{A-a}{'''}} \dots n)^{T-t}.$$

This polynomial has the same form as the previous one, but it can be of
the same or a different nature. It is of the same nature if the same formula
can be used to compute the number of its terms; this is true if the relations
among the various exponents in the latter polynomial are the same as those
in the former polynomial. It has a different nature if a different formula
must be used.

(135.) We have seen in (84 ff) that the expression for the number of terms
of the polynomial

$$([(u^A, x'^{\overset{A}{'}})^B, (u^A, y''^{\overset{A}{''}})^{\overset{B}{'}}, (x'^{\overset{A}{'}}, y''^{\overset{A}{''}})^{\overset{B}{''}}]^C, z'''^{\overset{A}{'''}} \dots n)^T$$

can have eight different forms, resulting in eight different expressions for the degree D of the final equation. Likewise, D takes as many different expressions as there are possible expressions for the number of terms of polynomials of the same form as the equations we just described in (133).

(136.) In general, we always consider one of the equations, multiplied by a polynomial of the same form, as we have done up to now: The resulting product or product equation always has the same form. Using the same approaches as we have up to now, and applying them *verbatim*, we see likewise that the expression of the number of remaining terms in the polynomial multiplier and in the product equation is always an exact differential of order $n - 1$, after cancelling as many terms as possible and not introducing new ones, using the $n - 1$ remaining equations. And finally, the resulting value of D is an exact differential of order n; consequently, it does not contain any of the exponents of the polynomial multiplier; rather, it is a function of the various exponents of the given equations.

(137.) We now see that the question of finding the value of D in all these equations reduces to finding an expression for the number of terms of an arbitrary polynomial of the form of current interest. It is then sufficient to differentiate this expression, using a process we have now seen more than enough.

(138.) But the different values of D that result from the various possible expressions for the number of terms of the polynomial multipler are not all admissible in all cases: We see from (117) that we must examine the sign of the sum of terms that multiply the same exponent in each value of D and check whether it is positive. If this examination is conclusive for all equations, the value of D is admissible; if it is not, then the value of D is inadequate; nevertheless there will always be a value of D for which the examination is conclusive, and when several possible values are admissible, they will be equal to one another.

(139.) Since there is great similarity between our current object and what we did previously, we could stop pursuing this branch of incomplete equations. However, we must not leave it before having, at least, given an idea of the different forms of the terms to be summed when looking for the number of terms in polynomials in this class, and how to compute their sum. Besides, we must also honor a commitment made in (67), that is, to give the value of the degree of the final equation in all equations of the form $(u^a \ldots n)^t = 0$, when the exponents a, $a_{\prime}$, $a_{\prime\prime}$ etc. are not subject to any other condition than

$$a + a_{\prime} + a_{\prime\prime} + a_{\prime\prime\prime} + \ldots > t,$$ including all exponents a, $a_{\prime}$, $a_{\prime\prime}$ etc. This condition must be true, otherwise the equation may not exist.

We now begin by solving the latter problem.

Problem XXIII

(140.) *We ask for the value of* $N(u^A \ldots n)^T$, *when the exponents* A, $A_{\prime}$, $A_{\prime\prime}$ *are arbitrary.*

This value is easy to find using (41), but there is value in looking for it using the method we have just used, which we will always use for questions of this form.

Let us first assume that there is only one exponent A, that is, that all other unknowns reach degree T.

Let us assume the polynomial is ordered with respect to the unknown corresponding to this exponent, say u, and let s be the exponent of u in an arbitrary term. Each term has the form $u^s(x\ldots n-1)^{T-s}$, from $s=0$ to $s=A$. We must therefore compute the sum of $N(x\ldots n-1)^{T-s}$, from $s=0$ to $s=A$. But this sum is $N(u\ldots n)^T - N(u\ldots n)^{T-A-1}$.

(141.) Let us assume now that only $n-2$ unknowns reach the degree T and that the two others, u and x, do not exceed the degrees A and A', respectively.

I order the polynomial $(u^A, \overset{A}{x'}, y, z\ldots n)^T$ with respect to x; each term is of the form $x^s(u^A, y, z\ldots n-1)^{T-s}$ from $s=0$ to $s=A$, or until $s+A=T$, depending on whether $\underset{\prime}{A} < T-A$ or $\underset{\prime}{A} > T-A$. We therefore encounter two cases.

First case

$\underset{\prime}{A} < T - A, \text{ or } \underset{\prime}{A} + A < T.$

In that case, the form $x^s(u^A, y, z\ldots n-1)^{T-s}$ occurs throughout the polynomial. We therefore only have to compute the sum of $N(u^A, y, z\ldots n-1)^{T-s}$ from $s=0$ to $s=\underset{\prime}{A}$. However we just saw that $N(u^A, y, z\ldots n-1)^{T-s} = N(u\ldots n-1)^{T-s} - N(u\ldots n-1)^{T-A-s-1}$. Summing this quantity, we have

$$N(u^A, \overset{A}{x'}, y, z\ldots n)^T = N(u\ldots n)^T - N(u\ldots n)^{T-A-1}$$
$$-N(u\ldots n)^{T-\underset{\prime}{A}-1} + N(u\ldots n)^{T-A-\underset{\prime}{A}-2}$$

when $\underset{\prime}{A} < T - A$,

Second case

$\underset{\prime}{A} > T - A, \text{ or } A + \underset{\prime}{A} > T.$

In this case, the form $x^s(u^A, y, z\ldots n-1)^{T-s}$ occurs only from $s=0$ to $s=T-A$; past $s=T-A$, it is $x^s(u, y, z\ldots n-1)^{T-s}$ or $x^s(u\ldots n-1)^{T-s}$ until $s=\underset{\prime}{A}$. We therefore must compute the sum of (i) $N(u^A, y, z\ldots n-1)^{T-s}$ from $s=0$ to $s=T-A$; (ii) $N(u\ldots n-1)^{T-s}$ from $s=T-A$ excluded to $s=\underset{\prime}{A}$. Therefore we obtain

$$N(u^A, \overset{A}{x'}, \ldots n)^T = N(u\ldots n)^T - N(u\ldots n)^{T-A-1} - N(u\ldots n)^{T-\underset{\prime}{A}-1}$$

when $A + A' > T$.

(142.) Let us now assume that only $n-3$ unknowns reach the degree T and that the three others do not exceed the degrees A, $\underset{\prime}{A}$ and $\underset{\prime\prime}{A}$, respectively.

Let us order the polynomial $(u^A, x^{\overset{A}{\prime}}, y^{\overset{A}{\prime\prime}}, z \ldots n)^T$ with respect to y. Each term will be of the form $y^s(u^A, x^{\overset{A}{\prime}}, z \ldots n-1)^{T-s}$ from $s = 0$ to $s = \underset{\prime\prime}{A}$, or until $s + \underset{\prime}{A} = T$, or until $s + A = T$. In other words, until s equals the smallest of the three quantities $\underset{\prime\prime}{A}$, $T - \underset{\prime}{A}$, $T - A$. The six following cases arise:

$$
\begin{array}{c|c}
\underset{\prime\prime}{A} < T - \underset{\prime}{A} < T - A & T - \underset{\prime}{A} < T - A < \underset{\prime\prime}{A} \\[4pt]
\underset{\prime\prime}{A} < T - A < T - \underset{\prime}{A} & T - A < \underset{\prime\prime}{A} < T - \underset{\prime}{A} \\[4pt]
T - \underset{\prime}{A} < \underset{\prime\prime}{A} < T - A & T - A < T - \underset{\prime}{A} < \underset{\prime\prime}{A}.
\end{array}
$$

First case

$$\underset{\prime\prime}{A} < T - \underset{\prime}{A} < T - A.$$

In this case, the form $y^s(u^A, x^{\overset{A}{\prime}} \ldots n-1)^{T-s}$ occurs over the whole polynomial: We must therefore compute the sum of $N(u^A, x^{\overset{A}{\prime}} \ldots n-1)^{T-s}$ from $s = 0$ to $s = \underset{\prime\prime}{A}$.

From (141) we have $N(u^A, x^{\overset{A}{\prime}} \ldots n-1)^{T-s} = N(u \ldots n-1)^{T-s}$

$-N(u \ldots n-1)^{T-A-s-1} - N(u \ldots n-1)^{\overset{T-A-s-1}{\prime}} + N(u \ldots n-1)^{\overset{T-A-\underset{\prime}{A}-s-2}{}},$

if $A + \underset{\prime}{A} > T - s$.

We have $N(u^A, x^{\overset{A}{\prime}} \ldots n-1)^{T-s} = N(u \ldots n-1)^{T-s}$

$-N(u \ldots n-1)^{T-A-s-1} - N(u \ldots n-1)^{\overset{T-\underset{\prime}{A}-s-1}{}}$ if $A + \underset{\prime}{A} > T - s$.

Since s ends at $\underset{\prime\prime}{A}$, this case can be divided into two other subcases, which are

$$A + \underset{\prime}{A} < T - \underset{\prime\prime}{A}; \quad A + \underset{\prime}{A} > T - \underset{\prime\prime}{A}.$$

Since s begins at 0, two other subcases may also arise: $A + \underset{\prime}{A} < T$ and $A + \underset{\prime}{A} > T$. The second case may arise only with the first of the other two cases. Thus the three following cases may occur:

$$
\begin{aligned}
&A + \underset{\prime}{A} < T; \quad A + \underset{\prime}{A} < T - \underset{\prime\prime}{A}; \\
&A + \underset{\prime}{A} < T; \quad A + \underset{\prime}{A} > T - \underset{\prime\prime}{A}; \\
&A + \underset{\prime}{A} > T; \quad A + \underset{\prime}{A} > T - \underset{\prime\prime}{A}.
\end{aligned}
$$

In the first case, we compute the sum of the first expression from $s = 0$ to $s = \underset{\prime\prime}{A}$.

In the second case, we must compute (i) the first expression from $s = 0$ to $s = T - A - \underset{\prime}{A}$; (ii) the second expression from $s = T - A - \underset{\prime}{A}$ excluded until $s = \underset{\prime\prime}{A}$.

In the third case, we only have to compute the sum of the second expression from $s = 0$ to $s = \underset{\prime\prime}{A}$.

Therefore, we have

$$N(u^A, x^{\overset{A}{\prime}}, y^{\overset{A}{\prime\prime}}, z \ldots n)^T = N(u \ldots n)^T - N(u \ldots n)^{T-A-1}$$
$$-N(u \ldots n)^{T-\underset{\prime}{A}-1} - N(u \ldots n)^{T-\underset{\prime\prime}{A}-1} + N(u \ldots n)^{T-A-\underset{\prime}{A}-2}$$
$$+N(u \ldots n)^{T-A-\underset{\prime\prime}{A}-2} + N(u \ldots n)^{T-\underset{\prime}{A}-\underset{\prime\prime}{A}-2} - N(u \ldots n)^{T-A-\underset{\prime}{A}-\underset{\prime\prime}{A}-3}$$

when $A + \underset{\prime}{A} < T$ and $A + \underset{\prime}{A} < T - \underset{\prime\prime}{A}$.

If $A + \underset{\prime}{A} < T$ and $A + \underset{\prime}{A} > T - \underset{\prime\prime}{A}$, we have

$$N(u^A, x^{\overset{A}{\prime}}, y^{\overset{A}{\prime\prime}}, z \ldots n)^T = N(u \ldots n)^T - N(u \ldots n)^{T-A-1}$$
$$-N(u \ldots n)^{T-\underset{\prime}{A}-1} - N(u \ldots n)^{T-\underset{\prime\prime}{A}-1} + N(u \ldots n)^{T-A-\underset{\prime}{A}-2}$$
$$+N(u \ldots n)^{T-A-\underset{\prime\prime}{A}-2} + N(u \ldots n)^{T-\underset{\prime}{A}-\underset{\prime\prime}{A}-2}.$$

If $A + \underset{\prime}{A} > T$ and $A + \underset{\prime}{A} > T - \underset{\prime\prime}{A}$, we have

$$N(u^A, x^{\overset{A}{\prime}}, y^{\overset{A}{\prime\prime}}, z \ldots n)^T = N(u \ldots n)^T - N(u \ldots n)^{T-A-1}$$
$$-N(u \ldots n)^{T-\underset{\prime}{A}-1} - N(u \ldots n)^{T-\underset{\prime\prime}{A}-1} + N(u \ldots n)^{T-A-\underset{\prime\prime}{A}-2}$$
$$+N(u \ldots n)^{T-\underset{\prime}{A}-\underset{\prime\prime}{A}-2}.$$

Second case
$$\underset{\prime\prime}{A} < T - A < T - \underset{\prime}{A}.$$

This second case is the same as the first, modulo a swap of the variables A and $\underset{\prime}{A}$. This swap brings no change to the number of terms and therefore produces nothing new.

Third case
$$T - \underset{\prime}{A} < \underset{\prime\prime}{A} < T - A.$$

In this case, the form $y^s(u^A, x^{\overset{A}{\prime}}, z \ldots n)^{T-s}$ occurs only until $s = T - \underset{\prime}{A}$. Past $s = T - \underset{\prime}{A}$, it is $y^s(u^A \ldots n - 1)^{T-s}$ until $s = \underset{\prime\prime}{A}$. We therefore must compute the sum of

1. $N(u^A, x'^{A}, z \ldots n-1)^{T-s}$, from $s=0$ to $s = T - \underset{\prime}{A}$;

2. $N(u^A \ldots n-1)^{T-s}$, from $s = T - \underset{\prime}{A}$ excluded to $s = \underset{\prime\prime}{A}$.

We have

$$N(u^A, x'^{A}, z \ldots n-1)^{T-s} = N(u \ldots n-1)^{T-s} - N(u \ldots n-1)^{T-A-s-1}$$
$$-N(u \ldots n-1)^{T-\underset{\prime}{A}-s-1} + N(u \ldots n-1)^{T-A-\underset{\prime}{A}-s-2}$$

if $A + \underset{\prime}{A} < T - s$, and

$$N(u^A, x'^{A}, z \ldots n-1)^{T-s} = N(u \ldots n-1)^{T-s} - N(u \ldots n-1)^{T-A-s-1}$$
$$-N(u \ldots n-1)^{T-\underset{\prime}{A}-s-1}$$

if $A + \underset{\prime}{A} > T - s$. Then $N(u^A \ldots n-1)^{T-s} = N(u \ldots n-1)^{T-s} - N(u \ldots n-1)^{T-A-s-1}$.

Since s begins at 0, we may have $A + \underset{\prime}{A} < T$ or $A + \underset{\prime}{A} > T$.

In the first case, we must compute the sum of

1. the first expression from $s = 0$ to $s = T - A - \underset{\prime}{A}$;

2. the second expression from $s = T - \underset{\prime}{A} - A$ excluded to $s = T - \underset{\prime}{A}$;

3. the third expression from $s = T - \underset{\prime}{A}$ excluded to $s = \underset{\prime\prime}{A}$.

In the second case, we must compute the sum of

1. the second expression from $s = 0$ to $s = T - \underset{\prime}{A}$;

2. the third expression from $s = T - \underset{\prime}{A}$ excluded to $s = \underset{\prime\prime}{A}$.

Therefore, if $A + \underset{\prime}{A} < T$, we have

$$N(u^A, x'^{A}, y''^{A}, z \ldots n)^{T} = N(u \ldots n)^{T} - N(u \ldots n)^{T-A-1}$$
$$-N(u \ldots n)^{T-\underset{\prime}{A}-1} - N(u \ldots n)^{T-\underset{\prime\prime}{A}-1} + N(u \ldots n)^{T-A-\underset{\prime}{A}-2}$$
$$+N(u \ldots n)^{T-A-\underset{\prime\prime}{A}-2}.$$

If $A + \underset{\prime}{A} > T$, we have

$$N(u^A, x'^{A}, y''^{A}, z \ldots n)^{T} = N(u \ldots n)^{T} - N(u \ldots n)^{T-A-1}$$
$$-N(u \ldots n)^{T-\underset{\prime}{A}-1} - N(u \ldots n)^{T-\underset{\prime\prime}{A}-1} + N(u \ldots n)^{T-A-\underset{\prime\prime}{A}-2}.$$

Fourth case
$$T - \underset{\prime}{A} < T - A < \underset{\prime\prime}{A}.$$

In this case, the form $y^s(u^A, x^{\overset{A}{\prime}}, z\ldots n)^{T-s}$ occurs only from $s = 0$ to $s = T - \underset{\prime}{A}$.

Past $s = T - \underset{\prime}{A}$, it is $y^s(u^A, x, z\ldots n-1)^{T-s}$ until $s = T - A$.

Past $s = T - A$, it is $y^s(u, x, z\ldots n-1)^{T-s}$ or $y^s(u\ldots n-1)^{T-s}$ until $s = \underset{\prime\prime}{A}$.

Therefore, we must compute the sum of

1. $N(u^A, x^{\overset{A}{\prime}}, z\ldots n-1)^{T-s}$ from $s = 0$ to $s = T - \underset{\prime}{A}$;

2. $N(u^A\ldots n-1)^{T-s}$ from $s = T - \underset{\prime}{A}$ excluded to $s = T - A$;

3. $N(u\ldots n-1)^{T-s}$ from $s = T - A$ excluded to $s = \underset{\prime\prime}{A}$.

We have
$$N(u^A, x^{\overset{A}{\prime}}, z\ldots n-1)^{T-s} = N(u\ldots n-1)^{T-s} - N(u\ldots n-1)^{T-A-s-1}$$
$$-N(u\ldots n-1)^{T-\underset{\prime}{A}-s-1} + N(u\ldots n-1)^{T-A-\underset{\prime}{A}-s-2}$$
if $A + \underset{\prime}{A} < T - s$, and
$$N(u^A, x^{\overset{A}{\prime}}, z\ldots n-1)^{T-s} = N(u\ldots n-1)^{T-s} - N(u\ldots n-1)^{T-A-s-1}$$
$$-N(u\ldots n-1)^{T-\underset{\prime}{A}-s-1}$$
if $A + \underset{\prime}{A} > T - s$; and $N(u^A\ldots n-1)^{T-s} = N(u\ldots n-1)^{T-s} - N(u\ldots n-1)^{T-A-s-1}$ and $N(u\ldots n-1)^{T-s} = N(u\ldots n-1)^{T-s}$.
Therefore two cases can arise since s must begin at zero; these two cases are $A + \underset{\prime}{A} < T$ and $A + \underset{\prime}{A} > T$.

In the first case, we must compute the sum of

1. the first expression from $s = 0$ to $s = T - A - \underset{\prime}{A}$;

2. the second expression from $s = T - A - \underset{\prime}{A}$ excluded to $s = T - \underset{\prime}{A}$;

3. the third expression from $s = T - \underset{\prime}{A}$ excluded to $s = T - A$;

4. the fourth expression from $s = T - A$ excluded to $s = \underset{\prime\prime}{A}$.

In the second case, we must compute the sum of

1. the second expression from $s = 0$ to $s = T - \underset{\prime}{A}$;

2. the third expression from $s = T - \underset{\prime}{A}$ excluded to $s = T - A$;

3. the fourth expression from $s = T - A$ excluded to $s = \underset{\prime\prime}{A}$.

Therefore, if $A + \underset{\prime}{A} < T$, we have

$$N(u^A, x^{\overset{A}{\prime}}, y^{\overset{A}{\prime\prime}}, z \ldots n)^T = N(u \ldots n)^T - N(u \ldots n)^{T-A-1}$$
$$-N(u \ldots n)^{T-\underset{\prime}{A}-1} - N(u \ldots n)^{T-\underset{\prime\prime}{A}-1} + N(u \ldots n)^{T-A-\underset{\prime}{A}-2},$$

and if $A + \underset{\prime}{A} > T$, we have

$$N(u^A, x^{\overset{A}{\prime}}, y^{\overset{A}{\prime\prime}}, z \ldots n)^T = N(u \ldots n)^T - N(u \ldots n)^{T-A-1}$$
$$-N(u \ldots n)^{T-\underset{\prime}{A}-1} - N(u \ldots n)^{T-\underset{\prime\prime}{A}-1}.$$

Fifth case
$$T - A < \underset{\prime\prime}{A} < T - \underset{\prime}{A}.$$

Since this case differs from the third only by swapping A and $\underset{\prime}{A}$, we only need to swap these two quantities in the expression for the number of terms specific to the third case.

Sixth case
$$T - A < T - \underset{\prime}{A} < \underset{\prime\prime}{A}.$$

Since this case differs from the fourth case only by swapping $\underset{\prime}{A}$ and A, and since this swap does not change the expression for the number of terms relative to the fourth case, it follows that the sixth case does not provide any new expression.

(143.) Let us now bring together all different cases and the corresponding values for the number of terms. We see that the whole matter reduces to the following cases and expressions:

1. $A + \underset{\prime}{A} + \underset{\prime\prime}{A} < T$; $A + \underset{\prime}{A} < T$; $A + \underset{\prime\prime}{A} < T$; $\underset{\prime}{A} + \underset{\prime\prime}{A} < T$.

$$N(u^A, x^{\overset{A}{\prime}}, y^{\overset{A}{\prime\prime}}, z \ldots n)^T = N(u \ldots n)^T - N(u \ldots n)^{T-A-1}$$
$$-N(u \ldots n)^{T-\underset{\prime}{A}-1} - N(u \ldots n)^{T-\underset{\prime\prime}{A}-1} + N(u \ldots n)^{T-A-\underset{\prime}{A}-2}$$
$$+N(u \ldots n)^{T-A-\underset{\prime\prime}{A}-2} + N(u \ldots n)^{T-\underset{\prime}{A}-\underset{\prime\prime}{A}-2}$$
$$-N(u \ldots n)^{T-A-\underset{\prime}{A}-\underset{\prime\prime}{A}-3}.$$

2. $A + A_{'} + A_{''} > T;\ A + A_{'} < T;\ A + A_{''} < T;\ A_{'} + A_{''} < T.$

$$N(u^A, x^{\overset{A}{'}}, y^{\overset{A}{''}}, z\ldots n)^T = N(u\ldots n)^T - N(u\ldots n)^{T-A-1}$$
$$-N(u\ldots n)^{T-A_{'}-1} - N(u\ldots n)^{T-A_{''}-1} + N(u\ldots n)^{T-A-A_{'}-2}$$
$$+N(u\ldots n)^{T-A-A_{''}-2} + N(u\ldots n)^{T-A_{'}-A_{''}-2}.$$

3. $A + A_{'} + A_{''} > T;\ A + A_{'} > T;\ A + A_{''} < T;\ A_{'} + A_{''} < T.$

$$N(u^A, x^{\overset{A}{'}}, y^{\overset{A}{''}}, z\ldots n)^T = N(u\ldots n)^T - N(u\ldots n)^{T-A-1}$$
$$-N(u\ldots n)^{T-A_{'}-1} - N(u\ldots n)^{T-A_{''}-1} + N(u\ldots n)^{T-A-A_{''}-2}$$
$$+N(u\ldots n)^{T-A_{'}-A_{''}-2}.$$

4. $A + A_{'} + A_{''} > T;\ A + A_{'} < T;\ A + A_{''} < T;\ A_{'} + A_{''} > T.$

$$N(u^A, x^{\overset{A}{'}}, y^{\overset{A}{''}}, z\ldots n)^T = N(u\ldots n)^T - N(u\ldots n)^{T-A-1}$$
$$-N(u\ldots n)^{T-A_{'}-1} - N(u\ldots n)^{T-A_{''}-1} + N(u\ldots n)^{T-A-A_{'}-2}$$
$$+N(u\ldots n)^{T-A-A_{''}-2}.$$

5. $A + A_{'} + A_{''} > T;\ A + A_{'} < T;\ A + A_{''} > T;\ A_{'} + A_{''} < T.$

$$N(u^A, x^{\overset{A}{'}}, y^{\overset{A}{''}}, z\ldots n)^T = N(u\ldots n)^T - N(u\ldots n)^{T-A-1}$$
$$-N(u\ldots n)^{T-A_{'}-1} - N(u\ldots n)^{T-A_{''}-1} + N(u\ldots n)^{T-A-A_{'}-2}$$
$$+N(u\ldots n)^{T-A_{'}-A_{''}-2}.$$

6. $A + A_{'} + A_{''} > T;\ A + A_{'} > T;\ A + A_{''} > T;\ A_{'} + A_{''} < T.$

$$N(u^A, x^{\overset{A}{'}}, y^{\overset{A}{''}}, z\ldots n)^T = N(u\ldots n)^T - N(u\ldots n)^{T-A-1}$$
$$-N(u\ldots n)^{T-A_{'}-1} - N(u\ldots n)^{T-A_{''}-1} + N(u\ldots n)^{T-A-A_{'}-A_{''}-2}.$$

7. $A + A_{'} + A_{''} > T;\ A + A_{'} > T;\ A + A_{''} < T;\ A_{'} + A_{''} > T.$

$$N(u^A, x^{\overset{A}{'}}, y^{\overset{A}{''}}, z\ldots n)^T = N(u\ldots n)^T - N(u\ldots n)^{T-A-1}$$
$$-N(u\ldots n)^{T-A_{'}-1} - N(u\ldots n)^{T-A_{''}-1} + N(u\ldots n)^{T-A-A_{''}-2}.$$

8. $A + A_{'} + A_{''} > T;\ A + A_{'} < T;\ A + A_{''} > T;\ A_{'} + A_{''} > T.$

$$N(u^A, x^{\overset{A}{'}}, y^{\overset{A}{''}}, z\ldots n)^T = N(u\ldots n)^T - N(u\ldots n)^{T-A-1}$$
$$-N(u\ldots n)^{T-A_{'}-1} - N(u\ldots n)^{T-A_{''}-1} + N(u\ldots n)^{T-A-A_{'}-2}.$$

94

BOOK ONE

9. $A + A_\prime + A_{\prime\prime} > T;\ A + A_\prime > T;\ A + A_{\prime\prime} > T;\ A_\prime + A_{\prime\prime} > T.$

$$N(u^A, x^{A_\prime}, y^{A_{\prime\prime}}, z\ldots n)^T = N(u\ldots n)^T - N(u\ldots n)^{T-A-1}$$
$$-N(u\ldots n)^{T-A_\prime-1} - N(u\ldots n)^{T-A_{\prime\prime}-1}.$$

(144.) From these examples, we easily see how to find the value of $N(u^A \ldots n)^T$ for four, five, etc. different exponents, and for all cases that may arise using the same method. Therefore, we will not pursue these calculations further, in which only sums of quantities of the form $N(u\ldots n-1)^{P-s}$ appear. But our preceding work easily shows what the value of $N(u^A \ldots n)^T$ is for all cases. Here is the rule that we may infer by inspecting the preceding expressions and which can be confirmed via very easy arguments.

(145.) We compute all sums of 2, 3, 4, etc. exponents A, $A_\prime$, $A_{\prime\prime}$, $A_{\prime\prime\prime}$, etc.

and we compare the results with T. Any expression greater than T does not enter in the expression for the number of terms sought. The converse occurs for any combination less than T. This combination, when augmented with the number of quantities A, $A_\prime$, $A_{\prime\prime}$, etc., and deducted from T, gives the exponent of $N(u\ldots n)$ in the term it must contribute to the general expression. The sign of this term will be $+$ or $-$, depending on whether the number of quantities A, $A_\prime$, etc. entering in its exponent is even or odd.

For example, assume we ask for the value of $N(u^A, x^{A_\prime}, y^{A_{\prime\prime}}, z^{A_{\prime\prime\prime}}, u' \ldots n)^T$ when $A + A_\prime < T;\ A + A_{\prime\prime} < T;\ A + A_{\prime\prime\prime} < T;\ A_\prime + A_{\prime\prime} < T,\ A_\prime + A_{\prime\prime\prime} < T;$ $A_{\prime\prime} + A_{\prime\prime\prime} < T;\ A + A_\prime + A_{\prime\prime} < T;\ A + A_\prime + A_{\prime\prime\prime} < T;\ A + A_{\prime\prime} + A_{\prime\prime\prime} > T;\ A_\prime + A_{\prime\prime} + A_{\prime\prime\prime} > T;$ $A + A_\prime + A_{\prime\prime} + A_{\prime\prime\prime} > T.$
We find

$$N(u^A, x^{A_\prime}, y^{A_{\prime\prime}}, z^{A_{\prime\prime\prime}}, u' \ldots n)^T = N(u\ldots n)^T - N(u\ldots n)^{T-A-1}$$
$$-N(u\ldots n)^{T-A_\prime-1} - N(u\ldots n)^{T-A_{\prime\prime}-1} - N(u\ldots n)^{T-A_{\prime\prime\prime}-1}$$
$$+N(u\ldots n)^{T-A-A_\prime-2} + N(u\ldots n)^{T-A-A_{\prime\prime}-2} + N(u\ldots n)^{T-A-A_{\prime\prime\prime}-2}$$
$$+N(u\ldots n)^{T-A_\prime-A_{\prime\prime}-2} + N(u\ldots n)^{T-A_\prime-A_{\prime\prime\prime}-2} + N(u\ldots n)^{T-A_{\prime\prime}-A_{\prime\prime\prime}-2}$$
$$-N(u\ldots n)^{T-A-A_\prime-A_{\prime\prime}-3} - N(u\ldots n)^{T-A-A_\prime-A_{\prime\prime\prime}-3}.$$

General method to determine the degree of the final
equation for all cases of equations
of the form $(u^a \ldots n)^t = 0$

(146.) Until now, we have computed the differential of $N(u^{A+a} \ldots n)^{T+t}$, or simply $N(u^A \ldots n)^T$, by successively varying T by the increments t, t', t'', etc., A by the increments a, a', a'', etc., $A_\prime$ by the increments $a_\prime, a'_\prime, a''_\prime$, etc., and so on; it is quite easy to compute the different values of D that yield the degree of the final equation and the conditions for which those values of D are admissible.

From these conditions, we find that for three equations in three unknowns, we have the first form, as follows

First Form

$$A + \underset{\prime}{A} + \underset{\prime\prime}{A} < T; \quad A + \underset{\prime}{A} < T; \quad A + \underset{\prime\prime}{A} < T; \quad \underset{\prime}{A} + \underset{\prime\prime}{A} < T.$$

$$D = tt't'' - (t-a).(t'-a').(t''-a'') - (t-\underset{\prime}{a}).(t'-\underset{\prime}{a}').(t''-\underset{\prime}{a}'')$$
$$-(t-\underset{\prime\prime}{a}).(t'-\underset{\prime\prime}{a}').(t''-\underset{\prime\prime}{a}'')$$
$$+(t-a-\underset{\prime}{a}).(t'-a'-\underset{\prime}{a}').(t''-a''-\underset{\prime}{a}'')$$
$$+(t-a-\underset{\prime\prime}{a}).(t'-a'-\underset{\prime\prime}{a}').(t''-a''-\underset{\prime\prime}{a}'')$$
$$+(t-\underset{\prime}{a}-\underset{\prime\prime}{a}).(t'-\underset{\prime}{a}'-\underset{\prime\prime}{a}').(t''-\underset{\prime}{a}''-\underset{\prime\prime}{a}'')$$
$$-(t-a-\underset{\prime}{a}-\underset{\prime\prime}{a}).(t'-a'-\underset{\prime}{a}'-\underset{\prime\prime}{a}').(t''-a''-\underset{\prime}{a}''-\underset{\prime\prime}{a}'').$$

Conditions

$$\left.\begin{array}{r} t't'' - (t'-a').(t''-a'') - (t'-\underset{\prime}{a}').(t''-\underset{\prime}{a}'') \\ -(t'-\underset{\prime\prime}{a}').(t''-\underset{\prime\prime}{a}'') + (t'-a'-\underset{\prime}{a}').(t''-a''-\underset{\prime}{a}'') \\ +(t'-a'-\underset{\prime\prime}{a}').(t''-a''-\underset{\prime\prime}{a}'') \\ +(t'-\underset{\prime}{a}'-\underset{\prime\prime}{a}').(t''-\underset{\prime}{a}''-\underset{\prime\prime}{a}'') \\ -(t'-a'-\underset{\prime}{a}'-\underset{\prime\prime}{a}').(t''-a''-\underset{\prime}{a}''-\underset{\prime\prime}{a}'') \end{array}\right\} > 0,$$

$$\left.\begin{array}{r} (t'-a').(t''-a'') - (t'-a'-\underset{\prime}{a}').(t''-a''-\underset{\prime}{a}'')- \\ (t'-a'-\underset{\prime\prime}{a}').(t''-a''-\underset{\prime\prime}{a}'') \\ +(t'-a'-\underset{\prime}{a}'-\underset{\prime\prime}{a}').(t''-a''-\underset{\prime}{a}''-\underset{\prime\prime}{a}'') \end{array}\right\} > 0,$$

$$\left.\begin{array}{r} (t'-\underset{\prime}{a}').(t''-\underset{\prime}{a}'') - (t'-a'-\underset{\prime}{a}').(t''-a''-\underset{\prime}{a}'') \\ -(t'-\underset{\prime}{a}'-\underset{\prime\prime}{a}').(t''-\underset{\prime}{a}''-\underset{\prime\prime}{a}'') \\ +(t'-a'-\underset{\prime}{a}'-\underset{\prime\prime}{a}').(t''-a''-\underset{\prime}{a}''-\underset{\prime\prime}{a}'') \end{array}\right\} > 0,$$

$$\left.\begin{array}{r} (t'-\underset{\prime\prime}{a}').(t''-\underset{\prime\prime}{a}'') - (t'-a'-\underset{\prime\prime}{a}').(t''-a''-\underset{\prime\prime}{a}'') \\ -(t'-\underset{\prime}{a}'-\underset{\prime\prime}{a}').(t''-\underset{\prime}{a}''-\underset{\prime\prime}{a}'') \\ +(t'-a'-\underset{\prime}{a}'-\underset{\prime\prime}{a}').(t''-a''-\underset{\prime}{a}''-\underset{\prime\prime}{a}'') \end{array}\right\} > 0.$$

Second form
$$A + A_{/} + A_{//} > T;\ A + A_{/} < T;\ A + A_{//} < T;\ A_{/} + A_{//} < T.$$

$$D = tt't'' - (t - a).(t' - a').(t'' - a'') - (t - a_{/}).(t' - a'_{/}).(t'' - a''_{/})$$
$$-(t - a_{//}).(t' - a'_{//}).(t'' - a''_{//})$$
$$+(t - a - a_{/}).(t' - a' - a'_{/}).(t'' - a'' - a''_{/})$$
$$+(t - a - a_{//}).(t' - a' - a'_{//}).(t'' - a'' - a''_{//})$$
$$+(t - a_{/} - a_{//}).(t' - a'_{/} - a'_{//}).(t'' - a''_{/} - a''_{//}).$$

Conditions

$$\left.\begin{array}{r}
t't'' - (t' - a').(t'' - a'') \\
-(t' - a'_{/}).(t'' - a''_{/}) - (t' - a'_{//}).(t'' - a''_{//}) \\
+(t' - a' - a'_{/}).(t'' - a'' - a''_{/}) + (t' - a' - a'_{//}).(t'' - a'' - a''_{//}) \\
+(t' - a'_{/} - a'_{//}).(t'' - a''_{/} - a''_{//})
\end{array}\right\} > 0,$$

$$\left.\begin{array}{r}
(t' - a').(t'' - a'') - (t' - a' - a'_{/}).(t'' - a'' - a''_{/}) \\
-(t' - a' - a'_{//}).(t'' - a'' - a''_{//})
\end{array}\right\} > 0,$$

$$\left.\begin{array}{r}
(t' - a'_{/}).(t'' - a''_{/}) - (t' - a' - a'_{/}).(t'' - a'' - a''_{//}) \\
-(t' - a'_{/} - a'_{//}).(t'' - a''_{/} - a''_{//})
\end{array}\right\} > 0,$$

$$\left.\begin{array}{r}
(t' - a'_{//}).(t'' - a''_{//}) - (t' - a' - a'_{//}).(t'' - a'' - a''_{//}) \\
-(t' - a'_{/} - a'_{//}).(t'' - a''_{/} - a''_{//})
\end{array}\right\} > 0.$$

Third form
$$A + A_{/} + A_{//} > T;\ A + A_{/} > T;\ A + A_{//} < T;\ A_{/} + A_{//} < T.$$

$$D = tt't'' - (t - a).(t' - a').(t'' - a'')$$
$$-(t - a_{/}).(t' - a'_{/}).(t'' - a''_{/}) - (t - a_{//}).(t' - a'_{//}).(t'' - a''_{//})$$
$$+(t - a - a_{//}).(t' - a' - a'_{//}).(t'' - a'' - a''_{//})$$
$$+(t - a_{/} - a_{//}).(t' - a'_{/} - a'_{//}).(t'' - a''_{/} - a''_{//}).$$

Conditions

$$\left.\begin{array}{c} t't'' - (t' - a').(t'' - a'') - (t' - \underset{/}{a}').(t'' - \underset{/}{a}'') \\ -(t' - \underset{//}{a}').(t'' - \underset{//}{a}'') + (t' - a' - \underset{/}{a}').(t'' - a'' - \underset{/}{a}'') \\ +(t' - \underset{/}{a}' - \underset{//}{a}').(t'' - \underset{/}{a}'' - \underset{//}{a}'') \end{array}\right\} > 0,$$

$$(t' - a').(t'' - a'') - (t' - a' - \underset{//}{a}').(t'' - a'' - \underset{//}{a}'') > 0,$$

$$(t' - \underset{/}{a}').(t'' - \underset{/}{a}'') - (t' - \underset{/}{a}' - \underset{//}{a}').(t'' - \underset{/}{a}'' - \underset{//}{a}'') > 0,$$

$$\left.\begin{array}{c} (t' - \underset{//}{a}').(t'' - \underset{//}{a}'') - (t' - a' - \underset{//}{a}').(t'' - a'' - \underset{//}{a}'') \\ -(t' - \underset{/}{a}' - \underset{//}{a}').(t'' - \underset{/}{a}'' - \underset{//}{a}'') \end{array}\right\} > 0.$$

Fourth form

$$A + \underset{/}{A} + \underset{//}{A} > T; \quad A + \underset{/}{A} < T; \quad A + \underset{//}{A} < T; \quad \underset{/}{A} + \underset{//}{A} > T.$$

$$D = tt't'' - (t - a).(t' - a').(t'' - a'')$$
$$-(t - \underset{/}{a}).(t' - \underset{/}{a}').(t'' - \underset{/}{a}'') - (t - \underset{//}{a}).(t' - \underset{//}{a}').(t'' - \underset{//}{a}'')$$
$$+(t - a - \underset{/}{a}).(t' - a' - \underset{/}{a}').(t'' - a'' - \underset{/}{a}'')$$
$$+(t - a - \underset{//}{a}).(t' - a' - \underset{//}{a}').(t'' - a'' - \underset{//}{a}'').$$

Conditions

$$\left.\begin{array}{c} t't'' - (t' - a').(t'' - a'') - (t' - \underset{/}{a}').(t'' - \underset{/}{a}'') \\ -(t' - \underset{//}{a}').(t'' - \underset{//}{a}'') + (t' - a' - \underset{/}{a}').(t'' - a'' - \underset{/}{a}'') \\ +(t' - a' - \underset{//}{a}').(t'' - a'' - \underset{//}{a}'') \end{array}\right\} > 0,$$

$$\left.\begin{array}{c} (t' - a').(t'' - a'') - (t' - a' - \underset{/}{a}').(t'' - a'' - \underset{/}{a}'') \\ -(t' - a' - \underset{//}{a}').(t'' - a'' - \underset{//}{a}'') \end{array}\right\} > 0,$$

$$(t' - \underset{/}{a}').(t'' - a'') - (t' - a' - \underset{/}{a}').(t'' - a'' - \underset{/}{a}'') > 0,$$

$$(t' - \underset{//}{a}').(t'' - \underset{//}{a}'') - (t' - a' - \underset{//}{a}').(t'' - a'' - \underset{//}{a}'') > 0.$$

Fifth form

$$A + A_{\prime} + A_{\prime\prime} > T; \quad A + A_{\prime} < T; \quad A + A_{\prime\prime} > T; \quad A_{\prime} + A_{\prime\prime} < T.$$

$$D = tt't'' - (t - a).(t' - a').(t'' - a'')$$
$$- (t - a_{\prime}).(t' - a'_{\prime}).(t'' - a''_{\prime}) - (t - a_{\prime\prime}).(t' - a'_{\prime\prime}).(t'' - a''_{\prime\prime})$$
$$+ (t - a - a_{\prime}).(t' - a' - a'_{\prime}).(t'' - a'' - a''_{\prime})$$
$$+ (t - a_{\prime} - a_{\prime\prime}).(t' - a'_{\prime} - a'_{\prime\prime}).(t'' - a''_{\prime} - a''_{\prime\prime}).$$

Conditions

$$\left. \begin{array}{r} t't'' - (t' - a').(t'' - a'') - (t' - a'_{\prime}).(t'' - a''_{\prime}) \\ -(t' - a'_{\prime\prime}).(t'' - a''_{\prime\prime}) + (t' - a' - a'_{\prime}).(t'' - a'' - a''_{\prime}) \\ + (t' - a'_{\prime} - a'_{\prime\prime}).(t'' - a''_{\prime} - a''_{\prime\prime}) \end{array} \right\} > 0,$$

$$(t' - a').(t'' - a'') - (t' - a' - a'_{\prime}).(t'' - a'' - a''_{\prime}) > 0,$$

$$\left. \begin{array}{r} (t' - a'_{\prime}).(t'' - a''_{\prime}) - (t' - a' - a'_{\prime}).(t'' - a'' - a''_{\prime}) \\ - (t' - a'_{\prime} - a'_{\prime\prime}).(t'' - a''_{\prime} - a''_{\prime\prime}) \end{array} \right\} > 0,$$

$$(t' - a'_{\prime\prime}).(t'' - a''_{\prime\prime}) - (t' - a'_{\prime} - a'_{\prime\prime}).(t'' - a''_{\prime} - a''_{\prime\prime}) > 0.$$

Sixth form

$$A + A_{\prime} + A_{\prime\prime} > T; \quad A + A_{\prime} > T; \quad A + A_{\prime\prime} > T; \quad A_{\prime} + A_{\prime\prime} < T.$$

$$D = tt't'' - (t - a).(t' - a').(t'' - a'') - (t - a_{\prime}).(t' - a'_{\prime}).(t'' - a''_{\prime})$$
$$- (t - a_{\prime\prime}).(t' - a'_{\prime\prime}).(t'' - a''_{\prime\prime})$$
$$+ (t - a_{\prime} - a_{\prime\prime}).(t' - a'_{\prime} - a'_{\prime\prime}).(t'' - a''_{\prime} - a''_{\prime\prime}).$$

Conditions

$$\left. \begin{array}{r} t't'' - (t' - a').(t'' - a'') \\ -(t' - a'_{\prime}).(t'' - a''_{\prime}) - (t' - a'_{\prime\prime}).(t'' - a''_{\prime\prime}) \\ + (t' - a'_{\prime} - a'_{\prime\prime}).(t'' - a''_{\prime} - a''_{\prime\prime}) \end{array} \right\} > 0,$$

$$(t' - a').(t'' - a'') > 0,$$

$$(t' - a'_{\prime}).(t'' - a''_{\prime}) - (t' - a'_{\prime} - a'_{\prime\prime}).(t'' - a''_{\prime} - a''_{\prime\prime}) > 0,$$

$$(t' - a'_{\prime\prime}).(t'' - a''_{\prime\prime}) - (t' - a'_{\prime} - a'_{\prime\prime}).(t'' - a''_{\prime} - a''_{\prime\prime}) > 0.$$

Seventh form

$$A + A_{/} + A_{//} > T;\ A + A_{/} > T;\ A + A_{//} < T;\ A_{/} + A_{//} > T.$$

$$D = tt't'' - (t-a).(t'-a').(t''-a'')$$
$$-(t-a_{/}).(t'-a_{/}').(t''-a_{/}'') - (t-a_{//}).(t'-a_{//}').(t''-a_{//}'')$$
$$+(t-a-a_{//}).(t'-a'-a_{//}').(t''-a''-a_{//}'').$$

Conditions

$$\left.\begin{array}{l} t't'' - (t'-a').(t''-a'') - (t'-a_{/}').(t''-a_{/}'') \\[4pt] -(t'-a_{//}').(t''-a_{//}'') + (t'-a'-a_{//}').(t''-a''-a_{//}'') \end{array}\right\} > 0,$$

$$(t'-a').(t''-a'') - (t'-a'-a_{/}').(t''-a''-a_{/}'') > 0,$$

$$(t'-a_{/}').(t''-a_{/}'') > 0,$$

$$(t'-a_{//}').(t''-a_{//}'') - (t'-a'-a_{//}').(t''-a''-a_{//}'') > 0.$$

Eighth form

$$A + A_{/} + A_{//} > T;\ A + A_{/} < T;\ A + A_{//} > T;\ A_{/} + A_{//} > T.$$

$$D = tt't'' - (t-a).(t'-a').(t''-a'')$$
$$-(t-a_{/}).(t'-a_{/}').(t''-a_{/}'') - (t-a_{//}).(t'-a_{//}').(t''-a_{//}'')$$
$$+(t-a-a_{/}).(t'-a'-a_{/}').(t''-a''-a_{/}'').$$

Conditions

$$\left.\begin{array}{l} t't'' - (t'-a').(t''-a'') - (t'-a_{/}').(t''-a_{/}'') \\[4pt] -(t'-a_{//}').(t''-a_{//}'') + (t'-a'-a_{/}').(t''-a''-a_{/}'') \end{array}\right\} > 0,$$

$$(t'-a').(t''-a'') - (t'-a'-a_{/}').(t''-a''-a_{/}'') > 0,$$

$$(t'-a_{/}').(t''-a_{/}'') - (t'-a'-a_{/}').(t''-a''-a_{/}'') > 0,$$

$$(t'-a_{//}').(t''-a_{//}'') > 0.$$

Ninth form

$$A + \underset{\prime}{A} + \underset{\prime\prime}{A} > T;\ A + \underset{\prime}{A} > T;\ A + \underset{\prime\prime}{A} > T;\ \underset{\prime}{A} + \underset{\prime\prime}{A} > T.$$

$$D = tt't'' - (t-a).(t'-a').(t''-a'') - (t-\underset{\prime}{a}).(t'-\underset{\prime}{a}').(t''-\underset{\prime}{a}'')$$
$$-(t-\underset{\prime\prime}{a}).(t'-\underset{\prime\prime}{a}').(t''-\underset{\prime\prime}{a}'').$$

Conditions

$$\left.\begin{array}{c} t't'' - (t'-a').(t''-a'') \\ -(t'-\underset{\prime}{a}').(t''-\underset{\prime}{a}'') - (t'-\underset{\prime\prime}{a}').(t''-\underset{\prime\prime}{a}'') \end{array}\right\} > 0,$$

$$(t'-a').(t''-a'') > 0,$$
$$(t'-\underset{\prime}{a}').(t''-\underset{\prime}{a}'') > 0,$$
$$(t'-\underset{\prime\prime}{a}').(t''-\underset{\prime\prime}{a}'') > 0.$$

For an arbitrary number of unknowns, we can easily determine all values of D for equations of the form $(u^a \ldots n)^t = 0$, given arbitrary values for a, $\underset{\prime}{a}$, $\underset{\prime\prime}{a}$ etc.. We can also find the conditions for which each possible value of D is admissible. Using the argument in (145), we can compute all of the different forms taken by the value of $N(u^A \ldots n)^T$. Using differentiations, this value immediately gives the value of D and the conditions for which it is admissible. Only the length of the computations may vary.

Remarks

(147.)

1. In the case of three equations in three unknowns, we could have obtained the forms given in (146) from those provided in (120 ff), since they are special cases of these, except the first one, which we will discuss later. For a larger number of unknowns, the equations of the form $(u^A \ldots n)^t = 0$ are also special cases of the equations described in (133). However, the number of possible forms increases very fast with the number of unknowns and the expressions become incresingly complex: We have limited ourselves to the example of three equations in three unknowns to show how easy it is to find the values of D for an arbitrary number of unknowns and equations of the form $(u^A \ldots n)^t = 0$: We can simply derive these values from the more general forms we just discussed.

(148.)

2. The first of the nine forms presented in (146) cannot be derived from any of the eight presented in (120 ff). The reason is that there can be neither equations nor polynomials of that form. For example, consider the case of three unknowns. If we had $a + \underset{\prime}{a} + \underset{\prime\prime}{a} < t$, it is clear

that these three unknowns cannot reach the dimension t altogether, which contradicts the assumption: They can only reach the dimension $a + a\prime + a\prime\prime$ and then they belong to the forms given in (120 ff). From (83 ff), we have expressly excluded the case $A + A\prime + A\prime\prime < T$, and therefore it cannot be found in the eight forms given in (120 ff).

To conclude what is to be said about equations similar to those considered up until now, we will sketch the way to determine the number of terms of polynomials of this type. Indeed, we have reduced the search for the degree of the final equation to that question.

General considerations about the number of terms of other polynomials that are similar to those we have examined

(149.) Looking for the degree of the final equation resulting from polynomial equations that are analogous to those considered up until now reduces to searching for the number of terms in each polynomial. Before moving on to polynomials of another form, it is useful to discuss the precautions one must take to account for all possible forms of the number of terms while avoiding erroneous forms. We will also say a word about the various quantities we will have to sum and about the computation of their summation.

(150.) Consider a polynomial containing n unknowns, each of which cannot exceed a given degree (this degree may be different or the same for each equation). We assume that four of these unknowns are such that their pairwise combinations cannot exceed given dimensions. Likewise, the combinations of three unknowns cannot exceed given dimensions, and the combinations of four variables cannot exceed a given dimension either. Finally, we assume that the other variables, when combined among each other as well as with the first variables, achieve all possible dimensions until that of the polynomial. We represent such a polynomial by the following expression:

$$([(u^A, x'^A)^B, (u^A, y''^A)^{B'}, (x'^A, y''^A)^{B''}]^C,$$
$$[(u^A, x'^A)^B, (u^A, z'''^A)^{B'''}, (x'^A, z'''^A)^{B}_{IV}]^{C'}, \ldots$$
$$\ldots [(u^A, y''^A)^{B'}, (u^A, z'''^A)^{B'''}, (y''^A, z'''^A)^{B}_V]^{C''},$$
$$[(x'^A, y''^A)^{B''}, (x'^A, z'''^A)^{B}_{IV}, (y''^A, z'''^A)^{B}_V]^{C'''}, r_{IV} \ldots n)^T.$$

In order to show how we determine the number of terms, we will begin by assuming that the exponents A_{IV}, A_V, etc. of the unknowns other than u, x, y, z, are all equal to T, as we have done previously in (84). Indeed, from (77), it is easy to derive the other cases once we have the number of terms in this case.

Let us now assume that the polynomial is ordered in ascending powers of any one of the four letters u, x, y, z. Let us choose z, for example. Each term has the form

$$z^s([(u^A, x'^A)^B, (u^A, y''^A)^{B'}, (x'^A, y''^A)^{B''}]^C, r \ldots n - 1)^{T-s}$$

from $s = 0$ to s equal to the smallest of the seven quantities

$$B_V - A_{\prime\prime};\ B_{IV} - A_{\prime};\ B_{\prime\prime\prime} - A;\ C_{\prime\prime\prime} - B_{\prime\prime};\ C_{\prime\prime} - B_{\prime};\ C_{\prime} - B;\ E - C.$$

From all possible cases that may arise, we select the following, to illustrate several precautions we need to take:

$$B_V - A_{\prime\prime} < B_{IV} - A_{\prime} < B_{\prime\prime\prime} - A < C_{\prime\prime\prime} - B_{\prime\prime} < C_{\prime\prime} - B_{\prime} < C_{\prime} - B < E - C.$$

It follows that from $s = 0$, the form of each term is

$$z^s\big([(u^A, x_\prime^{\,A})^B, (u^A, y_{\prime\prime}^{\,A})_\prime^{\,B}, (x_\prime^{\,A}, y_{\prime\prime}^{\,A})_{\prime\prime}^{\,B}]^C, r\ldots n-1\big)^{T-s},$$

until $s = B_V - A_{\prime\prime}$.

Past $s = B_V - A_{\prime\prime}$, the form is

$$z^s\big([(u^A, x_\prime^{\,A})^B, (u^A, y_V^{\,B-s})_\prime^{\,B}, (x_\prime^{\,A}, y_V^{\,B-s})_{\prime\prime}^{\,B}]^C, r\ldots n-1\big)^{T-s},$$

until $s = B_{IV} - A_{\prime}$.

Past $s = B_{IV} - A_{\prime}$, the form is

$$z^s\big([(u^A, x_{IV}^{\,B-s})^B, (u^A, y_V^{\,B-s})_\prime^{\,B}, (x_{IV}^{\,B-s}, y_V^{\,B-s})_{\prime\prime}^{\,B}]^C,$$
$$r\ldots n-1\big)^{T-s},$$

until $s = B_{\prime\prime\prime} - A$.

Past $s = B_{\prime\prime\prime} - A$, the form is

$$z^s\big([(u_{\prime\prime\prime}^{\,B-s}, x_{IV}^{\,B-s})^B, (u_{\prime\prime\prime}^{\,B-s}, y_V^{\,B-s})_\prime^{\,B}, (x_{IV}^{\,B-s}, y_V^{\,B-s})_{\prime\prime}^{\,B}]^C,$$
$$r\ldots n-1\big)^{T-s},$$

until $s = C_{\prime\prime\prime} - B_{\prime\prime}$.

Past $s = C_{\prime\prime\prime} - B_{\prime\prime}$, the form is

$$z^s\big([(u_{\prime\prime\prime}^{\,B-s}, x_{IV}^{\,B-s})^B, (u_{\prime\prime\prime}^{\,B-s}, y_V^{\,B-s})_\prime^{\,B}, (x_{IV}^{\,B-s}, y_V^{\,B-s})_{\prime\prime\prime}^{\,C-s}]^C,$$
$$r\ldots n-1\big)^{T-s},$$

until $s = C_{\prime\prime} - B_{\prime}$.

Past $s = C_{\prime\prime} - B_{\prime}$, the form is

$$z^s\big([(u_{\prime\prime\prime}^{\,B-s}, x_{IV}^{\,B-s})^B, (u_{\prime\prime\prime}^{\,B-s}, y_V^{\,B-s})_{\prime\prime}^{\,C-s}, (x_{IV}^{\,B-s}, y_V^{\,B-s})_{\prime\prime\prime}^{\,C-s}]^C,$$
$$r\ldots n-1\big)^{T-s},$$

until $s = C_{\prime} - B$.

Past $s = C_{\prime} - B$, the form is

$$z^s\big([(u_{\prime\prime\prime}^{\,B-s}, x_{IV}^{\,B-s})_\prime^{\,C-s}, (u_{\prime\prime\prime}^{\,B-s}, y_V^{\,B-s})_{\prime\prime}^{\,C-s}, (x_{IV}^{\,B-s}, y_V^{\,B-s})_{\prime\prime\prime}^{\,C-s}]^C,$$
$$r\ldots n-1\big)^{T-s},$$

until $s = E - C$.

Past $s = E - C$, the form is

$$z^s([[(u'''^{B-s}, x\overset{IV}{}^{B-s})'^{C-s}, (u'''^{B-s}, y\overset{V}{}^{B-s})''^{C-s}, (x\overset{IV}{}^{B-s}, y\overset{V}{}^{B-s})'''^{C-s}]^{E-s},$$
$$r \ldots n - 1)^{T-s},$$

until $s = \underset{'''}{A}$, which is the highest value s can reach.

The task of finding the number of terms of these eight polynomials still remains, from what was said in (84 ff). We must also sum these eight expressions, over the values of s for which each one holds.

However, the range of s over which each polynomial occurs does not determine the proper form to use to compute the number of its terms. For example, the third polynomial occurs from $s = \underset{IV}{B} - \underset{'}{A}$ to $s = \underset{'''}{B} - A$. But the expression for the number of its terms, which belongs to any one of the eight forms given in (92 ff) as soon as this polynomial occurs, can then belong to any one of these eight forms before s becomes equal to $\underset{'''}{B} - A$: It can consecutively belong to many of these eight forms before $s = \underset{'''}{B} - A$. The same thing can happen to the other polynomials. It is also possible that the expression for the number of terms belongs to forms derived from the eight presented expressions (92 ff), by virtue of what was said in (101).

Indeed, let us assume, for example, that A, $\underset{'}{A}$, $\underset{''}{A}$; B, $\underset{'}{B}$, $\underset{''}{B}$; C be such that the expressions of the number of terms of the first and second polynomials belong to the sixth form, which requires $C - B > \underset{'}{B} - A$; $C - B > \underset{''}{B} - \underset{'}{A}$;

$$C - \underset{'}{B} > \underset{''}{B} - \underset{''}{A}.$$

Considering the third polynomial, the expression of the number of terms will belong to the sixth form, as long as $C - B > \underset{'}{B} - A$; $C - B > \underset{''}{B} - \underset{IV}{B} + s$;

$$C - \underset{'}{B} > \underset{''}{B} - \underset{V}{B} + s.$$

But as soon as s (which grows continuously) changes any of these inequalities, we reach another form. We can easily imagine that changes in these forms will occur even more frequently in the polynomials following the third polynomial. In fact, not only will we have to look at the eight forms presented in (92 ff), but also those that can be derived from (101).

To find the different forms we will encounter consecutively, we note that the parameters of the problem are always sufficient to answer this question.

For example, let us assume that the inequalities among the quantities A, $\underset{'}{A}$, $\underset{''}{A}$; B, $\underset{'}{B}$, $\underset{''}{B}$, and C are such that the expression of the number of terms of the first of our eight polynomials belongs to the sixth form. We therefore have $C - B > \underset{'}{B} - A$; $C - B > \underset{''}{B} - \underset{'}{A}$; $C - \underset{'}{B} > \underset{''}{B} - \underset{''}{A}$, where the first of our eight polynomials satisfies the general conditions given in (83).

The second of these eight polynomials still belongs to the same form as long as $C - B > \underset{'}{B} - A$; $C - B > \underset{''}{B} - \underset{'}{A}$; $C - \underset{'}{B} > \underset{''}{B} - \underset{V}{B} - s$. Indeed, what was $A_{''}$ in the first polynomial has now become $B_V - s$. But as soon as $C - B_{'} < B_{''} - B_V + s$, the expression for the number of terms cannot be taken from the sixth form anymore; instead, it must be taken from the fifth form. Therefore, we must determine whether we can have $C - B_{'} < B_{''} - B_V + s$ before having $s = B_{IV} - A_{'}$, that is, before reaching the third polynomial. For this to happen, we must have $C - B_{'} - B_{''} + B_V < B_{IV} - A_{'}$. Therefore, if $C - B_{'} - B_{''} + B_V > B_{IV} - A_{'}$, the expression for the number of terms of the second polynomial belongs to the sixth form from $s = B_V - A_{''}$ to $s = B_{IV} - A_{'}$, that is, over the entire range of values for which this polynomial occurs. But if instead we have $C - B_{'} - B_{''} + B_V < B_{IV} - A_{'}$, the expression for the number of terms of the second polynomial belongs to the sixth form only from $s = B_V - A_{''}$ to $s = C - B_{'} - B_{''} + B_V$. Past this term, it belongs to the fifth form until $s = B_{IV} - A_{'}$.

Moreover, we must have $C - B_{'} - B_{''} + B_V > B_V - A_{''}$ for this second case to occur, that is, $C > B_{'} + B_{''} - A_{''}$. This condition holds by hypothesis, since it is nothing but $C - B_{'} > B_{''} - A_{''}$.

Let us now consider the third polynomial. We will assume that the first of the two cases in the argument above holds for the second polynomial. Then the expression for the number of terms of the third polynomial will belong to the sixth form as long as $C - B_{'} > B_{''} - A$; $C - B_{'} > B_{''} - B_{IV} + s$; $C - B_{'} > B_{''} - B_V + s$. Therefore it will no longer belong to this form under two possible circumstances: First, when $C - B < B_{''} - B_{IV} + s$; second, when $C - B_{'} < B_{''} - B_V + s$. For these conditions to occur, we must have s smaller than $B_{'''} - A$. We must therefore have $C - B - B_{''} + B_{IV} < B_{'''} - A$, and $C - B_{'} - B_{''} + B_V < B_{'''} - A$. Hence, four cases arise:

$$C - B - B_{''} + B_{IV} < B_{'''} - A; \quad C - B_{'} - B_{''} + B_V < B_{'''} - A;$$
$$C - B - B_{''} + B_{IV} < B_{'''} - A; \quad C - B_{'} - B_{''} + B_V > B_{'''} - A;$$
$$C - B - B_{''} + B_{IV} > B_{'''} - A; \quad C - B_{'} - B_{''} + B_V < B_{'''} - A;$$
$$C - B - B_{''} + B_{IV} > B_{'''} - A; \quad C - B_{'} - B_{''} + B_V > B_{'''} - A.$$

In the last case, the expression for the number of terms keeps belonging to the sixth form throughout the domain over which the third polynomial applies.

In the third case, it takes this form only from $s = B_I - A$ to $s = C - B_I - B_{II} - B_V$. Afterwards, it takes the fifth form from $s = C - B_I - B_{II} + B_V$ to $s = B_{III} - A$.

In the second case, the expression of the number of terms takes the sixth form only until $s = C - B - B_{II} + B_{IV}$. Past that term, it takes the fourth form until $s = B_{III} - A$.

In the first case, the expression of the number of terms takes the sixth form until s equals the smallest of the quantities $C - B_I - B_{II} + B_V$ and $C - B - B_{II} - B_{IV}$.

This leads to the following two cases:

$$C - B_I - B_{II} + B_V > C - B - B_{II} + B_{IV}$$

or

$$B + B_V > B_I + B_{IV}$$

and

$$C - B_I - B_{II} + B_V < C - B - B_{II} + B_{IV}$$

or

$$B + B_V < B_I + B_{IV}.$$

Assume $B + B_V < B_I + B_{IV}$: Past $s = C - B_I - B_{II} + B_V$, the expression for the number of terms takes the fifth form, until $s = C - B - B_{II} + B_{IV}$. Past $s = C - B - B_{II} + B_{IV}$, it takes the third form until $s = B_{III} - A$. The same developments occur when $B + B_V > B_I + B_{IV}$.

(151.) These developments are enough to see what to do for any polynomials, as long as we assume that these polynomials satisfy the conditions described in (83), as we have implicitly done until now.

But these conditions do not necessarily hold: Therefore, it makes sense to identify the characteristics of polynomials for which these conditions must hold and those for which they are not necessary.

We first remark that the first of our eight polynomials must necessarily satisfy the conditions mentioned in (83); otherwise, the polynomial of interest would not belong to the class we have assumed up to now.

Second, the second polynomial must also satisfy the same conditions; however, this is not as obvious. Assume this polynomial does not satisfy one of these conditions. For example, assume that $A_I + B_V - s < B_{II}$ holds for some s less than $B_{IV} - A_I$. Then, clearly, past $s = A_I + B_V - B_{II}$ the two unknowns x and y can reach the dimension $A_I + B_V - s$ together. Then, z together with x and y cannot reach a dimension higher than $A_I + B_V$. However, the assumptions $A_{'} + B_{V} - B_{''} < B_{IV} - A_{'}$ and $B_{IV} - A_{'} < B_{'} - A_{''} < C_{'''} - B_{''}$ give $A_{'} + B_{V} - B_{''} < C_{'''} - B_{''}$ or $A_{'} + B_{V} < C_{'''}$. Therefore, we cannot assume that $A_{'} + B_{V} - s < B_{''}$ before $s = B_{IV} - A_{'}$ without reaching a contradiction with the fact that x, y and z, taken together, must be able to reach the dimension $C_{'''}$. Likewise, we can see that it is impossible for the second polynomial not to satisfy any other condition of (83).

(152.) We will also remark that, while we discover whether the partial polynomial of interest must satisfy the conditions (83) or not, we also identify the conditions of existence for the main polynomial. For example, we have seen earlier that it was not possible to have $A_{'} + B_{V} < C_{'''}$. Thus, we can conclude that $A_{'} + B_{V} > C_{'''}$ is one of the conditions for the existence of the main polynomial, which is of interest throughout this work. Likewise, we see that we must have $A_{'} + B_{V} - s > B_{'}$. By replacing s with its highest value in the same second polynomial, we get $A_{'} + B_{V} - B_{IV} + A_{'} > B_{'}$ or $A_{'} + B_{V} - B_{IV} > B_{'} - A_{'}$, and consequently $A_{'} + B_{V} - B_{IV} > B_{'} - A_{''}$ or $A_{'} + A_{''} > B_{'}$, which is another condition of existence of the main polynomial.

We see likewise that the third polynomial must satisfy all the conditions (83) over all its domain, that is, from $s = B_{IV} - A_{'}$ to $s = B_{'''} - A$. For example, we always have $B_{IV} - s + B_{V} - s > B_{''}$. Indeed, if we assumed $B_{IV} + B_{V} - 2s < B_{''}$ before $s = B_{'''} - A$, then x, y and z together could reach the dimension $B_{IV} + B_{V} - s$ only, as soon as s exceeds $\dfrac{B_{IV} + B_{V} - B_{''}}{2}$; thus when s reaches $B_{'''} - A$, they could only reach the dimension $B_{IV} + B_{V} - B_{'''} + A$ together. But since $(B_{IV} + B_{V} - B_{''})/2 < B_{'''} - A$, we have $B_{IV} + B_{V} - B_{''} - B_{'''} + A < B_{'''} - A < C_{''} - B_{''}$. We would therefore get $B_{IV} + B_{V} - B_{'''} + A < C_{'''}$, which implies that x, y and z together could not reach the dimension $C_{'''}$. Therefore, they could never reach this dimension for any value of s since $B_{IV} + B_{V} - s$ only decreases as s increases.

Therefore, the condition $(B_{IV} + B_{V} - B_{''})/2 > B_{'''} - A$ or $B_{IV} + B_{V} - B_{''} > 2(B_{'''} - A)$ is also essential for the existence of the main polynomial.

Likewise, we see that the fourth partial polynomial must satisfy the conditions (83) over all its domain, and we easily find new corresponding conditions for the existence of the main polynomial.

The story is different for the fifth polynomial. Following the same argument, we must conclude that $B_{'''} - s + B_{IV} - s > B$ and $B_{'''} - s + B_{V} - s > B_{'}$ must hold over the whole domain of this polynomial and that the same is true for many other conditions in (83). But, for example, the condition $B_{IV} - s + B_V - s > C_{III} - s$ must not necessarily hold. Indeed, the condition involving C_{III}, that is, the condition that x, y and z reach the dimension C_{III} together is explicit. Therefore, the relation between $B_{IV} - s + B_V - s$ and $C_{III} - s$ must not necessarily hold anymore.

Two cases therefore arise: The first, where $B_{IV} - s + B_V - s > C_{III} - s$ until $s = B_{III} - A$ at least, and $B_{IV} - s + B_V - s < C_{III} - s$ before $s = B_{III} - A$. In the first case, the fifth polynomial must still satisfy all the conditions (83). But in the second case, the condition that would lead to $B' + B_{III} + C_{III} - s > 2C$ changes into $B' + B_{III} + B_{IV} + B_V - 2s > 2C$. Thus, the condition $B' + B_{III} + B_{IV} + B_V - 2(B_{III} - A) > 2C$ becomes one of the main conditions for the existence of the main polynomial, one of whose characteristics is $B_{IV} - s + B_V - s < C_{III} - s$, that is, $B_{IV} + B_V - B_{III} + A < C_{III}$. Thus, for the fifth polynomial, we have either $B_{IV} + B_V - B_{III} + A > C_{III}$ or $B_{IV} + B_V - B_{III} + A < C_{III}$. In the first case, $B' + B_{III} + C_{III} - B_{III} + A > 2C$ is one of the essential conditions for the existence of the main polynomial; in the second case, the condition $B' + B_{III} + B_{IV} + B_V - 2(B_{III} - A) > 2C$ will be essential for the existence of the main polynomial.

Likewise, we see that we must have $B_{III} + B_{IV} - 2s > B$ throughout the domain of the sixth polynomial. However, the conditions $B_{III} - s + B_V - s > C_{II} - s$ and $B_{IV} - s + B_V - s > C_{III} - s$ do not hold. This leads us to four new cases, whose characteristics are easy to determine, as well as the conditions that result from them, and decide the existence of the main polynomial.

In the seventh polynomial, none of the conditions $B_{III} - s + B_{IV} - s > C' - s$, $B_{III} - s + B_V - s > C_{II} - s$, $B_{IV} - s + B_V - s > C_{III} - s$ is essential over the domain of this polynomial. We therefore obtain all eight cases that can be provided by the comparison of those three inequalities. Thus, we determine the characteristics and conditions resulting for the existence of the main polynomial using an argument similar to that above.

For example, consider the case when we simultaneously have $B + B_{IV} - s < C_{III}$; $B' + B_V - s < C_{III}$; $B_{IV} + B_V - s < C_{III}$; the characteristics of the polynomial are $B + B_{IV} - E + C < C_{III}$; $B' + B_{III} + B_V - E + C < C_{III}$; $B_{IV} + B_V - E + C < C_{III}$. One of the essential conditions for the existence of the main polynomial is $B_{III} - s + B_{IV} - s + B_{III} - s + B_V - s + B_{IV} - s + B_V - s > 2C$ or $B_{III} + B_{IV} + B_V - 3s > C$, that is, $\underset{\prime\prime\prime}{B} + \underset{IV}{B} + \underset{V}{B} - 3(E - C) > C$.

Considering the eighth polynomial, we will be able to make all assumptions that are compatible with $s < \underset{\prime\prime\prime}{A}$.

We therefore see that as early as the fifth polynomial, the expression for the number of terms does not immediately belong to any of the eight forms presented in (92 ff). But we will always be able to deduct it from one of these forms, by remembering what was said in (101).

(153.) It therefore remains to discuss the nature of the terms to be summed and about the way to compute their sum.

Following our argument, and thinking about the different combinations of the exponents A, $\underset{\prime}{A}$, $\underset{\prime\prime}{A}$; B, $\underset{\prime}{B}$, $\underset{\prime\prime}{B}$; C; T, in the eight forms given in (92 ff) and about those that can be inferred from the cases mentioned in (101), we see that, in addition to the terms of the form $N(u\ldots n-1)^{P-s}$ and $N(u\ldots n-1)^{P+s}$, $N(u)^{Q+Rs} \times N(u\ldots n-1)^{P\pm s}$, that we have already taught to sum in (69 ff), we also encounter the new following forms: $N(u\ldots n-1)^{P\pm 2s}$, $N(u\ldots n-1)^{P\pm 3s}$, etc., $N(u\ldots n-1)^{\frac{P\pm s}{2}}$, etc., $N(u)^{Q+Rs} \times N(u\ldots n-2)^{P\pm 2s}$, $N(u)^{Q+Rs} \times N(u\ldots n-2)^{\frac{P\pm s}{2}}$, etc. In other similar polynomials, we will encounter, in general, terms of the form

$$N(u\ldots p)^{A'+B's} \times N(u\ldots q)^{\frac{P+Qs}{k}}.$$

(154.) Our object is not to give a detailed theory about computing the sum of such quantities. It is only to show the way to do so and we will limit ourselves to showing how to compute the sum of $N(u\ldots n-1)^{P+2s}$, $N(u\ldots n-1)^{P-2s}$, $N(u\ldots n-1)^{\frac{P-s}{2}}$, $N(u\ldots n-1)^{\frac{P+s}{2}}$. Concerning $N(u)^{Q+Rs} \times N(u\ldots n-2)^{\frac{P\pm s}{2}}$, or even $N(u\ldots 2)^{Q+Rs} \times N(u\ldots n-3)^{\frac{P\pm s}{2}}$, or $N(u\ldots 3)^{Q+Rs} \times N(u\ldots n-4)^{\frac{P\pm s}{2}}$, etc., we will always be able to recast the computation of their sum to that of $N(u\ldots n-1)^{\frac{P\pm s}{2}}$, by means of the following example.

(155.) Assume we have, for example, $N(u\ldots 2)^{Q+3s} \times N(u\ldots n-2)^{\frac{P-s}{2}}$. We know (35) that $N(u\ldots 2)^{Q+3s} = \frac{(Q+3s+1).(Q+3s+2)}{2}$. We assume that the latter takes the form

$$A + B.\frac{P-s}{2} + C.\left(\frac{P-s}{2} - 1\right).\left(\frac{P-s}{2}\right).$$

By performing the required multiplications and comparing terms to terms the coefficients in s on each side of the equality, we will easily get A, B and C. Considering these quantities to be known, the quantity $N(u\ldots 2)^{Q+3s} \times N(u\ldots n-2)^{\frac{P-s}{2}}$ becomes

$$A \times N(u\ldots n-2)^{\frac{P-s}{2}} + B.\left(\frac{P-s}{2}\right).N(u\ldots n-2)^{\frac{P-s}{2}}$$
$$+ C.\left(\frac{P-s}{2} - 1\right).\left(\frac{P-s}{2}\right).N(u\ldots n-2)^{\frac{P-s}{2}}.$$

However,

$$\frac{P-s}{2}N(u\ldots n-2)^{\frac{P-s}{2}} = (n-1) \times N(u\ldots n-1)^{\frac{P-s}{2}-1}$$
$$= (n-1) \times N(u\ldots n-1)^{\frac{P-2-s}{2}},$$

which is always of the form $N(u\ldots n-1)^{\frac{P-s}{2}}$. Thus, the sum can be computed using the same methods.

Likewise

$$\left(\tfrac{P-s}{2}-1\right).\left(\tfrac{P-s}{2}\right).N(u\ldots n-2)^{\frac{P-s}{2}}$$
$$= (n-1)n \times N(u\ldots n)^{\frac{P-s}{2}-2} = (n-1)n \times N(u\ldots n)^{\frac{P-4-s}{2}},$$

which is again of the form $N(u\ldots n)^{\frac{P-s}{2}}$.

We therefore see how we are always able to recast $N(u\ldots p)^{A'+B's} \times N(u\ldots q)^{\frac{P+Qs}{k}}$ to the form $N(u\ldots q)^{\frac{P+Qs}{k}}$.

We only have to consider terms of the form $N(u\ldots n-1)^{P+Qs}$ and $N(u\ldots n-1)^{\frac{P+Qs}{k}}$. Let us illustrate how to do so by focusing our attention on $N(u\ldots n-1)^{P-2s}$, $N(u\ldots n-1)^{P+2s}$, $N(u\ldots n-1)^{\frac{P-s}{2}}$, and $N(u\ldots n-1)^{\frac{P+s}{2}}$. This will show how to proceed for any other value of Q and k.

(156.) To compute the sum of the quantities of the form[3] $N(u\ldots n-1)^{P-2s}$, I differentiate $N(u\ldots n)^{P-2s-1}$, by varying s by the quantity -1, and I obtain

$$N(u\ldots n)^{P-2s-1} - N(u\ldots n)^{P-2s+1}$$

$$= \frac{\left[\begin{array}{l}(P-2s).(P-2s+1).(P-2s+2)\ldots(P-2s+n-1)\\ -(P-2s+2).(P-2s+3).(P-2s+4)\ldots(P-2s+n+1)\end{array}\right]}{1.2.3\ldots n}$$

$$= \frac{(P-2s+2).(P-2s+3)\ldots(P-2s+n-1)}{1.2.3\ldots(n-2)}$$
$$\times \frac{(P-2s).(P-2s+1)-(P-2s+n).(P-2s+n+1)}{(n-1)n}$$

$$= \frac{(P-2s+2).(P-2s+3)\ldots(P-2s+n-1)}{1.2.3\ldots(n-2)}$$
$$\times \left(\frac{-2n(P-2s+1)-n.(n-1)}{n(n-1)}\right)$$

$$= \frac{-2(P-2s+1).(P-2s+2).(P-2s+3)\ldots(P-2s+n-1)}{1.2.3\ldots(n-2).(n-1)}$$
$$- \frac{(P-2s+2).(P-2s+3)\ldots(P-2s+n-1)}{1.2.3\ldots(n-2)}$$

$$= -2N(u\ldots n-1)^{P-2s} - N(u\ldots n-2)^{P-2s+1}.$$

Therefore,

$$dN(u\ldots n)^{P-2s-1} = -2N(u\ldots n-1)^{P-2s} - N(u\ldots n-2)^{P-2s+1}.$$

Thus

$$\int N(u\ldots n-1)^{P-2s} = -\frac{1}{2}N(u\ldots n)^{P-2s-1} - \frac{1}{2}\int N(u\ldots n-2)^{P-2s+1}.$$

For the same reason

$$\int N(u\ldots n-2)^{P-2s+1} = -\frac{1}{2}N(u\ldots n-1)^{P-2s} - \frac{1}{2}\int N(u\ldots n-3)^{P-2s+2},$$

---

[3]I will not insist enough on the fact that $P-2s$ is a positive integer throughout this work; the same is true of $\frac{P\pm s}{2}$ and in general of $\frac{P+Qs}{k}$ in $N(u\ldots n-1)^{\frac{P+Qs}{k}}$.

$$\int N(u\ldots n-3)^{P-2s+2} = -\frac{1}{2}N(u\ldots n-2)^{P-2s+1} - \frac{1}{2}\int N(u\ldots n-4)^{P-2s+3},$$

and so on. Thus,

$$\int N(u\ldots n-1)^{P-2s} = -\frac{1}{2}N(u\ldots n)^{P-2s-1} + \frac{1}{4}N(u\ldots n-1)^{P-2s}$$
$$-\frac{1}{8}N(u\ldots n-2)^{P-2s+1} + \frac{1}{16}N(u\ldots n-3)^{P-2s+2} - \text{ etc.} + C.$$

Thus, if we ask for the sum of this quantity from $s = K$ included to $s = L$ included, with $L > K$, we obtain

$$\int N(u\ldots n-1)^{P-2s} = -\frac{1}{2}N(u\ldots n)^{P-2L-1} + \frac{1}{2}N(u\ldots n)^{P-2K+1}$$
$$+\frac{1}{4}N(u\ldots n-1)^{P-2L} - \frac{1}{4}N(u\ldots n-1)^{P-2K+2} - \frac{1}{8}N(u\ldots n-2)^{P-2L+1}$$
$$+\frac{1}{8}N(u\ldots n-2)^{P-2K+3} + \frac{1}{16}N(u\ldots n-3)^{P-2L+2}$$
$$-\frac{1}{16}N(u\ldots n-3)^{P-2K+4} - \text{ etc.}$$

We can continue this sequence until $n = 0$ included, and observe that, in this case, $N(u\ldots n)^R = 1$ for any R.

(157.) In order to sum quantities of the form $N(u\ldots n-1)^{P+2s}$ from $s = R$ included to $s = L$ included, with $L > R$, we find

$$\int N(u\ldots n-1)^{P+2s} = \frac{1}{2}N(u\ldots n)^{P+2L+1} - \frac{1}{2}N(u\ldots n)^{P+2K-1}$$
$$-\frac{1}{4}N(u\ldots n-1)^{P+2L+2} + \frac{1}{4}N(u\ldots n-1)^{P+2K} + \frac{1}{8}N(u\ldots n-2)^{P+2L+3}$$
$$-\frac{1}{8}N(u\ldots n-2)^{P+2K+1} - \frac{1}{16}N(u\ldots n-3)^{P+2L+4}$$
$$+\frac{1}{16}N(u\ldots n-3)^{P+2K+2} + \text{ etc.}$$

by differentiating $N(u\ldots n)^{P+2s+1}$ and operating and reasoning, as we have previously done.

(158.) If we differentiate $N(u\ldots n)^{P+3s+2}$ using a similar process, we find

$$dN(u\ldots n)^{P+3s+2} = 3N(u\ldots n-1)^{P+3s} + 3N(u\ldots n-2)^{P+3s+1}$$
$$+N(u\ldots n-3)^{P+3s+2},$$

leading us to conclude

$$\int N(u\ldots n-1)^{P+3s} = \frac{1}{3}N(u\ldots n)^{P+3s+2} - \int N(u\ldots n-2)^{P+3s+1}$$
$$-\frac{1}{3}\int N(u\ldots n-3)^{P+3s+2}.$$

Likewise,

$$\int N(u\ldots n-2)^{P+3s+1} = \frac{1}{3}N(u\ldots n-1)^{P+3s+3} - \int N(u\ldots n-3)^{P+3s+2}$$
$$-\frac{1}{3}\int N(u\ldots n-4)^{P+3s+3},$$

$$\int N(u\ldots n-3)^{P+3s+2} = \frac{1}{3}N(u\ldots n-2)^{P+3s+4} - \int N(u\ldots n-4)^{P+3s+3}$$
$$-\frac{1}{3}\int N(u\ldots n-5)^{P+3s+4},$$

and so on. From there, it is easy to conclude about the value of $\int N(u\ldots n-1)^{P+3s}$.

(159.) We now see how to proceed to obtain the value of $\int N(u\ldots n-1)^{P-3s}$ and in general to obtain that of $\int N(u\ldots n-1)^{P+Qs}$.

(160.) Let us now focus on quantities of the form $N(u\ldots n-1)^{\frac{P+s}{2}}$.

From (156), we remark that whenever quantities of this form arise, $P + s$ can always take on two possible values, generally represented by $P + r + s$, where r is equal to 0 or 1, depending upon whether $P + s$ is even or odd. We will therefore assume we must compute the sum of $N(u \ldots n - 1)^{\frac{P+r+s}{2}}$. Since s varies by units of 1, we must consider the parts $N(u \ldots n-1)^{\frac{P+s}{2}}$ and $N(u \ldots n-1)^{\frac{P+r+s}{2}}$ separately to compute the value of $\int N(u \ldots n-1)^{\frac{P+r+s}{2}}$. The first part expresses all quantities $N(u \ldots n-1)^{\frac{P+r+3s}{2}}$ in which $P + s$ is even, and the second expresses all those where $P + s$ is odd.

Therefore, we must compute the sum of $N(u \ldots n - 1)^{\frac{P+s}{2}}$, when s varies by increments of 2, and likewise to compute the sum of $N(u \ldots n - 1)^{\frac{P+r+s}{2}}$ when s varies by increments of 2. Bringing those two forms back together, we obtain the total value of $\int N(u \ldots n - 1)^{\frac{P+r+s}{2}}$ when s varies by increments of 1.

It is easy to get $\int N(u \ldots n - 1)^{\frac{P+r+s}{2}}$ from $\int N(u \ldots n-1)^{\frac{P+s}{2}}$ by simply changing P into $P + r$. We will therefore concentrate on $\int N(u \ldots n-1)^{\frac{P+s}{2}}$ only.

In order to compute the sum of $N(u \ldots n - 1)^{\frac{P+s}{2}}$, when s varies in increments of 2, I notice that by defining $z = \frac{P+s}{2}$, z varies by increments of 1 when s varies by increments of 2. Thus, the question reduces to computing the sum of $N(u \ldots n - 1)^z$ when z varies by increments of 1. This sum is known to be $N(u \ldots n)^z$, that is, $N(u \ldots n)^{\frac{P+s}{2}}$. Likewise, we find

$$\int N(u \ldots n - 1)^{\frac{P+r+s}{2}} = N(u \ldots n)^{\frac{P+r+s}{2}}.$$

Therefore,

$$\int N(u \ldots n - 1)^{\frac{P+s}{2}} = N(u \ldots n)^{\frac{P+s}{2}} + N(u \ldots n)^{\frac{P+r+s}{2}} + C.$$

Therefore two cases arise when computing this sum from $s = K$ included to $s = L$ included, with $L > K$: $P + L$ is even or $P + L$ is odd. In the first case, the sum from s being an arbitrary number up to $s = L$ is $N(u \ldots n)^{\frac{P+L}{2}} + N(u \ldots n)^{\frac{P+L+r-1}{2}} + C$, that is, $2N(u \ldots n)^{\frac{P+L}{2}} + C$.

If instead $P + L$ is odd, the sum from an arbitrary number up to $s = L$ is

$$N(u \ldots n)^{\frac{P+r+L}{2}} + N(u \ldots n)^{\frac{P+L-1}{2}} + C;$$

that is,

$$N(u \ldots n)^{\frac{P+L+1}{2}} + N(u \ldots n)^{\frac{P+L-1}{2}} + C.$$

Following the same argument, the sum from any number to $s = K - 1$ is $2N(u \ldots n)^{\frac{P+K-1}{2}} + C$ if $P + K - 1$ is even. It is $N(u \ldots n)^{\frac{P+K}{2}} + N(u \ldots n)^{\frac{P+K-2}{2}}$ if $P + K - 1$ is odd.

Therefore, four cases can arise when computing the sum:

If $P + L$ and $P + K - 1$ are both even, we have

$$\int N(u \ldots n - 1)^{\frac{P+s}{2}} = 2N(u \ldots n)^{\frac{P+L}{2}} - 2N(u \ldots n)^{\frac{P+K-1}{2}}.$$

If $P + L$ is even and $P + K - 1$ is odd, we have

$$\int N(u \ldots n - 1)^{\frac{P+s}{2}} = 2N(u \ldots n)^{\frac{P+L}{2}} - N(u \ldots n)^{\frac{P+K}{2}} - 2N(u \ldots n)^{\frac{P+K-2}{2}}.$$

If $P + L$ is odd and $P + K - 1$ is even, we obtain

$$\int N(u \ldots n-1)^{\frac{P+s}{2}} = N(u \ldots n)^{\frac{P+L+1}{2}} + N(u \ldots n)^{\frac{P+L-1}{2}}$$
$$-2N(u \ldots n)^{\frac{P+K-1}{2}}.$$

Finally, if $P + L$ and $P + K - 1$ are both odd, we have

$$\int N(u \ldots n-1)^{\frac{P+s}{2}} = N(u \ldots n)^{\frac{P+L+1}{2}} + N(u \ldots n)^{\frac{P+L-1}{2}}$$
$$-N(u \ldots n)^{\frac{P+K}{2}} - N(u \ldots n)^{\frac{P+K-2}{2}}.$$

It is quite easy to see how to compute the sum of $N(u \ldots n-1)^{\frac{P-s}{2}}$. We will not spend time on it.

(161.) From (160), if we had to compute the sum of $N(u \ldots n-1)^{\frac{P+3s}{2}}$, we would compute the sum of $N(u \ldots n-1)^{\frac{P+3s}{2}}$ where s varies by increments of 2 for even values of $P + 3s$, and the sum of $N(u \ldots n-1)^{\frac{P+r+3s}{2}}$ where s varies by increments of 2 for the odd values of $P + 3s$.

Let us examine how to compute the sum of $N(u \ldots n-1)^{\frac{P+3s}{2}}$ where s varies by increments of 2. Defining $z = \frac{P+3s}{2}$, it is clear that when s varies by increments of 2, z varies by increments of 3. We therefore have to compute the sum of $N(u \ldots n-1)^z$ when z varies by increments of 3. In other terms, writing $z = Q+3z'$, the problem is to compute the sum of $N(u \ldots n-1)^{Q+3z'}$ when z' varies by increments of 1. This is easy from what was said in (158).

(162.) The constant Q we have introduced serves the following purposes: Since we have defined $\frac{P+3s}{2} = z$ and $z = Q+3z'$, we have $\frac{P+3s}{2} = Q+3z'$, where z' is a positive integer. From there we find $z' = \frac{P+3s-2Q}{6}$. We must therefore adjust Q so that $P + 3s - 2Q$ can be divided by 6 when we reach the extreme values of s, $K - 1$ and L; this task is easy.

(163.) We therefore see how to compute the sum of $N(u \ldots n-1)^{\frac{P+Qs}{2}}$ and in general $N(u \ldots n-1)^{\frac{P+Qs}{k}}$.

Indeed, consider, for example, $N(u \ldots n-1)^{\frac{P+s}{3}}$. Since $\frac{P+s}{3}$ must be an integer, this expression really is $\frac{P+r+s}{3}$, where r is 0, 1 or 2, depending on whether $P + s$ exceeds a multiple of 3 by 0, 1 or 2. We therefore must compute the sum involving $\frac{P+s}{3}$ when s varies by increments of 3, then the sum involving $\frac{P+s+1}{3}$ when s varies by increments of 3, then finally the sum involving $\frac{P+s+2}{3}$ when s varies by increments of 3, and finally by bringing these sums together. However, defining $z = \frac{P+s}{3}$, the question reduces to computing the sum of $N(u \ldots n-1)^z$ when z varies by increments of 1. Then we substitute $P + 1$ and $P + 2$ instead of P in this sum.

We also see that the total sum will take several different possible values, depending on whether the quantities $P+L$, $P+L+1$, $P+L+2$, $P+K-1$, $P + K$, $P + K + 1$ exceed a multiple of 3 by an amount of 0, 1 or 2. But since we already gave an example in (160), we will not go into further detail.

Conclusion about first-order incomplete equations

(164.) First-order incomplete equations are those such that $p < n$ out of n unknowns satisfy the following conditions:

1. Each of these p unknowns connot exceed a given degree. This degree can be different or the same for each unknown.

2. Combinations of two of these cannot reach beyond a given dimension. These dimensions may be the same or different for each pair of variables.

3. Combinations of three of these unknowns cannot reach beyond a given dimension. These dimensions may be the same or different for each 3-uple of variables.

4. Combinations of four of these unknowns cannot reach beyond a given dimension. These dimensions may be the same or different for each 4-uple of variables.

5. This continues until the combination of p of these variables, which cannot reach beyond a given dimension.

6. Finally, the other unknowns can reach, alone or in combination with any of the previous variables, all possible dimensions up to the degree of the polynomial.

Since the number of these equations is the same as the number of unknowns they contain, it is always possible to determine the degree of the final equation resulting from the elimination of $n-1$ of these unknowns.

Indeed, it is easy to see that: (i) The most general form one might adopt for the polynomial multipler is the very same form of these equations; (ii) the form of each of the polynomials whose number of terms determines how many terms can be cancelled in the polynomial multiplier and in the product equation have the same form as that of these equations.

Moreover, we will make sure that all these different polynomials share the same nature, following the same argument as that already used in (105).

We have indicated how to compute the number of terms of any first-order polynomial. Since this allows us to also compute the number of terms that may be cancelled both in the polynomial multiplier and in the product equation, we will always be able to compute the number of the remaining terms. Thus, we will also be able to determine the degree of the final equation independently from the computation of the polynomial multiplier itself.

But, remembering our observations (117) about incomplete first-order equations with only three unknowns, and remembering we had found eight different expressions for the number of terms of this type of polynomials and therefore for the degree of the final equations, we must expect a prodigious number of possible expressions as the complexity of the polynomials grows. We can get an idea of this by looking at the discussion presented in (150 ff).

However, looking again at (117) we can conclude that it is always possible to find which expression for the degree of the final equation is the proper one, and the conditions characterizing all possible cases these equations may satisfy.

We also see that computing all possible expressions for the final equation resulting from an arbitrary number of equations would represent a prodigious amount of work, even for four unknowns. However, if all equations share the same nature, then we never have to look at all possible expressions to determine the degree of the final equation. This expression results directly from the differentiation of the expression for the number of terms of a polynomial with the same nature as these equations. When the equations do not share the same nature, we can safely determine the degree of the final equation only by inspecting all possible forms it can take and all attendant admissibility conditions. This is a similar exercise as that presented in (118). It is, however, infinitely more lengthy.

By the way, this complexity is inherent to the problem at hand: It is impossible to reduce the number of possible expressions present in our method as it is impossible to reduce the number of possible combinations of four letters below 24. We believe we have done everything possible by giving a way to know all possible expressions and, among all these expressions, the proper answer to the question. To expect more than that would be to expect the impossible.

SECTION III
About incomplete polynomials and second-, third-,
fourth-, etc., order incomplete equations

(165.) No matter how general the polynomials we have dealt with in the previous section are, they do not yet cover all possible polynomials and all possible equations. Their form is not as general as necessary for us to claim that we can determine the lowest degree of the final equation for any algebraic equation.

It is not enough to specify the highest dimensions reached by the unknowns, taken either alone or in combinations of 2, 3, etc., to capture all the variations that may influence the degree of the final equation. While these have a very significant influence on the degree of the final equation, there are many other equations for which these dimensions would only give an upper bound to the degree of the final equation.

In addition to the kind of polynomials considered until now, we can imagine similar polynomials, but whose properties would hold only over a certain number of consecutive dimensions of the equation, and which would be succeeded by similar patterns, which would hold again over another finite number of consecutive dimensions of the equation, and so on.

(166.) We now try to explain our concept of incomplete polynomials of a given order. Beginning with a first-order incomplete polynomial, we remove a number of its terms from a given dimension of this polynomial to its highest dimension. This polynomial then looks like an incomplete first-order polynomial when considered from that dimension until the highest degree, but its coefficients are different from those suggested by the lower dimensional terms of this polynomial.

For example, consider the polynomial $(x^A, y^{\overset{A}{\prime}})^T$. Assume that past the dimension $\underset{\prime}{T} < T$, we suppress all terms where x exceeds the degree $A' < A$, and all terms where y exceeds the degree $\underset{\prime}{A'} < \underset{\prime}{A}$; we get a second-order polynomial in two unknowns. We represent this polynomial as

$$\begin{pmatrix} x^{A'}, y^{\overset{A'}{\prime}} \\ x^A, y^{\overset{A}{\prime}} \end{pmatrix}^{\overset{T}{\underset{\prime}{T}}}.$$

Likewise, consider the polynomial

$$[(x^A, y^{\overset{A}{\prime}})^B, (x^A, z^{\overset{A}{\prime\prime}})^{\overset{B}{\prime}}, (y^{\overset{A}{\prime}}, z^{\overset{A}{\prime\prime}})^{\overset{B}{\prime\prime}}]^T.$$

Assume we suppress all terms where x, y and z exceed the degrees A', $\underset{\prime}{A'}$, $\underset{\prime\prime}{A'}$ (smaller than A, $\underset{\prime}{A}$, $\underset{\prime\prime}{A}$, respectively) for all terms whose dimension is greater than $\underset{\prime}{T} < T$. Assume, moreover, that we suppress all terms where pairwise combinations of x, y and z exceed the degrees B', $\underset{\prime}{B'}$, $\underset{\prime\prime}{B'}$ (smaller than B,

B, B, respectively) for all terms whose dimension is greater than $T < T$.
The result is an incomplete, second-order polynomial in three unknowns, which we represent as

$$
\left[
\left(
\begin{matrix}
x^{A'},\, y^{A'} \\
x^{A},\, y^{A}
\end{matrix}
\right)^{B'}_{B},\;
\left(
\begin{matrix}
x^{A'},\, z^{A'} \\
x^{A},\, z^{A}
\end{matrix}
\right)^{B'}_{B},\;
\left(
\begin{matrix}
y^{A'},\, z^{A'} \\
y^{A},\, z^{A}
\end{matrix}
\right)^{B'}_{B}
\right]^{T}_{T}.
$$

Assume that we suppress all terms where the degree of x exceeds $A'' < A'$ and the degree of y exceeds $A'' < A'$ in terms whose dimension T is smaller than T and greater than T in the polynomial

$$
\left(
\begin{matrix}
x^{A},\, y^{A'} \\
x^{A},\, y^{A}
\end{matrix}
\right)^{T}_{T}.
$$

We thus get a third-order, incomplete polynomial, which we will represent by

$$
\left(
\begin{matrix}
x^{A''}\, y^{A''} \\
x^{A'},\, y^{A'} \\
x^{A},\, y^{A}
\end{matrix}
\right)^{T}_{T}{}^{T'}_{T''}
$$

or $\left(x^{A'',A',A},\, y^{A'',A',A}\right)^{T,T',T''}$.

And in general, we will represent an incomplete polynomial of arbitrary order as $\left(u^{A,\bar{A},\bar{\bar{A}},\bar{\bar{\bar{A}}},\text{ etc.} \ldots n}\right)^{T,\bar{T},\bar{\bar{T}},\bar{\bar{\bar{T}}},\text{ etc.}}$, assuming the polynomial is incomplete with respect to variables taken alone. In this expression, A, $\bar{A}$, $\bar{\bar{A}}$, $\bar{\bar{\bar{A}}}$, etc. represent the various largest exponents of the same unknown in the intervals between the dimensions $T, \bar{T}, \bar{\bar{T}}, \bar{\bar{\bar{T}}}$, etc. If this polynomial is also incomplete with respect to unknowns taken in pairs and combinations of 3, we will use the notation

$$
\left[
\left(x^{A,\bar{A}},\, y^{A,\bar{A}}\right)^{B,\bar{B}},\;
\left(x^{A,\bar{A}},\, z^{A,\bar{A}}\right)^{B,\bar{B}},\;
\left(y^{A,\bar{A}},\, z^{A,\bar{A}}\right)^{B,\bar{B}}
\right]^{T,\bar{T}}
$$

instead of the notation

$$
\left[
\left(x^{A'}, y^{A'}_{\prime} \right)^{\substack{B' \\ B}} ,
\left(x^{A}_{\prime}, z^{A'}_{\prime\prime} \right)^{\substack{B'_{\prime} \\ B}} ,
\left(y^{A'}_{\prime}, z^{A'}_{\prime\prime} \right)^{\substack{B'_{\prime\prime} \\ B_{\prime\prime}}}
\right]^{\substack{T \\ T_{\prime}}} .
$$

From the above examples we see what we mean by polynomials of various orders.

(167.) Polynomials whose order exceeds the first degree are far different from first-order polynomials. In incomplete first-order equations, the polynomial multiplier and the product equation are always polynomials whose order is the same as these equations. The same is true of all polynomials whose number of terms contribute to giving the expression for the degree of the final equation.

In incomplete equations whose order exceeds 1, the polynomial multiplier, the product equation, and all polynomials whose number of terms contribute to the expression of the degree of the final equation are polynomials whose orders are different.

(168.) Following this observation, we can anticipate that the form of the polynomial multiplier is not as easily determined as for incomplete equations of order 1. It is not always true that the chosen polynomial multipliers always lead to an expression for the degree of the final equation that can be written as an exact differential, whose order equals that of the number of equations. When this occurs, the form of the polynomial multiplier does not provide information about the degree of the final equation. Thus we could determine the degree of the multiplier polynomial only by picking a polynomial multiplier whose order is indefinite. But we must remark that looking for this polynomial based solely on the requirement to satisfy the condition that the degree of the final equation be an exact differential equal to the number of equations would lead to an infinite amount of work. We would have to consider an infinite number of cases and subdivisions into different subcases as they arise from the various inequalities that exist among the exponents of the equations.

We therefore should not expect a general treatment of the question, as we were able to do with incomplete, first-order equations. Even if we were not concerned about the amount of work we just mentioned, we would not be able to display the prodigious number of quantities involved.

(169.) We will therefore limit ourselves to present the method, by using only the simplest possible polynomial multiplier. We will also limit our investigation to equations in two and three unknowns. We will already see that the polynomial is not sufficient to give a general expression for the degree of the final equation in all possible cases; consequently we will have to use a polynomial multipler with a larger number of undetermined exponents in cases other than those we will present.

As for equations in two unknowns, the simplest possible polynomial will always work.

By the way, solving for the degree of the final equation in the general case of arbitrary order equations is an immense task, though this is not necessarily true for specific equations. Only looking for a general expression would require all this work. We will see in the Second Book on the elimination method that we can always reach the final equation whose order is as low as possible for specific systems of equations.

About the number of terms in incomplete
polynomials of arbitrary order

(170.) In order to limit notational complexity, we will use the expression $(u^{A,\bar{A},\bar{\bar{A}},\bar{\bar{\bar{A}}}}, \text{etc.} \ldots n)^{T,\bar{T},\bar{\bar{T}},\bar{\bar{\bar{T}}}}$, etc. only to represent a polynomial with arbitrary order, irrespective of whether the degree of variables is bounded for single variables only or for the combination of any two such variables, etc.

PROBLEM XXIV
We ask for the value of $N(u^{A,\bar{A},\bar{\bar{A}},\bar{\bar{\bar{A}}}}, \text{etc.} \ldots n)^{T,\bar{T},\bar{\bar{T}},\bar{\bar{\bar{T}}}}$, etc.

(171.) Let us assume first that we only have to deal with $(u^{A,\bar{A}} \ldots n)^{T,\bar{T}}$. If we compare this polynomial with $(u^{A} \ldots n)^{T}$, we see that, from T to $\bar{T}$, the following number of terms is missing:

$$[N(u^{\bar{A}} \ldots n)^{T} - N(u^{\bar{A}} \ldots n)^{\bar{T}}] - [N(u^{A} \ldots n)^{T} - N(u^{A} \ldots n)^{\bar{T}}]$$
$$= ddN(u^{\bar{A}} \ldots n)^{T} \ldots \left(\begin{array}{cc} T & \bar{A} \\ T - \bar{T}, 0 & 0, \bar{A} - A \end{array} \right).$$

Thus

$$N(u^{A,\bar{A}} \ldots n)^{T} = N(u^{\bar{A}} \ldots n)^{T}$$
$$-ddN(u^{\bar{A}} \ldots n)^{T} \ldots \left(\begin{array}{cc} T & \bar{A} \\ T - \bar{T}, 0 & 0, \bar{A} - A \end{array} \right).$$

Consider now $N(u^{A,\bar{A},\bar{\bar{A}}} \ldots n)^{T,\bar{T},\bar{\bar{T}}}$. Comparing again with the polynomial $(u^{\bar{A}} \ldots n)^{T}$, we see that:

1. From T to $\bar{T}$, the number of missing terms is

$$ddN(u^{\bar{\bar{A}}} \ldots n)^{T} \ldots \left(\begin{array}{cc} T & \bar{\bar{A}} \\ T - \bar{T}, 0 & 0, \bar{\bar{A}} - A \end{array} \right).$$

2. From $\bar{T}$ to $\bar{\bar{T}}$, the number of missing terms is

$$ddN(u^{\bar{\bar{A}}} \ldots n)^{\bar{T}} \ldots \left(\begin{array}{cc} \bar{T} & \bar{\bar{A}} \\ \bar{T} - \bar{\bar{T}}, 0 & 0, \bar{\bar{A}} - \bar{A} \end{array} \right).$$

Thus

$$N(u^{A,\bar{A},\bar{\bar{A}}}\ldots n)^{T,\bar{T},\bar{\bar{T}}} = N(u^{\bar{A}}\ldots n)^T$$
$$-ddN(u^{\bar{A}}\ldots n)^T\ldots \begin{pmatrix} T & \bar{A} \\ T-\bar{T},0 & 0,\bar{A}-A \end{pmatrix}$$
$$-ddN(u^{\bar{A}}\ldots n)^{\bar{T}}\ldots \begin{pmatrix} \bar{T} & \bar{A} \\ \bar{T}-\bar{\bar{T}},0 & 0,\bar{A}-\bar{A} \end{pmatrix}.$$

Following the same argument, we find

$$N(u^{A,\bar{A},\bar{\bar{A}},\bar{\bar{\bar{A}}}}\ldots n)^{T,\bar{T},\bar{\bar{T}},\bar{\bar{\bar{T}}}} = N(u^{\bar{A}}\ldots n)^T$$
$$-ddN(u^{\bar{\bar{A}}}\ldots n)^T\ldots \begin{pmatrix} T & \bar{\bar{A}} \\ T-\bar{T},0 & 0,\bar{\bar{A}}-A \end{pmatrix}$$
$$-ddN(u^{\bar{\bar{A}}}\ldots n)^{\bar{T}}\ldots \begin{pmatrix} \bar{T} & \bar{\bar{A}} \\ \bar{T}-\bar{\bar{T}},0 & 0,\bar{\bar{A}}-\bar{A} \end{pmatrix}$$
$$-ddN(u^{\bar{\bar{A}}}\ldots n)^{\bar{\bar{T}}}\ldots \begin{pmatrix} \bar{\bar{T}} & \bar{\bar{A}} \\ \bar{\bar{T}}-\bar{\bar{\bar{T}}},0 & 0,\bar{\bar{A}}-\bar{\bar{A}} \end{pmatrix},$$

and so on.

About the form of the polynomial multiplier and of the polynomials whose number of terms impact the degree of the final equation resulting from a given number of incomplete equations with arbitrary order

(172.) Consider the polynomial multiplier, the product equation, and the polynomials whose number of terms determine the number of terms that may be cancelled both from the polynomial multiplier and in the product equation. Since all these polynomials and equations have different orders, it would require quite a bit more work than Section II, if we wanted to conclude about the form of the polynomials whose number of terms enters into the expression of the final equation from an inderminate polynomial multiplier.

But thinking further about the polynomials whose number of terms expresses the number of terms that may be cancelled in the polynomial multiplier and in the product equation of Section II, we believe we can find a general approach to determine the principal characteristics of such polynomials for incomplete equations of arbitrary order.

Let us recall the following ideas in order to make our point.

(173.) Consider an arbitrary number n of incomplete, first-order equations, represented by $(u^a\ldots n)^t = 0$, $(u^{a'}\ldots n)^{t'} = 0$, $(u^{a''}\ldots n)^{t''} = 0$, etc. and let us write as $(u^A\ldots n)^T$ the polynomial multiplier of the first equation.

Using the second equation only, we have seen that we can cancel

$$N(u^{A-a'},\ldots n)^{T-t'}$$

terms from the polynomial multiplier, and $N(u^{A+a-a'},\ldots n)^{T+t-t'}$ terms in the product equation.

Using the second and third equations alone, we have seen we can cancel

$$N(u^{A-a'}\ldots n)^{T-t'} - N(u^{A-a'-a''}\ldots n)^{T-t'-t''} + N(u^{A-a''}\ldots n)^{T-t''}$$

terms in the polynomial multiplier and

$$N(u^{A+a-a'} \ldots n)^{T+t-t'} - N(u^{A+a-a'-a''} \ldots n)^{T+t-t'-t''}$$
$$+N(u^{A+a-a''} \ldots n)^{T+t-t''}$$

terms in the product equation. Similar expressions follow (see 60 ff).

Let us define $A - a' - a'' = A'$ and $T - t' - t'' = T'$. Then, the polynomial multiplier is $(u^{A'+a'+a''} \ldots n)^{T'+t'+t''}$, and the product equation is $(u^{A'+a+a'+a''} \ldots n)^{T+t+t'+t''}$. Using the second and third equations alone, we can cancel

$$N(u^{A'+a''} \ldots n)^{T'+t''} - N(u^{A'} \ldots n)^{T'} + N(u^{A'+a'} \ldots n)^{T'+t'}$$

terms in the polynomial multiplier and

$$N(u^{A'+a+a''} \ldots n)^{T'+t+t''} - N(u^{A'+a} \ldots n)^{T'+t} + N(u^{A'+a+a'} \ldots n)^{T'+t+t'}$$

terms in the product equation.

(174.) If we successively multiply an arbitrary first-order polynomial $(u^A \ldots n)^T$ by all the proposed first-order equations, we obtain a polynomial $(u^{A'+a+a'+a''+a'''+ \text{ etc.}} \ldots n)^{T'+t+t'+t''+t'''+ \text{ etc.}}$ that we can regard as the generator of all other polynomials that may arise in the question being asked.

Indeed:

1. The form of this polynomial is that of the product equation.

2. The polynomial $(u^{A+a''+a'''+ \text{ etc.}} \ldots n)^{T+t''+t'''+ \text{ etc.}}$ is the form of the polynomial multiplier.

3. $N(u^{A+a''+a'''+ \text{ etc.}} \ldots n)^{T+t''+t'''+ \text{ etc.}}$ is the number of terms that may be cancelled in the polynomial multiplier using the second equation alone. Likewise, $N(u^{A'+a'+a''+ \text{ etc.}} \ldots n)^{T+t''+t'''+ \text{ etc.}}$ is the number of terms that may be cancelled in the polynomial multiplier using the third equation alone. The quantity

$$N(u^{A+a''+a'''+ \text{ etc.}} \ldots n)^{T+t''+t'''}, \text{ etc.}$$
$$-N(u^{A+a'''}, \text{ etc.} \ldots n)^{T+t'''}, \text{ etc.}$$
$$+N(u^{A+a'+a'''}, \text{ etc.} \ldots n)^{T+t'+t'''}, \text{ etc.}$$

is the number of terms that can be cancelled using the second and third equations.

4. Likewise we see the number of terms that can be cancelled in the product equation is

$$N(u^{A+a+a''+a'''}, \text{ etc.} \ldots n)^{T+t+t''+t'''}, \text{ etc.}$$
$$-N(u^{A+a+a'''}, \text{ etc.} \ldots n)^{T+t+t'''}, \text{ etc.}$$
$$+N(u^{A+a+a'+a'''}, \text{ etc.} \ldots n)^{T+t+t'+t'''}, \text{ etc}.$$

Thus we see, in general, that we can always find the expression for the number of terms that can be cancelled using an arbitrary number of equations.

We propose to follow these observations to deal with higher order incomplete equations. In order not to disrupt the flow of our discourse, we will first introduce a few useful notions about the reduction of the differentials we will encounter.

Useful notions for the reduction of differentials that enter in the expression of the number of terms of a polynomial with arbitrary order

(175.) The expression of the number of terms in an incomplete polynomial whose order is greater than 1 contains, as we have seen in (171), second differences. The variations of these differences are composites of the variety of exponents of the given equations when they are used to seek the degree of the final equation. We are interested in unweaving, among these second differences, which ones are those whose clustering can lead to exact differentials whose order equals that of the number of equations. To do so, it is necessary to decompose these second differences into other second differences whose variation is as close to them as possible. The following notions are introduced in order to facilitate this process.

(176.) Let $F(u)$ be an arbitrary function of u. Consider the quantity $d[F(u)] \ldots \left(\begin{array}{c} u \\ a+b \end{array} \right)$. We want to decompose it into two differentials, where the variation is a in the first differential and b in the second. We get

$$d[F(u)] \ldots \left(\begin{array}{c} u \\ a+b \end{array} \right) = d[F(u)] \ldots \left(\begin{array}{c} u \\ a \end{array} \right) + d[F(u+a)] \ldots \left(\begin{array}{c} u+a \\ b \end{array} \right).$$

Indeed, the variation of $F(u)$, when u becomes $u+a+b$, is composed of the variation of $F(u)$ when u becomes $u+a$ added to the variation of $F(u+a)$ when $u+a$ becomes $u+a+b$.

(177.) For the same reason

$$dd[F(u)] \ldots \left(\begin{array}{c} u \\ a+b, c+d \end{array} \right)$$
$$= dd[F(u)] \ldots \left(\begin{array}{c} u \\ a, c+d \end{array} \right) + dd[F(u+a)] \ldots \left(\begin{array}{c} u+a \\ b, c+d \end{array} \right)$$
$$= dd[F(u)] \ldots \left(\begin{array}{c} u \\ a, c \end{array} \right) + dd[F(u+c)] \ldots \left(\begin{array}{c} u+c \\ a, d \end{array} \right)$$
$$+ dd[F(u+a)] \ldots \left(\begin{array}{c} u+a \\ b, c \end{array} \right) + dd[F(u+a+c)] \ldots \left(\begin{array}{c} u+a+c \\ b, d \end{array} \right).$$

(178.) If, conversely, we had $d[F(u)] \ldots \left(\begin{array}{c} u \\ a-b \end{array} \right)$ we would get

$$d[F(u)] \ldots \left(\begin{array}{c} u \\ a-b \end{array} \right) = d[F(u)] \ldots \left(\begin{array}{c} u \\ a \end{array} \right) + d[F(u+a)] \ldots \left(\begin{array}{c} u+a \\ -b \end{array} \right).$$

(179.) And if we had $d[F(u)] \ldots \left(\begin{array}{c} u \\ -a-b \end{array} \right)$ we would get

$$d[F(u)] \ldots \left(\begin{array}{c} u \\ -a-b \end{array} \right) = d[F(u)] \ldots \left(\begin{array}{c} u \\ -a \end{array} \right) + d[F(u-a)] \ldots \left(\begin{array}{c} u-a \\ -b \end{array} \right).$$

(180.) We will apply these principles to quantities of the form

$$ddN(u^{A+\bar{a}+\bar{a}'}\ldots n)^{T+t+t'}\ldots\left[\begin{array}{cc:cc} T+t+t' & & A+\bar{a}+\bar{a}' \\ -(t'-\bar{t}'),0 & & 0,-(\bar{a}-a+\bar{a}'-a') \end{array}\right],$$

in which the two variations of $T+t+t'$ are $-(t'-\bar{t}')$ and 0. In addition, those of $A+\bar{a}+\bar{a}'$ are 0 and $-(\bar{a}-a+\bar{a}'-a')$. Our objective will be to reduce these differentials to others where there are no other variations than $-(t'-\bar{t}')$, 0, $-(\bar{a}'-a')$, and $-(\bar{a}-a)$. We therefore have

$$ddN(u^{A+\bar{a}+\bar{a}'}\ldots n)^{T+t+t'}\ldots\left[\begin{array}{cc:cc} T+t+t' & & A+\bar{a}+\bar{a}' \\ -(t'-\bar{t}'),0 & & 0,-(\bar{a}-a+\bar{a}'-\bar{a}') \end{array}\right]$$

$$= ddN(u^{A+\bar{a}+\bar{a}'}\ldots n)^{T+t+t'}\ldots\left[\begin{array}{cc:cc} T+t+t' & & A+\bar{a}+\bar{a}' \\ -(t'-\bar{t}'),0 & & 0,-(\bar{a}-a) \end{array}\right]$$

$$+ ddN(u^{A+\bar{a}+\bar{a}'}\ldots n)^{T+t+t'}\ldots\left[\begin{array}{cc:cc} T+t+t' & & A+\bar{a}+\bar{a}' \\ -(t'-\bar{t}'),0 & & 0,-(\bar{a}'-a') \end{array}\right].$$

By the way, we will still present the variations in a positive form, for simplicity's sake, although all those encountered in the differentials turn out to be negative: Since their result must be a differential whose order equals the dimension of the differentiated quantity, this convention will make no difference in the expression of the final value (16).

PROBLEM XXV

Consider $(u^{a,\bar{a}}\ldots n)^{t,\bar{t}}=0$, $(u^{a',\bar{a}'}\ldots n)^{t',\bar{t}'}=0$, $(u^{a'',\bar{a}''}\ldots n)^{t'',\bar{t}''}=0$, etc. n incomplete, second-order equations, each containing the same number n of unknowns. We ask for the degree of the final equation.

(181.) Consider first the case of two equations only. Assuming we have multiplied the second equation by the polynomial $(u^A\ldots n)^T$ (leading to the second-order polynomial $(u^{A+a',A+\bar{a}'}\ldots n)^{T+t',T+\bar{t}'}$), let us multiply it by the first equation. This yields the fourth-order polynomial

$$(u^{A+a+a',A+\bar{a}'+a,A+a'+\bar{a},A+\bar{a}'+\bar{a}}\ldots n)^{T+t+t',T+\bar{t}'+t,T+t'+\bar{t},T+\bar{t}'+\bar{t}}.$$

The quantities

$$A+a+a',\ A+\bar{a}'+a,\ A+a'+\bar{a},\ A+\bar{a}'+\bar{a}$$

and

$$T+t+t',\ T+\bar{t}'+t,\ T+t'+\bar{t},\ T+\bar{t}'+\bar{t}$$

are ordered such that the first is smallest and the last is largest in the first group, and the converse holds in the second group. The relative order of the intermediate quantities does not matter. This being done, it is easy to see that:

1. We can always cancel $N(u^A\ldots n)^T$ terms in the polynomial

$$(u^{A+a',A+\bar{a}'}\ldots n)^{T+t',T+\bar{t}'}$$

using the second equation.

2. Likewise, we can cancel $N(u^{A+a,A+\bar{a}}\ldots n)^{T+t,T+\bar{t}}$ terms in the polynomial

$$(u^{A+a'+a,A+\bar{a}'+a,A+a'+\bar{a},A+\bar{a}'+\bar{a}}\ldots n)^{T+t+t',T+\bar{t}'+t,T+t'+\bar{t},T+\bar{t}'+\bar{t}}.$$

Pick an arbitrary polynomial of the form $(u^{A+a',A+\bar{a}'}\ldots n)^{T+t',T+\bar{t}'}$. If we multiply the first equation by this polynomial, we will therefore always be able to reduce the number of terms in the product equation to a number of terms expressed by

$$N(u^{A+a'+a,A+\bar{a}'+a,A+a'+\bar{a},A+\bar{a}'+\bar{a}}\ldots n)^{T+t'+t,T+\bar{t}'+t,T+t'+\bar{t},T+\bar{t}'+\bar{t}}$$
$$-N(u^{A+a,A+\bar{a}}\ldots n)^{T+t,T+\bar{t}} - N(u^{A+a',A+\bar{a}'}\ldots n)^{T+t',T+\bar{t}'} + N(u^A\ldots n)^T.$$

(182.) Let us now assume that there are three equations and let us pick a polynomial of the form

$$(u^{A+a'+a'',A+\bar{a}+a'',A+a'+\bar{a}'',A+\bar{a}'+\bar{a}''}\ldots n)^{T+t'+t'',T+\bar{t}'+t'',T+t'+\bar{t}'',T+\bar{t}'+\bar{t}''}.$$

In this form the varieties of exponents of u and those of the exponents of the polynomial can follow each other in many different ways.

Assume we multiply the first equation by this polynomial. We get a product equation in which the sequence of the exponents of A and that of the varieties of T are

$$A + a + a' + a'', \; A + a + \bar{a}' + a'', \; A + a + a' + \bar{a}'', \; A + a + a' + \bar{a}'',$$
$$A + \bar{a} + a' + a'', \; A + \bar{a} + \bar{a}' + a'', \; A + \bar{a} + a' + \bar{a}'', \; A + \bar{a} + \bar{a}' + \bar{a}''.$$

$$T + t + t' + t'', \; T + t + \bar{t}' + t'', \; T + t + t' + \bar{t}'', \; T + t + \bar{t}' + \bar{t}'',$$
$$T + \bar{t} + t' + t'', \; T + \bar{t} + \bar{t}' + t'', \; T + \bar{t} + t' + \bar{t}'', \; T + \bar{t} + \bar{t}' + \bar{t}''.$$

If, for brevity's sake, we represent the number of terms of this polynomial or this product equation by N', we see that

1. We can always reduce the number of terms of the multiplier polynomial to a number equal to

$$N(u^{A+a'+a'',A+\bar{a}'+a'',A+a'+\bar{a}'',A+\bar{a}'+\bar{a}''}\ldots n)^{T+t'+t'',T+\bar{t}'+t'',T+t'+\bar{t}'',T+\bar{t}'+\bar{t}''}$$
$$-N(u^{A+a',A+\bar{a}'}\ldots n)^{T+t',T+\bar{t}'} - N(u^{A+a'',A+\bar{a}''}\ldots n)^{T+t'',T+\bar{t}''}$$
$$+N(u^A\ldots n)^T,$$

   using the last two equations.

2. Consequently, using the same two equations and the coefficients of the polynomial multiplier, we can reduce the size of the product equation to a number of terms given by

$$N' - N(u^{A+a+a',A+a+\bar{a}',A+a'+\bar{a},A+\bar{a}+\bar{a}'})^{T+t+t',T+t+\bar{t}',T+\bar{t}+t',T+\bar{t}+\bar{t}'}$$
$$-N(u^{A+a'+a'',A+a+a'',A+a+a'',A+\bar{a}+\bar{a}''})^{T+t+t'',T+t+\bar{t}'',T+\bar{t}+t'',T+\bar{t}+\bar{t}''}$$
$$+N(u^{A+a,A+\bar{a}}\ldots n)^{T+t,T+\bar{t}}$$
$$-N(u^{A+a'+a'',A+\bar{a}'+a'',A+a'+\bar{a}'',A+\bar{a}'+\bar{a}''})^{T+t'+t'',T+\bar{t}'+t'',T+t'+\bar{t}'',T+\bar{t}'+\bar{t}''}$$
$$+N(u^{A+a',A+\bar{a}'})^{T+t',T+\bar{t}'} + N(u^{A+a'',A+\bar{a}''})^{T+t'',T+\bar{t}''} - N(u^A)^T.$$

   It is now easy to see how we can compute the number of remaining terms for an arbitrary number of equations.

(183.) Therefore, if we can obtain a general expression of the degree of the final equation for these kind of equations, the expression for the number of remaining terms must be an exact differential of order n.

If this does not happen, the form we have picked for the polynomial multiplier is not the appropriate one in general and this polynomial multiplier has a different order.

(184.) The form we have chosen for the polynomial multiplier works for incomplete, arbitrary order equations in two unknowns. This is not true for a larger number of unknowns. It can happen only in specific cases.

Let us first see how it is generally appropriate for incomplete, arbitrary order equations in two unknowns. We will then see a few cases when it occurs for incomplete equations with a larger number of unknowns, and we will conclude by showing how to obtain this form in other cases.

(185.) Consider two equations in two unknowns. Let us pick the polynomial

$$(u^{A+a',A+\bar{a}'}\ldots 2)^{T+t',T+\bar{t}'}$$

as our polynomial multiplier.

The resulting product equation is then

$$(u^{A+a+a',A+\bar{a}'+a,A+a'+\bar{a},A+\bar{a}'+\bar{a}}\ldots 2)^{T+t+t',T+\bar{t}'+t,T+t'+\bar{t},T+\bar{t}'+\bar{t}}.$$

This equation presents two cases concerning the exponent of u and two cases concerning the terms in the exponents of the dimensions of this equation:

$$A+a+\bar{a}' < A+\bar{a}+a', \quad A+a+\bar{a}' > A+\bar{a}+a',$$
$$T+t+\bar{t}' > T+\bar{t}+t', \quad T+t+\bar{t}' < T+\bar{t}+t',$$

which can be rewritten

$$\bar{a}-a > \bar{a}'-a', \quad \bar{a}-a < \bar{a}'-a',$$
$$t-\bar{t} > t'-\bar{t}', \quad t-\bar{t} < t'-\bar{t}'.$$

Since each of the latter two cases can occur along with the first two, the following four cases arise

$$t-\bar{t} > t'-\bar{t}'; \quad \bar{a}-a > \bar{a}'-a';$$
$$t-\bar{t} > t'-\bar{t}'; \quad \bar{a}-a < \bar{a}'-a';$$
$$t-\bar{t} < t'-\bar{t}'; \quad \bar{a}-a > \bar{a}'-a';$$
$$t-\bar{t} < t'-\bar{t}'; \quad \bar{a}-a < \bar{a}'-a'.$$

In the first case, the product equation is just as we have represented it. In the second case, it is

$$(u^{A+a+a',A+\bar{a}'+a,A+\bar{a}'+\bar{a}}\ldots 2)^{T+t+t',T+\bar{t}'+t,T+\bar{t}'+\bar{t}};$$

that is, it is a third-order polynomial because, past the dimension $T+t+\bar{t}'$, the largest exponent of u is larger than the third term $A+\bar{a}+a'$, and that term is not significant anymore.

In the third case, the form is

$$(u^{A+a+a',A+a'+\bar{a},A+\bar{a}'+\bar{a}}\ldots 2)^{T+t+t',T+t'+\bar{t},T+\bar{t}'+\bar{t}}.$$

Indeed, the exponent $A + \bar{a} + a'$ is larger than $A + a + \bar{a}'$ and enters as soon as the dimension $T + \bar{t} + t'$, which is larger than $T + t + \bar{t}'$. Thus it will cover the variety $A + a + \bar{a}'$, which will therefore not be a variety anymore.

In the fourth case, the form is

$$(u^{A+a+a',\,A+\bar{a}+a',\,A+a+\bar{a}',\,A+\bar{a}'+\bar{a}} \ldots 2)^{T+t+t',\,T+\bar{t}+t',\,T+t+\bar{t}',\,T+\bar{t}'+\bar{t}}.$$

Therefore, denoting by D the number of terms remaining in the final equation, we obtain, for the first case,

$$D = N(u^{A+a+a',\,A+\bar{a}'+a,\,A+a'+\bar{a},\,A+\bar{a}'+\bar{a}} \ldots 2)^{T+t+t',\,T+\bar{t}'+t,\,T+t'+\bar{t},\,T+\bar{t}'+\bar{t}}$$
$$-N(u^{A+a,A+\bar{a}} \ldots 2)^{T+t,T+\bar{t}} - N(u^{A+a',A+\bar{a}'} \ldots 2)^{T+t',T+\bar{t}'} + N(u^{A} \ldots 2)^{T}.$$

But from (171) we have

1.

$$N(u^{A+a+a',\,A+\bar{a}'+a,\,A+a'+\bar{a},\,A+\bar{a}'+\bar{a}} \ldots 2)^{T+t+t',\,T+\bar{t}'+t,\,T+t'+\bar{t},\,T+\bar{t}'+\bar{t}}$$
$$= N(u^{A+\bar{a}+\bar{a}'} \ldots 2)^{T+t+t'}$$
$$-ddN(u^{A+\bar{a}+\bar{a}'} \ldots 2)^{T+t+t'} \ldots \begin{pmatrix} T+t+t' & A+\bar{a}+\bar{a}' \\ t'-\bar{t}',0 & \vdots & 0,\bar{a}-a+\bar{a}'-a' \end{pmatrix}$$
$$-ddN(u^{A+\bar{a}+\bar{a}'} \ldots 2)^{T+t+\bar{t}'} \ldots \begin{pmatrix} T+t+\bar{t}' & A+\bar{a}+\bar{a}' \\ t-\bar{t}-t'+\bar{t}',0 & \vdots & 0,\bar{a}-a \end{pmatrix}$$
$$-ddN(u^{A+\bar{a}+\bar{a}'} \ldots 2)^{T+\bar{t}+t'} \ldots \begin{pmatrix} T+\bar{t}+t' & A+\bar{a}+\bar{a}' \\ t'-\bar{t}',0 & \vdots & 0,\bar{a}'-a' \end{pmatrix}.$$

2.

$$N(u^{A+a,A+\bar{a}} \ldots 2)^{T+t,T+\bar{t}} = N(u^{A+\bar{a}} \ldots 2)^{T+t}$$
$$-ddN(u^{A+\bar{a}} \ldots 2)^{T+t} \ldots \begin{pmatrix} T+t & A+\bar{a} \\ t-\bar{t},0 & \vdots & 0,\bar{a}-a \end{pmatrix}.$$

3.

$$N(u^{A+a',A+\bar{a}'} \ldots 2)^{T+t',T+\bar{t}'} = N(u^{A+\bar{a}'} \ldots 2)^{T+t'}$$
$$-ddN(u^{A+\bar{a}'} \ldots 2)^{T+t'} \ldots \begin{pmatrix} T+t' & A+\bar{a}' \\ t'-\bar{t}',0 & \vdots & 0,\bar{a}'-a' \end{pmatrix}.$$

Moreover from (180), we have

1.

$$ddN(u^{A+\bar{a}+\bar{a}'} \ldots 2)^{T+t+t'} \ldots \begin{pmatrix} T+t+t' & A+\bar{a}+\bar{a}' \\ (t'-\bar{t}'),0 & \vdots & 0,(\bar{a}-a+\bar{a}'-a') \end{pmatrix}$$
$$= ddN(u^{A+\bar{a}+\bar{a}'} \ldots 2)^{T+t+t'} \ldots \begin{pmatrix} T+t+t' & A+\bar{a}+\bar{a}' \\ (t'-\bar{t}'),0 & \vdots & 0,(\bar{a}-a) \end{pmatrix}$$
$$+ddN(u^{A+\bar{a}+\bar{a}'} \ldots 2)^{T+t+t'} \ldots \begin{pmatrix} T+t+t' & A+a+\bar{a} \\ (t'-\bar{t}'),0 & \vdots & 0,(\bar{a}'-a') \end{pmatrix}.$$

2.

$$ddN(u^{A+\bar{a}+\bar{a}'} \ldots 2)^{T+t+\bar{t}'} \ldots \begin{pmatrix} T+t+\bar{t}' & A+\bar{a}+\bar{a}' \\ (t-\bar{t}-t'+\bar{t}'),0 & \vdots & 0,(\bar{a}-a) \end{pmatrix}$$
$$= ddN(u^{A+\bar{a}+\bar{a}'} \ldots 2)^{T+t+\bar{t}'} \ldots \begin{pmatrix} T+t+\bar{t}' & A+\bar{a}+\bar{a}' \\ t-\bar{t},0 & \vdots & 0,(\bar{a}-a) \end{pmatrix}$$
$$-ddN(u^{A+\bar{a}+\bar{a}'} \ldots 2)^{T+t+\bar{t}'} \ldots \begin{pmatrix} T+\bar{t}+\bar{t}' & A+\bar{a}+\bar{a}' \\ t'-\bar{t}',0 & \vdots & 0,(\bar{a}-a) \end{pmatrix}.$$

Performing these substitutions, we get

$$D = N(u^{A+\bar{a}+\bar{a}'}\dots 2)^{T+t+t'} - N(u^{A+\bar{a}}\dots 2)^{T+t'}$$
$$- N(u^{A+\bar{a}'}\dots 2)^{T+t'} + N(u^A\dots 2)^T$$

$$-ddN(u^{A+\bar{a}+\bar{a}'}\dots 2)^{T+t+t'}\dots \begin{pmatrix} T+t+t' & A+\bar{a}+\bar{a}' \\ (t'-\bar{t}'),0 & 0,(\bar{a}-a) \end{pmatrix}$$

$$-ddN(u^{A+a+\bar{a}'}\dots 2)^{T+t+t'}\dots \begin{pmatrix} T+t+t' & A+a+\bar{a}' \\ (t'-\bar{t}'),0 & 0,(\bar{a}'-a') \end{pmatrix}$$

$$-ddN(u^{A+\bar{a}+\bar{a}'}\dots 2)^{T+t+\bar{t}'}\dots \begin{pmatrix} T+t+\bar{t}' & A+\bar{a}+\bar{a}' \\ (t-\bar{t}),0 & 0,(\bar{a}-a) \end{pmatrix}$$

$$+ddN(u^{A+\bar{a}+\bar{a}'}\dots 2)^{T+\bar{t}+\bar{t}'}\dots \begin{pmatrix} T+\bar{t}+\bar{t}' & A+\bar{a}+\bar{a}' \\ (t'-\bar{t}'),0 & 0,(\bar{a}-a) \end{pmatrix}$$

$$-ddN(u^{A+\bar{a}+\bar{a}'}\dots 2)^{T+\bar{t}+t'}\dots \begin{pmatrix} T+\bar{t}+t' & A+\bar{a}+\bar{a}' \\ (t'-\bar{t}'),0 & 0,(\bar{a}'-a') \end{pmatrix}$$

$$+ddN(u^{A+\bar{a}}\dots 2)^{T+t}\dots \begin{pmatrix} T+t & A+\bar{a} \\ t-\bar{t},0 & 0,\bar{a}-a \end{pmatrix}$$

$$+ddN(u^{A+\bar{a}'}\dots 2)^{T+t'}\dots \begin{pmatrix} T+t' & A+\bar{a}' \\ t'-\bar{t}',0 & 0,\bar{a}'-a' \end{pmatrix}.$$

At this point, if we remark that

1.

$$-ddN(u^{A+\bar{a}+\bar{a}'}\dots 2)^{T+t+t'}\dots \begin{pmatrix} T+t+t' & A+\bar{a}+\bar{a}' \\ (t'-\bar{t}'),0 & 0,(\bar{a}-a) \end{pmatrix}$$
$$+ddN(u^{A+\bar{a}+\bar{a}'}\dots 2)^{T+\bar{t}+\bar{t}'}\dots \begin{pmatrix} T+\bar{t}+\bar{t}' & A+\bar{a}+\bar{a}' \\ (t'-\bar{t}'),0 & 0,(\bar{a}-a) \end{pmatrix}$$
$$= -d^3N(u^{A+\bar{a}+\bar{a}'}\dots 2)^{T+t+t'}\dots \begin{pmatrix} T+t+t' & A+\bar{a}+\bar{a}' \\ (t'-\bar{t}'),0,t-\bar{t}+t'-\bar{t}' & 0,(\bar{a}-a),0 \end{pmatrix}$$
$$= 0.$$

2. Likewise

$$-ddN(u^{A+\bar{a}+\bar{a}'}\dots 2)^{T+t+\bar{t}'}\dots \begin{pmatrix} T+t+\bar{t}' & A+\bar{a}+\bar{a}' \\ (t-\bar{t}),0 & 0,-(\bar{a}-a) \end{pmatrix}$$
$$+ddN(u^{A+\bar{a}}\dots 2)^{T+t}\dots \begin{pmatrix} T+t & A+\bar{a} \\ t-\bar{t},0 & 0,\bar{a}-a \end{pmatrix}$$
$$= -d^3N(u^{A+\bar{a}+\bar{a}'}\dots 2)^{T+t+\bar{t}'}\dots \begin{pmatrix} T+t+t' & A+\bar{a}+\bar{a}' \\ (t-\bar{t}'),0,t' & 0,(\bar{a}-a),\bar{a}' \end{pmatrix}$$
$$= 0.$$

3. Also,

$$-ddN(u^{A+\bar{a}+\bar{a}'}\dots 2)^{T+\bar{t}+t'}\dots \begin{pmatrix} T+\bar{t}+t' & A+\bar{a}+\bar{a}' \\ (t'-\bar{t}'),0 & 0,(\bar{a}'-a') \end{pmatrix}$$
$$+ddN(u^{A+\bar{a}'}\dots 2)^{T+t'}\dots \begin{pmatrix} T+t' & A+\bar{a}' \\ t'-\bar{t}',0 & 0,\bar{a}'-a' \end{pmatrix}$$
$$= -d^3N(u^{A+\bar{a}+\bar{a}'}\dots 2)^{T+\bar{t}+t'}\dots \begin{pmatrix} T+\bar{t}+t' & A+\bar{a}+\bar{a}' \\ (t'-\bar{t}'),0,\bar{t} & 0,(\bar{a}'-a'),\bar{a} \end{pmatrix} = 0.$$

4. And finally

$$N(u^{A+\bar{a}+\bar{a}'}\ldots 2)^{T+t+t'} - N(u^{A+\bar{a}}\ldots 2)^{T+t'}$$
$$-N(u^{A+\bar{a}'}\ldots 2)^{T+t'} + N(u^{A}\ldots 2)^{T}$$
$$= ddN(u^{A+\bar{a}+\bar{a}'}\ldots 2)^{T+t+t'}\ldots\begin{pmatrix} T+t+t' & : & A+\bar{a}+\bar{a}' \\ t,t' & & \bar{a},\bar{a}' \end{pmatrix}.$$

We see that the value of D reduces to

$$D = ddN(u^{A+\bar{a}+\bar{a}'}\ldots 2)^{T+t+t'}\ldots\begin{pmatrix} T+t+t' & : & A+\bar{a}+\bar{a}' \\ t,t' & & \bar{a},\bar{a}' \end{pmatrix}$$
$$-ddN(u^{A+\bar{a}+\bar{a}'}\ldots 2)^{T+t+t'}\ldots\begin{pmatrix} T+t+t' & : & A+\bar{a}+\bar{a}' \\ (t'-\bar{t}'),0 & & 0,(\bar{a}'-a') \end{pmatrix}.$$

That is,

$$D = tt' - (t-\bar{a}).(t'-\bar{a}') - (t'-\bar{t}').(\bar{a}'-a').$$

(186.) Given the amount of detail we have spent on the calculations for the first case, it is probably superfluous to do the same for each of the remaining three cases. We will therefore only give the results. But prior to that, we observe that if $\underset{\prime}{a}$ and $\underset{\prime}{\bar{a}}$ are two varieties of the exponents of the second unknown in the first equation and $\underset{\prime}{a}'$ and $\underset{\prime}{\bar{a}}'$ represent similar quantities for the second equation, we will obtain the following two cases relative to the second unknown:

$$\underset{\prime}{\bar{a}} - \underset{\prime}{a} > \underset{\prime}{\bar{a}}' - \underset{\prime}{a}' \quad\text{and}\quad \underset{\prime}{\bar{a}} - \underset{\prime}{a} < \underset{\prime}{\bar{a}}' - \underset{\prime}{a}'.$$

Each of these two cases can arise alongside the four others. Therefore, the value D of the degree of the final equation in incomplete, second-order equations in two unknowns can take on eight values depending upon the eight relative orders among the exponents of the equations and the unknowns.

We will find these eight values in the table below, where we have introduced

$$D' = tt' - (t-\bar{a}).(t'-\bar{a}') - (t-\underset{\prime}{\bar{a}}).(t'-\underset{\prime}{\bar{a}}').$$

Table of all possible values of the degree of the final equations for all possible cases of incomplete, second-order equations in two unknowns

Case	Corresponding values for D
$\left\{\begin{array}{l} t-\bar{t} < t'-\bar{t}' \\ \bar{a}-a < \bar{a}'-a' \\ \underset{\prime}{\bar{a}}-\underset{\prime}{a} < \underset{\prime}{\bar{a}}'-\underset{\prime}{a}' \end{array}\right.$	$D = D' - (t-\bar{t}).(a'-a+\bar{a}-\underset{\prime}{a})$
$\left\{\begin{array}{l} t-\bar{t} < t'-\bar{t}' \\ \bar{a}-a > \bar{a}'-a' \\ \underset{\prime}{\bar{a}}-\underset{\prime}{a} < \underset{\prime}{\bar{a}}'-\underset{\prime}{a}' \end{array}\right.$	$D = D' - (t-\bar{t}).(\bar{a}'-a'+\underset{\prime}{\bar{a}}-\underset{\prime}{a})$

Case	Corresponding values for D
$\begin{cases} t - \bar{t} < t' - \bar{t}' \\ \bar{a} - a < \bar{a}' - a' \\ \underset{\prime}{\bar{a}} - \underset{\prime}{a} > \underset{\prime}{\bar{a}'} - \underset{\prime}{a'} \end{cases}$	$D = D' - (t - \bar{t}).(\bar{a} - a + \underset{\prime}{\bar{a}'} - \underset{\prime}{a'})$
$\begin{cases} t - \bar{t} < t' - \bar{t}' \\ \bar{a} - a > \bar{a}' - a' \\ \underset{\prime}{\bar{a}} - \underset{\prime}{a} > \underset{\prime}{\bar{a}'} - \underset{\prime}{a'} \end{cases}$	$D = D' - (t - \bar{t}).(\bar{a}' - a' + \underset{\prime}{\bar{a}'} - \underset{\prime}{a'})$
$\begin{cases} t - \bar{t} > t' - \bar{t}' \\ \bar{a} - a < \bar{a}' - a' \\ \underset{\prime}{\bar{a}} - \underset{\prime}{a} < \underset{\prime}{\bar{a}'} - \underset{\prime}{a'} \end{cases}$	$D = D' - (t' - \bar{t}').(\bar{a} - a + \underset{\prime}{\bar{a}} - \underset{\prime}{a})$
$\begin{cases} t - \bar{t} > t' - \bar{t}' \\ \bar{a} - a > \bar{a}' - a' \\ \underset{\prime}{\bar{a}} - \underset{\prime}{a} < \underset{\prime}{\bar{a}'} - \underset{\prime}{a'} \end{cases}$	$D = D' - (t' - \bar{t}').(\bar{a}' - a' + \underset{\prime}{\bar{a}} - \underset{\prime}{a})$
$\begin{cases} t - \bar{t} > t' - \bar{t}' \\ \bar{a} - a < \bar{a}' - a' \\ \underset{\prime}{\bar{a}} - \underset{\prime}{a} > \underset{\prime}{\bar{a}'} - \underset{\prime}{a'} \end{cases}$	$D = D' - (t' - \bar{t}').(\bar{a} - a + \underset{\prime}{\bar{a}'} - \underset{\prime}{a'})$
$\begin{cases} t - \bar{t} > t' - \bar{t}' \\ \bar{a} - a > \bar{a}' - a' \\ \underset{\prime}{\bar{a}} - \underset{\prime}{a} > \underset{\prime}{\bar{a}'} - \underset{\prime}{a'} \end{cases}$	$D = D' - (t' - \bar{t}').(\bar{a}' - a' + \underset{\prime}{\bar{a}'} - \underset{\prime}{a'}).$

(187.) Let us assume the two equations are third order and represented by

$$(u^{\underset{\prime}{a}, \bar{a}, \bar{\bar{a}}} \ldots 2)^{t, \bar{t}, \bar{\bar{t}}} = 0, \quad (u^{a', \bar{a}', \bar{\bar{a}}'} \ldots 2)^{t', \bar{t}', \bar{\bar{t}}'} = 0.$$

From what was said in (181 ff), we must take a polynomial multiplier of the form

$$(u^{A+a', A+\bar{a}', A+\bar{\bar{a}}'} \ldots 2)^{T+t', T+\bar{t}', T+\bar{\bar{t}}'}.$$

Then the product equation will have the two following series as varieties in the exponents of u and in the exponents of its dimensions:

$$A + a + a', \ A + \bar{a} + a', \ A + \bar{\bar{a}} + a', \ A + a + \bar{a}', \ A + \bar{a} + \bar{a}',$$
$$A + \bar{\bar{a}} + \bar{a}', \ A + a + \bar{\bar{a}}', \ A + \bar{a} + \bar{\bar{a}}', \ A + \bar{\bar{a}} + \bar{\bar{a}}',$$

$$T + t + t', \ T + \bar{t} + t', \ T + \bar{\bar{t}} + t', \ T + t + \bar{t}', \ T + \bar{t} + \bar{t}',$$
$$T + \bar{\bar{t}} + \bar{t}', \ T + t + \bar{\bar{t}}', \ T + \bar{t} + \bar{\bar{t}}', \ T + \bar{\bar{t}} + \bar{\bar{t}}'.$$

In these series, the order of the terms may vary depending on the relative values of

$$a, a'; \ \bar{a}, \bar{a}'; \ \bar{\bar{a}}, \bar{\bar{a}}'; \ t, t'; \ \bar{t}, \bar{t}'; \ \bar{\bar{t}}, \bar{\bar{t}}';$$

only the order of the first and the last quantities in each series remains invariant.

Limiting ourselves to a single case, we assume that the order of these quantities corresponds to the order in which they are written. That is, we assume $A + \bar{\bar{a}} + \bar{\bar{a}}'$ is the largest number; $A + \bar{a} + \bar{\bar{a}}'$ is larger than all its predecessors but smaller than its successor; $A + a + \bar{\bar{a}}'$ is larger than all its predecessors but smaller than all its successors; and so on. Let us assume the converse order holds for the quantities $T + \bar{\bar{t}} + \bar{\bar{t}}'$, $T + \bar{t} + \bar{\bar{t}}'$, $T + t + \bar{\bar{t}}'$, taken in the same order. All these conditions reduce to the following conditions:

$$\bar{\bar{a}}' - \bar{a}' > \bar{\bar{a}} - a; \; \bar{\bar{a}}' - a' > \bar{\bar{a}} - a$$
$$t' - \bar{t}' > t - \bar{\bar{t}}; t' - \bar{\bar{t}}' > t - \bar{t}.$$

The product equation then becomes an incomplete polynomial of order 9. The number of terms that can be cancelled using the second equation is the number of terms of the polynomial $(u^{A+a, A+\bar{a}, A+\bar{\bar{a}}} \ldots 2)^{T+t, T+\bar{t}, T+\bar{\bar{t}}}$.

And the number of terms that can be cancelled in the polynomial multiplier using the same second equation is the number of terms of the polynomial $(u^A \ldots 2)^T$.

Therefore, denoting by N' the number of terms in the product equation, by N'' the number of terms that can cancelled using the second equation, by N''' the number of terms in the polynomial multiplier, and by N^{IV} the number of terms that can be cancelled in the polynomial multiplier using the second equation, we can actually compute these values.

Dealing with these quantities as in (185) and substituting the results in the equation $D = N' - N'' - N''' + N^{IV}$, where D is the degree of the final equation, we find

$$D = ddN(u^{A+\bar{a}+\bar{\bar{a}}'})^{T+t+t'} \ldots \begin{bmatrix} T+t+t' & : & A+\bar{a}+\bar{\bar{a}}' \\ t,t' & & \bar{a}, \bar{\bar{a}}' \end{bmatrix}$$

$$-ddN(u^{A+\bar{a}+\bar{\bar{a}}'})^{T+t+t'} \ldots \begin{bmatrix} T+t+t' & : & A+\bar{a}+\bar{\bar{a}}' \\ t-\bar{t},0 & & 0,\bar{a}-a+\bar{\bar{a}}'-a' \end{bmatrix}$$

$$-ddN(u^{A+\bar{a}+\bar{\bar{a}}'})^{T+\bar{t}+t'} \ldots \begin{bmatrix} T+\bar{t}+t' & : & A+\bar{a}+\bar{\bar{a}}' \\ \bar{t}-\bar{\bar{t}},0 & & 0,\bar{a}-\bar{a}+\bar{\bar{a}}'-a' \end{bmatrix}$$

$$-ddN(u^{A+\bar{a}+\bar{\bar{a}}'})^{T+t+\bar{t}'} \ldots \begin{bmatrix} T+t+\bar{t}' & : & A+\bar{a}+\bar{\bar{a}}' \\ t-\bar{t},0 & & 0,\bar{a}-a+\bar{\bar{a}}'-\bar{a}' \end{bmatrix}$$

$$-ddN(u^{A+\bar{a}+\bar{\bar{a}}'})^{T+\bar{t}+t'} \ldots \begin{bmatrix} T+\bar{t}+t' & : & A+\bar{a}+\bar{\bar{a}}' \\ (\bar{t}-\bar{\bar{t}}),0 & & 0,\bar{a}-\bar{a}+\bar{\bar{a}}'-\bar{a}' \end{bmatrix}$$

$$+ddN(u^{A+\bar{a}+\bar{\bar{a}}'})^{T+\bar{t}+\bar{t}'} \ldots \begin{bmatrix} T+\bar{\bar{t}}+\bar{t}' & : & A+\bar{\bar{a}}+\bar{a}' \\ t-\bar{\bar{t}},0 & & 0,\bar{\bar{a}}'-a' \end{bmatrix},$$

that is,

$$D = tt' - (t-\bar{a}).(t'-\bar{\bar{a}}') - (t-\bar{t}).(\bar{a}-a+\bar{\bar{a}}'-a')$$
$$-(\bar{t}-\bar{\bar{t}}).(\bar{a}-\bar{a}+\bar{\bar{a}}'-a') - (t-\bar{t}).(\bar{a}-a+\bar{\bar{a}}'-\bar{a}')$$
$$-(t-\bar{t}).(\bar{a}-\bar{a}+\bar{\bar{a}}'-\bar{a}') + (t-\bar{\bar{t}}).(\bar{\bar{a}}'-a').$$

Rewriting $t - \bar{\bar{t}} = t - \bar{t} + \bar{t} - \bar{\bar{t}}$, the expression of D reduces to

$$D = tt' - (t-\bar{\bar{a}}).(t'-\bar{\bar{a}}') - 2(t-\bar{t}).(\bar{a}-a)$$
$$-2(t-\bar{\bar{t}}).(\bar{\bar{a}}-\bar{a}) - (2t-\bar{t}-\bar{\bar{t}}).(\bar{a}'-\bar{\bar{a}}').$$

We can write, likewise, the expression of D for all other sequences of the quantities

$$\bar{\bar{a}} - a, \ \bar{\bar{a}} - \bar{a}, \ \bar{a} - a, \ \bar{\bar{a}}' - a', \ \bar{\bar{a}}' - \bar{a}', \ \bar{a}' - a';$$
$$t - \bar{t}, \ t - \bar{\bar{t}}, \ \bar{t} - \bar{\bar{t}}, \ t' - \bar{t}', \ t' - \bar{\bar{t}}', \ \bar{t}' - \bar{\bar{t}}'.$$

This value can be obtained if the product equation is an incomplete polynomial of order 9 as we just saw or a polynomial of lower order, as often happens depending on the values of the quantities t relative to the quantities a.

From what was said in (186), we will have no more difficulties finding the value of D , with respect to the quantites similar to a, $\bar{a}$, $\bar{\bar{a}}$, a', $\bar{a}'$, $\bar{\bar{a}}'$, for the second unknown.

(188.) In general, the degree of the final equation turns out to always be a function of differentials of order no less than 2 for incomplete equations in two unknowns of arbitrary order. As a consequence, in any case we can write an explicit expression for the degree of the final equation as a function of the exponents of the given equations.

(189.) The same does not hold for equations involving a larger number of unknowns. The form we have indicated in (181 ff) does not always lead to an expression for the degree of the final equation, which is a function of differentials of order no less than the number of unknowns. For example, consider equations in three unknowns of the form $(u^{a,\bar{a}}, \ldots 3)^{t,\bar{t}} = 0$. The polynomial multiplier cannot, in general, be an incomplete polynomial of order less than that of the polynomial

$$\left(u^{A+a'+a'',A+a'+a^{\bar{}'},A+\bar{a}'+a'',A+\bar{a}'+a^{\bar{}'}} \ldots 3\right)^{T+t'+t'',T+t'+t^{\bar{}'},T+\bar{t}'+t'',T+\bar{t}'+t^{\bar{}'}}.$$

This is the polynomial given in (181 ff) that we must indeed use as a polynomial multiplier for the equation $(u^{a,\bar{a}} \ldots 3)^{t,\bar{t}} = 0$. But it may happen, as will be true in several cases, that this polynomial multiplier must be incomplete of higher order. It is nevertheless not always impossible to determine directly which form this polynomial must take in each case, and therefore the degree of the final equation as a function of the exponents of the equations and the unknowns. But this is an extremely complicated task. We will show a few cases where the form given in (181 ff) is sufficient. We will then give an idea about the general process to follow to determine the degree of the final equation in all cases. We will see in the Second Book that the elimination process we will present always leads to an equation whose degree is the lowest possible, without knowing this degree a priori.

(190.) We propose to determine the degree of the final equation for three equations in three unknowns of the form

$$(u^{a,\bar{a}} \ldots 3)^{t,\bar{t}} = 0, \ (u^{a',\bar{a}'} \ldots 3)^{t',\bar{t}'} = 0, \ (u^{a'',\bar{a}''} \ldots 3)^{t'',\bar{t}''} = 0.$$

Let us use

$$\left(u^{A+a'+a'',A+a'+a^{\bar{}''},A+\bar{a}'+a'',A+\bar{a}'+a^{\bar{}''}} \ldots 3\right)^{T+t'+t'',T+t'+t^{\bar{}''},T+\bar{t}'+t'',T+\bar{t}'+t^{\bar{}''}}$$

as a polynomial multiplier of the first equation. The order of the quantities can be different from what we see here, depending upon the relative values of these quantities.

The product equation is an eighth-order polynomial, in which the varieties of the exponents of u and of the exponents of the polynomial are as follows:

$$A + a + a' + a'', \ A + a + a' + \bar{a}'', \ A + a + \bar{a}' + a'', \ A + a + \bar{a}' + \bar{a}'',$$
$$A + \bar{a} + a' + a'', \ A + \bar{a} + a' + \bar{a}'', \ A + \bar{a} + \bar{a}' + a'', \ A + \bar{a} + \bar{a}' + \bar{a}'';$$

$$T + t + t' + t'', \ T + t + t' + \bar{t}'', \ T + t + \bar{t}' + t'', \ T + t + \bar{t}' + \bar{t}'',$$
$$T + \bar{t} + t' + t'', \ T + \bar{t} + t' + \bar{t}'', \ T + \bar{t} + \bar{t}' + t'', \ T + \bar{t} + \bar{t}' + \bar{t}''.$$

Let us denote by N' the number of terms in this equation, by N'' the number of terms that can be cancelled in this equation using the second equation, by N''' the number of terms that can be cancelled using the third equation, and by N^{IV} the number of terms that can be cancelled using the third equation, in the polynomial used to cancel N'' terms in the product equation.

Likewise, let us denote by $N'_{\prime}, N''_{\prime}, N'''_{\prime}, N^{IV}_{\prime}$ the quantities relative to the polynomial multiplier that are analog to N', N'', N''', N^{IV} for the product equation. The degree of the final equation can be written as

$$D = N' - N'' - N''' + N^{IV} - N'_{\prime} + N''_{\prime} + N'''_{\prime} - N^{IV}_{\prime}.$$

However, it is easy to obtain N' for arbitrary relative sequences of the exponents, from what was said in (171). Regarding $N'', N''', N^{IV}, N'_{\prime}$, etc.,

it is easy to see that N'' is the number of terms of an incomplete polynomial of order 4, whose varieties are as follows:

$$A + a + a'', \ A + a + \bar{a}'', \ A + \bar{a} + a'', \ A + \bar{a} + \bar{a}'';$$
$$T + t + t'', \ T + t + \bar{t}'', \ T + \bar{t} + t''; \ T + \bar{t} + \bar{t}''.$$

N''' is the number of terms of an incomplete polynomial of order 4 whose varieties are

$$A + a + a', \ A + a + \bar{a}', \ A + \bar{a} + a', \ A + \bar{a} + \bar{a}';$$
$$T + t + t', \ T + t + \bar{t}', \ T + \bar{t} + t', \ T + \bar{t} + \bar{t}'.$$

N^{IV} is the number of terms of an incomplete polynomial of order 2 whose varieties are

$$A + a, \ A + \bar{a};$$
$$T + t, \ T + \bar{t}.$$

$N'_{\prime}$ is the number of terms of an incomplete polynomial of order 4 whose varieties are

$$A + a' + a'', \ A + a' + \bar{a}'', \ A + \bar{a}' + a'', \ A + \bar{a}' + \bar{a}'';$$
$$T + t' + t'', \ T + t' + \bar{t}'', \ T + \bar{t}' + t'', \ T + \bar{t}' + \bar{t}''.$$

$N''_{\prime}$ is the number of terms of an incomplete polynomial of order 2 whose varieties are

$$A + a'', \ A + \bar{a}'';$$
$$T + t'', \ T + \bar{t}''.$$

$N'''_{\prime}$ is the number of terms of an incomplete polynomial of order 2 whose varieties are

$$A + a', \ A + \bar{a}';$$
$$T + t', \ T + \bar{t}'.$$

Finally, $N^{IV}_{\prime}$ is the number of terms of the incomplete first-order polynomial $(u^A \ldots 3)^T$.

We therefore easily obtain the expression for each of the quantities entering in the value of D for each case of the various relative values of the exponents.

Since the result of their substitution into the expression for D is not, in general, the composition of differentials whose order is no less than three, let us limit ourselves to studying one of the cases for which the result of such a substitution leads to such a case.

Let us assume, for example, that the three given equations satisfy $t - \bar{t} = t' - \bar{t}' = t'' - \bar{t}''$.

Then several among the varieties of the exponents of u have the same dimension in the product equation, and the sequence of these varieties is

$$A + a + a' + a'', \ A + a + a' + \bar{a}'', \ A + a + \bar{a}' + \bar{a}'', \ A + \bar{a} + \bar{a}' + \bar{a}'',$$
$$A + a + \bar{a}' + a'', \ A + \bar{a} + a' + \bar{a}'',$$
$$A + \bar{a} + a' + a'', \ A + \bar{a} + \bar{a}' + a''.$$

$$T + t + t' + t'', \ T + t + t' + \bar{t}'', \ T + t + \bar{t}' + \bar{t}'', \ T + \bar{t} + \bar{t}' + \bar{t}''.$$

That is, the product equation is then a fourth-order polynomial only, whose form is determined by the largest of the three exponents

$$A + a + a' + \bar{a}'', \ A + a + \bar{a}' + a'', \ A + \bar{a} + a' + a'',$$

and the largest of the three exponents

$$A + a + \bar{a}' + \bar{a}'', \ A + \bar{a} + a' + \bar{a}'', \ A + \bar{a} + \bar{a}' + \bar{a}''.$$

Thus the case when $t - \bar{t} = t' - \bar{t}' = t'' - \bar{t}''$ consists of the six following cases:

$$
\begin{array}{l|l}
\bar{a} - a > \bar{a}' - a' > \bar{a}'' - a'' & \bar{a}' - a' > \bar{a}'' - a'' > \bar{a} - a \\
\bar{a} - a > \bar{a}'' - a'' > \bar{a}' - a' & \bar{a}'' - a'' > \bar{a} - a > \bar{a}' - a' \\
\bar{a}' - a' > \bar{a} - a > \bar{a}'' - a'' & \bar{a}'' - a'' > \bar{a}' - a' > \bar{a} - a.
\end{array}
$$

Consider the first of these six cases: The product equation is a fourth-order polynomial whose varieties are

$$A + a + a' + a'', \ A + \bar{a} + a' + a'', \ A + \bar{a} + \bar{a}' + a'', \ A + \bar{a} + \bar{a}' + \bar{a}'',$$
$$T + t + t' + t'', \ T + t + t' + \bar{t}'', \ T + t + \bar{t}' + \bar{t}'', \ T + \bar{t} + \bar{t}' + \bar{t}''.$$

The polynomial whose number of terms is given by N'' is third-order and its varieties are

$$A + a + a'', \quad A + \bar{a} + a'', \quad A + \bar{a} + \bar{a}''.$$
$$T + t + t'', \quad T + t + \bar{t}'', \quad T + \bar{t} + \bar{t}''.$$

The polynomial whose number of terms is given by N''' is third-order and its varieties are

$$A + a + a', \quad A + \bar{a} + a', \quad A + \bar{a} + \bar{a}'.$$
$$T + t + t', \quad T + t + \bar{t}', \quad T + \bar{t} + \bar{t}'.$$

The polynomial whose number of terms is given by N^{IV} is second-order and its varieties are

$$A + a, \quad A + \bar{a}.$$
$$T + t, \quad T + \bar{t}.$$

The polynomial whose number of terms is given by $\underset{\prime}{N'}$ is third-order and its varieties are

$$A + a' + a'', \quad A + \bar{a}' + a'', \quad A + \bar{a}' + \bar{a}''.$$
$$T + t' + t'', \quad T + t' + \bar{t}'', \quad T + \bar{t}' + \bar{t}''.$$

The varieties of the polynomial whose number of terms is given by $\underset{\prime}{N''}$ are

$$A + a'', \quad A + \bar{a}''.$$
$$T + t'', \quad T + \bar{t}''.$$

The varieties of the polynomial whose number of terms is given by $\underset{\prime}{N'''}$ are

$$A + a', \quad A + \bar{a}'.$$
$$T + t', \quad T + \bar{t}'.$$

At this point, if we compute the values of N', N'', etc. using what was said in (171) and (180), and if we substitute them in the expression of D, we obtain

$$D = d^3 N(u^{A+\bar{a}+\bar{a}'+\bar{a}''})T+t+t'+t'' \cdots \left(\begin{array}{cc} T+t+t'+t'' & A+\bar{a}+\bar{a}'+\bar{a}'' \\ t, t', t'' & \bar{a}, \bar{a}', \bar{a}'' \end{array} \right)$$

$$-d^3 N(u^{A+\bar{a}+\bar{a}'+a''})T+t+t'+\bar{t}'' \cdots \left(\begin{array}{cc} T+t+t'+\bar{t}'' & A+\bar{a}+\bar{a}'+\bar{a}'' \\ t''-\bar{t}', 0, t'-\bar{t}'+\bar{t}'' & 0, \bar{a}'-a', \bar{a}'' \end{array} \right)$$

$$-d^3 N(u^{A+\bar{a}+a'+\bar{a}''})T+t+t'+\bar{t}'' \cdots \left(\begin{array}{cc} T+t+t'+\bar{t}'' & A+\bar{a}+a'+\bar{a}' \\ t''-\bar{t}'', 0, t' & 0, \bar{a}''-a'', a' \end{array} \right)$$

$$-d^3 N(u^{A+\bar{a}+\bar{a}'+a''})T+t+\bar{t}'+\bar{t}'' \cdots \left(\begin{array}{cc} T+t+\bar{t}'+\bar{t}'' & A+\bar{a}+\bar{a}'+\bar{a}'' \\ t''-\bar{t}'', 0, t-t'+\bar{t}' & 0, \bar{a}''-a'', \bar{a} \end{array} \right).$$

This expression reduces to

$$D = tt't'' - (t-\bar{a}).(t'-\bar{a}').(t''-\bar{a}'') - (t''-\bar{t}'').(\bar{a}'-a').(t'-\bar{t}'+\bar{t}''-\bar{a}'')$$
$$-(t''-\bar{t}'').(\bar{a}''-a'').(t-\bar{a}+\bar{t}'-a').$$

(191.) Likewise, we find the value of D corresponding to the five other cases.

(192.) The value of D can also be expressed as a function of the exponents of the equations and unknowns, using a polynomial multiplier such as described in (190):

1. when $\bar{a} - a = \bar{a}' - a' = \bar{a}'' - a''$,

2. when one of the three equations is incomplete but first-order only,

3. in a few other cases we will not pursue here.

Conclusion about incomplete equations of arbitrary order

(193.) From the preceding developments and concerning the following cases:

1. arbitrary number of complete equations of arbitrary order, containing the same number of unknowns;

2. incomplete first-order equations, whether they are incomplete relative to each unknown taken alone (see 140 ff), or incomplete relative to unknowns considered in combinations of two, three, four, etc.,

we conclude that we can use an incomplete, first-order polynomial whose form is as general as the proposed equations, denoted $(u^A \ldots n)^T$. Assuming we multiply this polynomial by all proposed equations successively, we always end up with a first-order polynomial whose form is that of what we have denoted the *product equation*.

Assume we also build all possible products of the polynomial $(u^A \ldots n)^T$ with the products of two, three, four, etc. of the proposed equations. The resulting products are also first-order polynomials whose number of terms contribute to the degree of the final equation.

It is always possible to pick the polynomial $(u^A \ldots n)^T$ such that the expression of the degree of the final equation is an exact differential whose order equals the number of unknowns. Consequently, this degree is a function of known quantities or of the exponents that determine the nature of these equations. In other words, incomplete first-order polynomials are sufficient to determine the degree of the final equation when considering complete or first-order, incomplete equations.

We now consider incomplete equations whose order is greater than 1. Assume that we use, likewise, a first-order polynomial $(u^A \ldots n)^T$, and that we multiply it successively, as described above, by all proposed equations. Assume this form is appropriate to determine the degree of the final equation. Then the total product and the partial products of this polynomial with products of two, three, four, etc. among the proposed equations is appropriate to represent all polynomials whose number of terms contribute to the expression of the degree of the final equation.

However, this simple form is enough to find a general expression for the degree of the final equation only in the case of equations in two unknowns. Concerning equations with a larger number of unknowns, it can lead to the degree of the final equation only when certain relations exist between the varieties and exponents of the equations and unknowns. In general, it is impossible to get an expression for the degree of the final equation as an exact differential of order equal to the number of unknowns.

Instead of the polynomial $(u^A \ldots n)^T$, we will have to use, in general, a polynomial $(u^{A,\bar{A},\bar{\bar{A}},\bar{\bar{\bar{A}}}, \text{etc.}} \ldots n)^{T,\bar{T},\bar{\bar{T}},\bar{\bar{\bar{T}}}, \text{etc}}$. Assuming we perform the same multiplications as described above, we can then make conclusions about the different expressions that the product equation and the polynomial multiplier can take, as well as the expressions of the other polynomials that we know enter in the expression of the degree of the final equation. From this general and unknown expression of the degree of the final equation, we could then compute the values of the quantities $\bar{\bar{A}} - A$, $\bar{\bar{A}} - A$, $\bar{A} - A$, etc., $T - \bar{\bar{T}}$, $T - \bar{\bar{T}}$, $T - \bar{T}$, etc., from the condition that the total number of terms composing this general expression becomes an exact differential whose order equals the number of unknowns.

But remember that using $(u^A \ldots n)^T$ leads to a product equation of order 16 in the case of three equations in three unknowns; this already yields a prodigious number of cases arising from that equation. When we are led to use the next form $(u^{A,\bar{A}} \ldots n)^{T,\bar{T}}$, we are led to a product equation of order 16, leading to an even greater number of cases to be discussed.

It should therefore be no surprise to the reader that we have decided not to pursue these matters any further.

No matter what, we see that we will always be able to find the degree of the final equation using this method, for any equation, since all equations belong to the set of incomplete equations of arbitrary order, and all polynomial multipliers belong to the set of arbitrary order, polynomial multipliers.

It will very often take a lot of work to determine the true form of the polynomial $(u^{A,\bar{A},\bar{\bar{A}},\bar{\bar{\bar{A}}}, \text{etc.}} \ldots n)^{T,\bar{T},\bar{\bar{T}},\bar{\bar{\bar{T}}}, \text{etc}}$, that is, to determine whether, for a given general case, this polynomial is $(u^A \ldots n)$, or $(u^{A,\bar{A}} \ldots n)^{T,\bar{T}}$, or $(u^{A,\bar{A},\bar{\bar{A}}} \ldots n)^{T,\bar{T},\bar{\bar{T}}}$, etc.

We will return to this subject in the rest of this book, and we will give an idea about how to compute an expression for the degree of the final equation using only the polynomial $(u^A \ldots n)$ introduced and described above. We will give reliable methods to determine the lowest degree of the final equation resulting from an arbitrary number of algrebraic equations, containing the same number of unknowns, and whose coefficients are arbitrary in any case. We will do even better, since we will give methods to determine the lowest degree of the final equation when specific relations hold among the coefficients of the equations.

Book Two

In which we give a process for reaching the final equation resulting from an arbitrary number of equations in the same number of unknowns, and in which we present many general properties of algebraic quantities and equations

GENERAL OBSERVATIONS

(194.) The method expressed in the first book for determining the degree of the final equation strongly indicates that the art of eliminating all variables but one reduces altogether to the elimination method in first-order equations, with an arbitrary number of unknowns. In appearance, it would seem that little remains to be said about this matter, since we know of methods that lead us quickly to the value of all unknowns in first-order equations. But even if these methods had all the perfection we propose to introduce, we would, by leaving our investigations here, leave aside more than one important topic and would neglect several important points in the General Theory of Equations.

Indeed:

1. We have seen in the first book that we must not consider all coefficients of the polynomial multiplier as equally qualified for elimination: We therefore must know which of these are really useful and what must or can be done about the others.

2. It is only for the sake of convenience that we have reduced the question of computing the degree of the final equation to multiplying only one of the proposed equations by a polynomial and to using the other equations to cancel all possible terms without introducing new ones in the product equation and in that polynomial. Indeed, the computations required to cancel these terms or to manipulate them in an arbitrary fashion are really about restating the elimination question by multiplying each equation by a specific polynomial multiplier, and by summing all these products to lead to the final equation by eliminating all other terms. It is therefore absolutely necessary to know these polynomial multipliers.

3. We should not content ourselves with having recast the elimination question in arbitrary degree equations to that of solving first-degree

equations: There must be as few of these equations as possible, and these should not hide information originally present in the problem. Indeed we must remark that if, sometimes, we can use a smaller number of indeterminate coefficients to get a simpler solution, this is usually done at the price of losing part of the information contained in the original question. We will give examples later.

When one particular system of equations is dealt with analytically and is underdetermined, we can determine the value of the superfluous coefficients by imposing arbitrary additional conditions. But if we assume them to be equal to zero for the sake of convenience, this assumption can help extract certain insightful patterns from the general expression of the coefficients. We will clarify this later.

4. It is not sufficient to ensure, via the methods given in the first book, that we have the lowest possible degree of the final equation when considering general and unrelated coefficients; it is also important to take advantage of specific relations existing among the coefficients that could lead to further lowering the degree of the final equation. Indeed, failure to do so would again lead to superfluous solutions to the problem.

The only known method that does not yield useless roots, the Method of Successive Eliminations, works only in the case of two equations in two unknowns, and still suffers some exceptions. This remark never seems to have occurred to any Analyst. This method avoids giving a degree higher than that arising from two generic equations of given degrees; however, it gives no idea about the symptoms by which we can recognize whether specific relations among coefficients can yield a final equation with lower degree. The resulting degree given by this method is independent of the specific values taken by the equations.

Our goal is to show how it is systematically possible to find the final equation with the lowest degree, whose roots are true solutions to the system of equations.

Finally, after having presented the most general method to solve the elimination problem, we will introduce how we can resolve it in the simplest possible way, that is, using the smallest possible number of coefficients, when we do not want to bother considering the specific relations among the coefficients that could lead to lowering the degree of the final equation. Since this and other knowledge rely on the Theory of elimination in first-degree equations, we will begin with perfecting this method to our desired level of satisfaction.

A new elimination method for first-order equations with an arbitrary number of unknowns

(195.) As analysis is making progress, work has focused on more complex problems, and it was quickly realized that all methods to solve equations did not share the same efficiency, although they all lead to the same result. Among those who worked towards reducing the complexity of these methods, Mr. Cramer gave first, in his analysis of curves, a rule to obtain the value of each unknown in this kind of equation, devoid of any superfluous quantity.

I then gave in the 1764 *Memoirs of the Academy of Sciences,* a rule that appeared to be much more practical, since, so to speak, it requires no more attention that that required to write a letter.

Since then, Mr. Vandermonde and Mr. de la Place have provided, in the second volume of the 1772 *Memoirs of the Academy,* new methods to easily construct the elimination formulas specific to this kind of equation.

When faced with the problem of applying these methods to the general elimination problem, I figured that all these methods were still quite insufficient from a practical standpoint.

(196.) For as long as the proposed equations contain all possible terms, these methods do not yield any superfluous term. But they do not take advantage of missing terms in the equations. The formulas always are exhaustive, and they must be written entirely before a long comparison process can yield simplifications due to missing terms. The formulas must be built for generic instances of the equations, and therefore we must perform the same amount of work as if these equations were generic.

But in our general problem of elimination, unknowns never appear all simultaneously in all equations. And since the number of these unknowns is very high, we see that general elimination formulas could yield much more superfluous work than useful work: We dare to say that this work would soon become impractical.

(197.) Instead of aiming at giving general elimination formulas for first-degree equations, we propose to give a rule that applies to general equations, but also takes advantage of simplifications offered by specific equations: This rule would proceed the same way for both, but would perform only the necessary computations to get the value of the unknowns. This rule applies indifferently to numerical or formal equations, without requiring any new formula. Unless I am mistaken, here is this rule.

General rule to compute the values of the unknowns, altogether or separately, in first-order equations, whether these equations are symbolic or numerical

(198.) Let u, x, y, z, etc. be the unknowns, whose number is assumed to be n; assume n is also the number of equations.

Let a, b, c, d, etc. be the respective coefficients of the unknowns in the first equation.

Let a', b', c', d', etc. be the coefficients of the same unknowns in the second equation.

Let a'', b'', c'', d'', etc. be the coefficients of the same unknowns in the third equation, and so on.

Let us assume that the constant term in each equation is also the factor of an unknown denoted t.

Form the product $uxyzt$ of all unknowns written in any order, but retain this order afterwards until the end of the operation.

Replace successively each unknown against its coefficient in the first equation, making sure to introduce a sign change for each even replacement:[4] This result is what we will describe as a *first line*.

In this *first line*, replace each unknown by its coefficient in the second equation, making sure to the flip sign for each even replacement. This yields a *second line*.

Replace each unknown by its coefficient in the third equation in this *second line* making sure again to flip signs for each even replacement. This constitutes the *third line*.

Continue the process until the last equation, included. The last *line* contains the value of the unknowns as follows.

Each unknown takes its value as the fraction whose numerator is the coefficient of that same unknown in the last or nth line and whose denominator is always the coefficient of the unknown t in the same nth line.

(199.) For example, consider the two equations

$$ax + by + c = 0,$$
$$a'x + b'y + c' = 0.$$

We ask for the value of x and y.

I introduce the unknown t in these two equations as follows:

$$ax + by + ct = 0,$$
$$a'x + b'y + c't = 0,$$

and I write the product xyt. In this product I replace x by a, then y by b, then t by c. Making sure to flip the sign of y, I obtain the first line:

$$ayt - bxt + cxy.$$

In this first line, I change x into a', y into b', t into c', and, observing the sign change rule, I obtain the following second line:

$$ab't - ac'y - a'bt + bc'x + a'cy - b'cx$$
$$\text{or } (ab' - a'b)t - (ac' - a'c)y + (bc' - b'c)x.$$

Therefore, by (198) I conclude that $x = \frac{bc' - b'c}{ab' - a'b}$, $y = \frac{-(ac' - a'c)}{ab' - a'b}$.

(200.) Consider the following three equations:

$$ax + by + cz + d = 0,$$
$$a'x + b'y + c'z + d' = 0,$$
$$a''x + b''y + c''z + d'' = 0.$$

I rewrite them as

$$ax + by + cz + dt = 0,$$
$$a'x + b'y + c'z + d't = 0,$$
$$a''x + b''y + c''z + d''t = 0.$$

---

[4]We assume all coefficients are preceded by the + sign. In the converse case, the rule given here still works by flipping signs.

I write the product $xyzt$.

I change x into a, y into b, z into c, t into d, and observing the sign rule I obtain the first line:

$$ayzt - bxzt + cxyt - dxyz.$$

I replace x by a', y by b', z by c', t by d', and observing the sign rule I obtain the second line:

$$(ab' - a'b)zt - (ac' - a'c)yt + (ad' - a'd)yz + (bc' - b'c)xt$$
$$- (bd' - b'd)xz + (cd' - c'd)xy.$$

I successively replace x by a'', y by b'', z by c'', t by d'', and observing the sign rule, I obtain the third line

$$[(ab' - a'b)c'' - (ac' - a'c)b'' + (bc' - b'c)a'']t$$
$$- [(ab' - a'b)d'' - (ad' - a'd)b'' + (bd' - b'd)a'']z$$
$$+ [(ac' - a'c)d' - (ad' - a'd)c'' + (cd' - c'd)a'']y$$
$$- [(bc' - b'c)d'' - (bd' - b'd)c'' + (cd' - c'd)b'']x,$$

from which I conclude by (198)

$$x = \frac{-[(bc' - b'c)d'' - (bd' - b'd)c'' + (cd' - c'd)b'']}{(ab' - a'b)c'' - (ac' - a'c)b'' + (bc' - b'c)a''},$$

$$y = \frac{+[(ac' - a'c)d'' - (ad' - a'd)c'' + (cd' - c'd)a'']}{(ab' - a'b)c'' - (ac' - a'c)b'' + (bc' - b'c)a''},$$

$$z = \frac{-[(ab' - a'b)d'' - (ad' - a'd)b'' + (bd' - b'd)a'']}{(ab' - a'b)c'' - (ac' - a'c)b'' + (bc' - b'c)a''}.$$

(201.) If we only wanted to obtain the value of x, for example, we would omit in the calculation all terms where neither x nor t are present.

This observation will be very useful to us later, because within the large number of indeterminate coefficients, we will only need to know a few of them.

(202.) If our object were to simply construct general elimination formulas, not only would it be sufficient to compute the value of a single unknown, but we would only need to have this value in the last line, because we know that the denominator can be easily extracted from the numerator, and the value of the numerator for any unknown can easily be extracted from the value of one of them.

Thus in the previous example, if I only wanted to obtain the value of x, I would need to compute the value of $xyzt$, ignoring all terms where x is absent. A little attention shows this is equivalent to computing the value of yzt.

We would get as follows

First line $bzt - cyt + dyz$

Second line $(bc' - b'c)t - (bd' - b'd)z + (cd' - c'd)y$

Third line $(bc' - b'c)d'' - (bd' - b'd)c'' + (cd' - c'd)b''.$

This is the numerator of the value of x, making sure to change all signs, because when computing yzt only, we must not forget that y was originally on an even location in $xyzt$.

(203.) Let us follow up by showing that our rule applies uniformly to the case when not all unknowns enter all equations, and also applies to numerical equations.

Assume the three following equations:

$$
\begin{aligned}
au + bx + e &= 0, \\
a'u + c'y + e' &= 0, \\
b''x + c''y + e'' &= 0.
\end{aligned}
$$

I therefore compute the value of $uxyz$ from (198) by introducing the new unknown z, and I get

First line $\quad axyz - buyz - euxy$

Second line $\quad -ac'xz + ae'xy - a'byz + bc'uz - be'uy - a'exy - ec'ux,$
$\quad\quad\quad\quad$ or $-ac'xz + (ae' - a'e)xy - a'byz + bc'uz - be'uy - ec'ux$

Third line $\quad -ac'b''z + ac'e''x + (ae' - a'e)b''y - (ae' - a'e)c''x$
$\quad\quad\quad\quad -a'bc''z + a'be''y - bc'e''u + be'c''u + b''ec'u$
$\quad\quad\quad\quad$ or $-(ac'b'' + a'bc'')z + [(ae' - a'e)b'' + a'be'']y$
$\quad\quad\quad\quad -[(ae' - a'e)c'' - ac'e'']x + [(e'c'' - e''c')b + b''c'e]u,$

from which we get

$$
u = \frac{-[(e'c'' - e''c')b + b''c'e]}{ac'b'' + a'bc''},
$$

$$
x = \frac{(ae' - a'e)c'' - ac'e''}{ac'b'' + a'bc''},
$$

$$
y = \frac{-[(ae' - a'e)b'' + a'be'']}{ac'b'' + a'bc''}.
$$

(204.) Consider the four following equations to illustrate an application to numerical equations:

$$
\begin{aligned}
2u + 3x - 8 &= 0, \\
3u + 2y - 9 &= 0, \\
4x + 3z - 20 &= 0, \\
2y + z - 10 &= 0.
\end{aligned}
$$

Having formed the product $uxyzt$ (198), I obtain

First line $\quad 2xyzt - 3uyzt - 8uxyz$

Second line $\quad -4xzt + 18xyz - 9yzt + 6uzt - 27uyz - 24xyz - 16uxz,$
$\quad\quad\quad\quad$ or $-4xzt - 6xyz - 9yzt + 6uzt - 27uyz - 16uxz$

Third line $\quad -16zt + 12xt + 80xz - 24yz - 18xy + 27yt$
$\quad\quad\quad\quad + 180yz - 18ut - 120uz - 81uy + 64uz - 48ux,$
or $\quad\quad\quad\quad -16zt + 12xt + 80xz + 156yz - 18xy + 27yt$
$\quad\quad\quad\quad - 18ut - 56uz - 81uy - 48ux$

Fourth line $\quad 38t + 152z + 114y + 76x + 38u.$

Thus from (198) we get $u = \frac{38}{38}$, $x = \frac{76}{38}$, $y = \frac{114}{38}$, $z = \frac{152}{38}$, that is, $u = 1$, $x = 2$, $y = 3$, $z = 4$.

(205.) If one of the lines becomes zero during the computation, it proves that the equation corresponding to that line is in fact not independent from the others and does not bring any new information to the problem. The number of equations is then not really equal to the number of unknowns, and this equation must be rejected.

For example, consider the three equations

$$
\begin{aligned}
2x + 3y + 5z + 6 &= 0, \\
3x + y + 2z + 5 &= 0, \\
10x + 8y + 14z + 22 &= 0.
\end{aligned}
$$

I get

First line	$2yzt - 3xzt + 5xyt - 6xyz$
Second line	$-7zt + 11yt - 8yz + xt - 9xz + 13xy$
Third line	$-98t + 154z + 88t - 242y - 64z + 112y + 10t$
	$- 22x - 90z + 126x + 130y - 104x,$

that is, zero.

Thus the third equation does not express more information than the other two: The problem is underdetermined, and expressed by the first two equations only.

(206.) If during the computations or at their end, one or more unknowns disappear, such that they are not present in the last line, then we must conclude that these unknowns are zero.

For example, consider the three equations

$$
\begin{aligned}
2x + 4y + 5z - 22 &= 0, \\
3x + 5y + 2z - 30 &= 0, \\
5x + 6y + 4z - 43 &= 0.
\end{aligned}
$$

I get

First line	$2yzt - 4xzt + 5xyt + 22xyz$
Second line	$-2zt + 11yt + 6yz - 17xt + 10xz - 106xy$
Third line	$-27t - 81y - 135x.$

From (198) we therefore get $x = \frac{-135}{-27}$, $y = \frac{-81}{-27}$, $z = \frac{0}{-27}$, that is, $x = 5$, $y = 3$, $z = 0$.

(207.) We can also extract a useful consequence from the rule we just gave (198), which we must not forget.

After computing the values of the unknowns, assume we must substitute them into an arbitrary expression where these unknowns do not exceed the first degree: We have an equivalent of this substitution by adding a new *line*, as if the quantity was a new equation, and by dividing this new line by the coefficient of the unknown t taken in the preceding line.

For example, if we ask for the value of the quantity $a''x + b''y + c''$ assuming the following two equations

$$ax + by + c = 0,$$
$$a'x + b'y + c' = 0.$$

I write the product xyt and then

First line $ayt - bxt + cxy$

Second line $(ab' - a'b)t - (ac' - a'c)y + (bc' - b'c)x$

Third line $(ab' - a'b)c'' - (ac' - a'c)b'' + (bc' - b'c)a''$.

Thus the result of the substitution is

$$\frac{(ab' - a'b)c'' - (ac' - a'c)b'' + (bc' - b'c)a''}{ab' - a'b}.$$

We will see later how we must proceed to compute the result of the same substitution in a quantity where the unknowns exceed the first degree.

(208.) Since the first-degree equations we will use later do not contain any known term, and since they contain one more unknown than there are equations, we will make a few remarks specific to these cases.

(209.) If we have an arbitrary number of first-degree equations with unspecified terms, and if the number of unknowns is one less than the number of equations, then the value of one of these unknowns is arbitrary. Moreover, the value of all the other unknowns is proportional to the value of that arbitrary value.

For example, consider the two equations

$$ax + by + cz = 0,$$
$$a'x + b'y + c'z = 0,$$

and assume we have given z an arbitray value, say 1; we then compute the corresponding values of x and y; to get the other values of x and y corresponding to any other value of z, we only have to multiply the value of x and y corresponding to $z = 1$ by the new value of z. Thus for $z = 1$, we get $x = \frac{bc' - b'c}{ab' - a'b}$, $y = \frac{-(ac' - a'c)}{ab' - a'b}$; therefore for $z = m$, we have $x = \frac{(bc' - b'c)}{ab' - a'b}.m$, $y = \frac{-(ac' - a'c)}{ab' - a'b}.m$.

(210.) From there it is easy to conclude that we can compute the value of the unknowns in such equations the following way.

Perform the computations as described in (198), considering one of the unknowns as if it had been introduced according to (198). Assign to the value of each unknown its coefficient in the last computed line. It will be one of the values of each of the unknowns. If we want other values, we multiply the value of each unknown that was just found by the same arbitrary number.

Thus we can compute all values of x, y and z in the two equations

$$ax + by + cz = 0,$$
$$a'x + b'y + c'z = 0.$$

I build the product xyz and then

First line $ayz - bxz + cxy$

Second line $(ab' - a'b)z - (ac' - a'c)y + (bc' - b'c)x.$

From these computations I conclude

$$z = ab' - a'b, \; y = -(ac' - a'c), \; x = bc' - b'c.$$

To obtain all the other corresponding values for x, y, z, I write $z = (ab' - a'b).m$, $y = -(ac' - a'c).m$, $x = (bc' - b'c).m$, where m is an arbitrary, positive or negative, integer or fractional number.

(211.) But since we will only need one arbitrary value later for each unknown, we will stop our effort and use the values arising immediately from what we will call the last line.

(212.) From this and what was said in (207), we can conclude that if we have as many equations (with no specifically known terms) as we have unknowns, the last line will also be the condition for which all these equations can hold together.

For example, if we have the three equations

$$\begin{aligned} ax + by + cz &= 0, \\ a'x + b'y + c'z &= 0, \\ a''x + b''y + c''z &= 0, \end{aligned}$$

I write

First line $ayz - bxz + cxy$

Second line $(ab' - a'b)z - (ac' - a'c)y + (bc' - b'c)x$

Third line $(ab' - a'b)c'' - (ac' - a'c)b'' + (bc' - b'c)a''.$

The condition equation is therefore

$$(ab' - a'b)c'' - (ac' - a'c)b'' + (bc' - b'c)a'' = 0.$$

Let us conclude with a remark that will be very useful later on, although it is quite simple.

(213.) The three above equations occur simultaneously if the condition equation is satisfied; but they can also occur in another case, when

$$x = 0, \; y = 0, \; z = 0.$$

This obvious solution is also a consequence of what we said in (206).

(214.) The general rule we just gave to compute the unknowns in first-order equations can still be improved upon, but we will talk about this when we have to deal with equations that require these improvements. We will then be able to present them more clearly.

A method to find functions of an arbitrary number of unknowns which are identically zero

(215.) When the number of quantities entering a computation is somewhat large, we know we do not want to expand all products that yield the result; rather, we group terms undergoing the same computations together as much as possible, and we only indicate the computations to be made. That method indeed simplifies computations, but prevents us from locating the terms that may cancel each other in the final result. Our goal here is to keep the advantages of such a method, without the drawback we just mentioned.

In a *Memoir* about elimination that was published in the 1764 *Memoirs of the Academy*, we have used such functions, but these functions were easily found: Since we had to consider no other sort of equations, we did not spend time to devise a method to handle more complex cases.

When we wanted to apply our elimination method, we found results we knew could be simplified, but without the help of the functions we are about to introduce, we had no other way than to expand all expressions to reach such a reduction. We found this job to be unrewarding. We therefore must make these types of functions known here, as well as the means to find them. The analyst may find these functions to be useful as well.

(216.) Let us assume we have n first-order equations in $n + 1$ unknowns. We also assume that no equation parameter is exactly known.

Let us imagine that we augment the number of equations by duplicating one of them; what we have called the last line in (198) will be not only be the condition for these $n + 1$ equations to have a solution, but also by (205) this condition will always be true; thus this function of the coefficients of the system will itself always be equal to zero.

This is a very easy way to find $n + 1$ functions of $n + 1$ parameters which are identically equal to zero.

(217.) For example, consider the two equations

$$ax + by + cz = 0,$$
$$a'x + b'y + c'z = 0.$$

Let us add a copy of the first equation to this system of two equations; that is, let us perform as if the system

$$ax + by + cz = 0,$$
$$a'x + b'y + c'z = 0,$$
$$ax + by + cz = 0$$

consisted of three different equations whose condition equation is what we are looking for. We write

First line	$ayz - bxz + cxy$
Second line	$(ab' - a'b)z - (ac' - a'c)y + (bc' - b'c)x$
Third line	$(ab' - a'b)c - (ac' - a'c)b + (bc' - b'c)a.$

Thus

$$(ab' - a'b)c - (ac' - a'c)b + (bc' - b'c)a = 0$$

is the condition equation.

But it is clear the third equation does not express anything different from the first one. Thus the latter quantity must equal zero. Therefore, given the series

$$a, \ b, \ c,$$
$$a', \ b', \ c',$$

we always have

$$(ab' - a'b)c - (ac' - a'c)b + (bc' - b'c)a = 0.$$

If we had repeated the second equation instead of the first, we would have obtained

$$(ab' - a'b)c' - (ac' - a'c)b' + (bc' - b'c)a' = 0.$$

(218.) Likewise, consider the three equations

$$
\begin{aligned}
ax + by + cz + dt &= 0, \\
a'x + b'y + c'z + d't &= 0, \\
a''x + b''y + c''z + d''t &= 0,
\end{aligned}
$$

and adjoin to them the equation

$$ax + by + cz + dt = 0.$$

We get

First line	$ayzt - bxzt + cxyt - dxyz$
Second line	$(ab' - a'b)zt - (ac' - a'c)yt + (ad' - a'd)yz$
	$+(bc' - b'c)xt - (bd' - b'd)xz + (cd' - c'd)xy$
Third line	$[(ab' - a'b)c'' - (ac' - a'c)b'' + (bc' - b'c)a'']t$
	$-[(ab' - a'b)d'' - (ad' - a'd)b'' + (bd' - b'd)a'')z$
	$+[(ac' - a'c)d'' - (ad' - a'd)c'' + (cd' - c'd)a'']y$
	$-[(bc' - b'c)d'' - (bd' - b'd)c'' + (cd' - c'd)b'']x.$

Finally, the fourth line will also be the (always true) condition equation

$$\left. \begin{aligned}
&[(ab' - a'b)c'' - (ac' - a'c)b'' + (bc' - b'c)a'']d \\
&-[(ab' - a'b)d'' - (ad' - a'd)b'' + (bd' - b'd)a'']c \\
&+[(ac' - a'c)d'' - (ad' - a'd)c'' + (cd' - c'd)a'']b \\
&-[(bc' - b'c)d'' - (bd' - b'd)c'' + (cd' - c'd)b'']a
\end{aligned} \right\} = 0.$$

Thus, given the series

$$
\begin{aligned}
&a, \ b, \ c, \ d, \\
&a', \ b', \ c', \ d', \\
&a'', \ b'', \ c'', \ d'',
\end{aligned}
$$

the three following functions of these twelve quantities are identically zero

$$\left.\begin{array}{l} [(ab' - a'b)c'' - (ac' - a'c)b'' + (bc' - b'c)a'']d \\ -[(ab' - a'b)d'' - (ad' - a'd)b'' + (bd' - b'd)a'']c \\ +[(ac' - a'c)d'' - (ad' - a'd)c'' + (cd' - c'd)a'']b \\ -[(bc' - b'c)d'' - (bd' - b'd)c'' + (cd' - c'd)b'']a \end{array}\right\} = 0,$$

$$\left.\begin{array}{l} [(ab' - a'b)c'' - (ac' - a'c)b'' + (bc' - b'c)a'']d' \\ -[(ab' - a'b)d'' - (ad' - a'd)b'' + (bd' - b'd)a'']c' \\ +[(ac' - a'c)d'' - (ad' - a'd)c'' + (cd' - c'd)a'']b' \\ -[(bc' - b'c)d'' - (bd' - b'd)c'' + (cd' - c'd)b'']a' \end{array}\right\} = 0,$$

$$\left.\begin{array}{l} [(ab' - a'b)c'' - (ac' - a'c)b'' + (bc' - b'c)a'']d'' \\ -[(ab' - a'b)d'' - (ad' - a'd)b'' + (bd' - b'd)a'']c'' \\ +[(ac' - a'c)d'' - (ad' - a'd)c'' + (cd' - c'd)a'']b'' \\ -[(bc' - b'c)d'' - (bd' - b'd)c'' + (cd' - c'd)b'']a'' \end{array}\right\} = 0.$$

It is easy but lengthy to extend these theorems. These are not the only ones we need to know.

(219.) Let us now consider the two series

$$a, \ b, \ c, \ d, \ e, \ f, \ \text{etc.}$$
$$a', \ b', \ c', \ d', \ e', \ f', \ \text{etc.}$$

We see that if we take combinations of any three of these variables, we get the following set of identities:

$$(ab' - a'b)c - (ac' - a'c)b + (bc' - b'c)a = 0,$$
$$(ab' - a'b)c' - (ac' - a'c)b' + (bc' - b'c)a' = 0,$$
$$(ab' - a'b)d - (ad' - a'd)b + (bd' - b'd)a = 0,$$
$$(ab' - a'b)d' - (ad' - a'd)b' + (bd' - b'd)a' = 0,$$
$$(ab' - a'b)e - (ae' - a'e)b + (be' - b'e)a = 0,$$
$$(ab' - a'b)e' - (ae' - a'e)b' + (be' - b'e)a' = 0,$$
etc.;
$$(bc' - b'c)d - (bd' - b'd)c + (cd' - c'd)b = 0,$$
$$(bc' - b'c)d' - (bd' - b'd)c' + (cd' - c'd)b' = 0,$$
$$(bc' - b'c)e - (be' - b'e)c + (ce' - c'e)b = 0,$$
$$(bc' - b'c)e' - (be' - b'e)c' + (ce' - c'e)b' = 0,$$
$$(bc' - b'c)f - (bf' - b'f)c + (cf' - c'f)b = 0,$$
$$(bc' - b'c)f' - (bf' - b'f)c' + (cf' - c'f)b' = 0,$$

and so on.

Let us now take any of these two equations, say the first two:

$$(ab' - a'b)c - (ac' - a'c)b + (bc' - b'c)a = 0,$$
$$(ab' - a'b)c' - (ac' - a'c)b' + (bc' - b'c)a' = 0.$$

Multiply the first equation by d' and the second by d. Substracting the second product from the first, we get

$$(ab' - a'b).(cd' - c'd) - (ac' - a'c).(bd' - b'd) + (bc' - b'c).(ad' - a'd) = 0.$$

(220.) Therefore, given the series

$$a, \ b, \ c, \ d,$$
$$a', \ b', \ c', \ d',$$

the identity

$$(ab' - a'b).(cd' - c'd) - (ac' - a'c).(bd' - b'd) + (bc' - b'c).(ad' - a'd) = 0$$

is always true.

Consider the two series

$$a, \ b, \ c, \ d, \ e, \ f, \ \text{etc.},$$
$$a', \ b', \ c', \ d', \ e', \ f', \ \text{etc.}$$

Taking combinations of any four of these variables, we easily get many functions of four variables which are identically equal to zero through this process.

(221.) Consider now three series:

$$a, \ b, \ c, \ d, \ e, \ f, \ \text{etc.},$$
$$a', \ b', \ c', \ d', \ e', \ f', \ \text{etc.},$$
$$a'', \ b'', \ c'', \ d'', \ e'', \ f'', \ \text{etc.}$$

According to (218), the three following identities will, for example, hold for the four quantities a, b, c, d:

$$\left.\begin{array}{l} [(ab' - a'b)c'' - (ac' - a'c)b'' + (bc' - b'c)a'']d \\ -[(ab' - a'b)d'' - (ad' - a'd)b'' + (bd' - b'd)a'']c \\ +[(ac' - a'c)d'' - (ad' - a'd)c'' + (cd' - c'd)a'']b \\ -[(bc' - b'c)d'' - (bd' - b'd)c'' + (cd' - c'd)b'']a \end{array}\right\} = 0,$$

$$\left.\begin{array}{l} [(ab' - a'b)c'' - (ac' - a'c)b'' + (bc' - b'c)a'']d' \\ -[(ab' - a'b)d'' - (ad' - a'd)b'' + (bd' - b'd)a'']c' \\ +[(ac' - a'c)d'' - (ad' - a'd)c'' + (cd' - c'd)a'']b' \\ -[(bc' - b'c)d'' - (bd' - b'd)c'' + (cd' - c'd)b'']a' \end{array}\right\} = 0,$$

$$\left.\begin{array}{l} [(ab' - a'b)c'' - (ac' - a'c)b'' + (bc' - b'c)a'']d'' \\ -[(ab' - a'b)d'' - (ad' - a'd)b'' + (bd' - b'd)a'']c'' \\ +[(ac' - a'c)d'' - (ad' - a'd)c'' + (cd' - c'd)a'']b'' \\ -[(bc' - b'c)d'' - (bd' - b'd)c'' + (cd' - c'd)b'']a'' \end{array}\right\} = 0.$$

Assume we multiply the first equation by e', the second by e, and we subtract the second product from the first.

Likewise, assume we multiply the first equation by e'' and the third by e, and that we subtract the second product from the first.

Finally, assume we multiply the second equation by e'' and the third equation by e' and that we subtract the second product from the first.

Then we get the three following identities:

$$\left.\begin{array}{l}
[(ab' - a'b)c'' - (ac' - a'c)b'' + (bc' - b'c)a''].(de' - d'e) \\
-[(ab' - a'b)d'' - (ad' - a'd)b'' + (bd' - b'd)a''].(ce' - c'e) \\
+[(ac' - a'c)d'' - (ad' - a'd)c'' + (cd' - c'd)a''].(be' - b'e) \\
-[(bc' - b'c)d'' - (bd' - b'd)c'' + (cd' - c'd)b''].(ae' - a'e)
\end{array}\right\} = 0,$$

$$\left.\begin{array}{l}
[(ab' - a'b)c'' - (ac' - a'c)b'' + (bc' - b'c)a''].(de'' - d''e) \\
-[(ab' - a'b)d'' - (ad' - a'd)b'' + (bd' - b'd)a''].(ce'' - c''e) \\
+[(ac' - a'c)d'' - (ad' - a'd)c'' + (cd' - c'd)a''].(be'' - b''e) \\
-[(bc' - b'c)d'' - (bd' - b'd)c'' + (cd' - c'd)b''].(ae'' - a''e)
\end{array}\right\} = 0,$$

$$\left.\begin{array}{l}
[(ab' - a'b)c'' - (ac' - a'c)b'' + (bc' - b'c)a''].(d'e'' - d''e') \\
-[(ab' - a'b)d'' - (ad' - a'd)b'' + (bd' - b'd)a''].(c'e'' - c''e') \\
+[(ac' - a'c)d'' - (ad' - a'd)c'' + (cd' - c'd)a''].(b'e'' - b''e') \\
-[(bc' - b'c)d'' - (bd' - b'd)c'' + (cd' - c'd)b''].(a'e'' - a''e')
\end{array}\right\} = 0.$$

We therefore see how we can find functions of five variables that are identically zero by combining any five terms from the three series.

(222.) Let us assume that, given the last three equations, we multiply the first one by f'', the second one by f', the third by f, and we then add the first and last equations and then subtract the second one from the sum. We get

$$\begin{array}{l}
[(ab' - a'b)c'' - (ac' - a'c)b'' + (bc' - b'c)a''] \\
\times [(de' - d'e)f'' - (de'' - d''e)f' + (d'e'' - d''e')f] \\
-[(ab' - a'b)d'' - (ad' - a'd)b'' + (bd' - b'd)a''] \\
\times [(ce' - c'e)f'' - (ce'' - c''e)f' + (c'e'' - c''e')f] \\
+[(ac' - a'c)d'' - (ad' - a'd)c'' + (cd' - c'd)a''] \\
\times [(be' - b'e)f'' - (be'' - b''e)f' + (b'e'' - b''e')f] \\
-[(bc' - b'c)d'' - (bd' - b'd)c'' + (cd' - c'd)b''] \\
\times [(ae' - a'e)f'' - (ae'' - a''e)f' + (a'e'' - a''e')f] \\
= 0.
\end{array}$$

We now see how we can find functions of six parameters that are identically zero, by taking combinations of any six terms.

Let us remark that the quantity $(ab' - a'b)c'' - (ac' - a'c)b'' + (bc' - b'c)a''$ can also be written as

$$(ab' - a'b)c'' - (ab'' - a''b)c' + (a'b'' - a''b')c,$$

such that we can write the preceding equation more elegantly as

$$\begin{array}{l}
[(ab' - a'b)c'' - (ab'' - a''b)c' + (a'b'' - a''b')c] \\
\times [(de' - d'e)f'' - (de'' - d''e)f' + (d'e'' - d''e')f] \\
-[(ab' - a'b)d'' - (ab'' - a''b)d' + (a'b'' - a''b')d] \\
\times [(ce' - c'e)f'' - (ce'' - c''e)f' + (c'e'' - c''e')f] \\
+[(ac' - a'c)d'' - (ac'' - a''b)d' + (a'c'' - a''c')d] \\
\times [(be' - b'e)f'' - (be'' - b''e)f' + (b'e'' - b''e')f] \\
-[(bc' - b'c)d'' - (bc'' - b''c)d' + (b'c'' - b''c')d] \\
\times [(ae' - a'e)f'' - (ae'' - a''e)f' + (a'e'' - a''e')f] \\
= 0.
\end{array}$$

(223.) This is enough to show how to find this kind of theorem. We see there is an infinity of other possible combinations leading to new functions identically equal to zero; these are easy to find.

About the form of the polynomial multiplier, or the polynomial multipliers, leading to the final equation

(224.) In the first book, we have seen that our approach to elimination consists of multiplying one of the equations by a polynomial. In this polynomial, we have eliminated all terms that may be cancelled using the other equations. Likewise, we have eliminated all terms of the product equation using the other equations. The number of remaining coefficients in the polynomial multiplier must be sufficient to cancel all remaining terms containing all unknowns but the one that was chosen to remain in the final equation.

From now on we consider an elimination method that appears to differ from this one, but which is in fact the same.

Let us assume that we multiply each of the given equations by its own polynomial multiplier and that we add all these products together. The result will be what we call the *sum equation*, which is to become the final equation through the cancellation of all terms corresponding to the unknowns to be eliminated.

We therefore have to:

1. determine the form of each of these polynomial multipliers;

2. determine the number of coefficients in each of these polynomials, which cannot be useful during the elimination process;

3. show whether the terms that must or may be rejected in each polynomial multiplier can be chosen arbitrarily;

4. determine the best use for these coefficients, if we do not need to reject them.

We first examine the first question.

(225.) The form of each multiplier is relatively easy to extract from all that was said in the first book. But the way we have considered this problem in (174) makes matters easier to explain, and this is how we will proceed.

(226.) We will assume that all equations share the same order and are incomplete; indeed, incomplete equations of lesser order are only specific cases of higher order, incomplete equations, such that we may assume they all have the same order as the highest order equation.

We will also assume these equations are incomplete for a similar reason: Complete equations are a subset of incomplete equations.

(227.) This being said, let us assume that, after having selected an incomplete, arbitrary first-order polynomial, we multiply it by one of the equations; that we multiply this product by the second equation; that we multiply this product again by the third equation; and so on. The final product is used to find the polynomial multipliers for each equation, as follows.

Assume we want to get the form of the polynomial multipier of the first equation: Eliminate all varieties of exponents of the final product related to the first equation, and you will get the form of its polynomial multiplier.

Likewise, assume we want to get the form of the polynomial multipier of the second equation: Eliminate all varieties of exponents of the final product related to the second equation, and you will get the form of its polynomial multiplier, and so on.

Let us clarify our statement with some examples.

(228.) Consider, for example, the two equations

$$(x^a, \overset{a}{y'})^b = 0; \quad (x^{a'}, \overset{a'}{y'})^{b'} = 0.$$

What we call the final product is

$$(x^{A+a+a'}, \overset{A+a+a'}{y'})^{B+b+b'},$$

as may be easily seen.

Thus, by suppressing all quantities related to the first equation from the varieties of the exponents $A + a + a'$, $\underset{'}{A} + \underset{'}{a} + \underset{'}{a'}$, $B + b + b'$ on the one hand and suppressing all quantites related to the second equation on the other hand, we get $(x^{A+a'}, \overset{A+a'}{y'})^{B+b'}$ as the form of the polynomial multiplier for the first equation and $(x^{A+a}, \overset{A+a}{y'})^{B+b}$ as the form of the polynomial multiplier for the second equation.

Consider now the three equations

$$[(x^a, \overset{a}{y'})^b, (x^a, \overset{a}{z''})^{\overset{b}{'}}, (\overset{a}{y'}, \overset{b}{z''})^{\overset{b}{''}}]^c = 0,$$
$$[(x^{a'}, \overset{a'}{y'})^{b'}, (x^{a'}, \overset{a'}{z''})^{\overset{b'}{'}}, (\overset{a'}{y'}, \overset{b'}{z''})^{\overset{b'}{''}}]^{c'} = 0,$$
$$[(x^{a''}, \overset{a''}{y'})^{b''}, (x^{a''}, \overset{a''}{z''})^{\overset{b''}{'}}, (\overset{a''}{y'}, \overset{b''}{z''})^{\overset{b''}{''}}]^{c''} = 0,$$

as seen in (82).

The form of the final product is

$$[(x^{A+a+a'+a''}, \overset{A+a+a'+a''}{y'})^{B+b+b'+b''},$$
$$(x^{A+a+a'+a''}, \overset{A+a+a'+a''}{z''})^{B+b+b'+b''},$$
$$(\overset{A+a+a'+a''}{y'}, \overset{A+a+a'+a''}{z''})^{B+b+b'+b''}]^{C+c+c'+c''}.$$

Suppressing successively all terms belonging to the first, second and third equations from the varieties of the exponents, we get

$$[(x^{A+a'+a''}, \overset{A+a'+a''}{y'})^{B+b'+b''},$$
$$(x^{A+a'+a''}, \overset{A+a'+a''}{z''})^{B+b'+b''},$$
$$(\overset{A+a'+a''}{y'}, \overset{A+a'+a''}{z''})^{B+b'+b''}]^{C+c'+c''}$$

as the form of the polynomial multiplier for the first equation,

$$[(x^{A+a+a''}, y'^{A+a+a''})^{B+b+b''},$$
$$(x^{A+a+a''}, z''^{A+a+a''})^{B+b+b''},$$
$$(y'^{A+a+a''}, z''^{A+a+a''})^{B+b+b''}]^{C+c+c''},$$

as the form of the polynomial multiplier for the second equation, and

$$[(x^{A+a+a'}, y'^{A+a+a'})^{B+b+b'}, (x^{A+a+a'}, z''^{A+a+a'})^{B+b+b'},$$
$$(y'^{A+a+a'}, z''^{A+a+a'})^{B+b+b'}]^{C+c+c'},$$

as the form of the multiplier polynomial for the third equation.

(229.) By the way, we do not always have to use a general, incomplete first-order polynomial in what we call the final product. It is enough that this polynomial contain the same varieties of exponents as the given equations, considered as incomplete and of first order.

Thus, when the equations are all incomplete of the form $(u^a, x'^a, y''^a \ldots n) = 0$, which is the simplest of all, we only have to use a polynomial of the form $(u^A, u'^A, y''^A \ldots n)^B$ to generate the final product.

About the requirement not to use all coefficients of the polynomial multipliers towards elimination

(230.) We have already said more than once that we must not consider all coefficients of the polynomial multipliers as useful towards elimination. In particular, what we have said in (45) about complete equations demonstrates our point. We think, however, that it is useful to reconsider this subject. It is all the more important that, if we used any one of the coefficients that must be rejected towards elimination, the resulting final equation would be wrong or would be trivial (e.g., all coefficients would disappear, thus bringing no new information).

Indeed, let us assume we have an arbitrary number of equations with the same number of unknowns. Multiply each equation by a polynomial where each term is undetermined. Add all these products and assume that the resulting sum, set equal to zero, is the final equation. To obtain this final equation, we must set each coefficient to zero in front of any variable or combination of variables different from that chosen to be in the final equation. Almost always, several coefficients will remain and their value is unconstrained. If these coefficients are used to destroy higher level terms in the hope of reducing the degree of the final equation, there will still be more coefficients than terms to be destroyed, such that the final equation could be reduced to be identically equal to zero for all values of the unknowns; such a conclusion is unacceptable.

Even if the number of such coefficients does not exceed the number of terms remaining after cancelling all terms that are allowed to be cancelled, they could still not be used to lower the degree of the final equation.

Indeed, intuition suggests that the final equation must feature a specific number of terms and corresponding degree, and that this degree cannot be arbitrarily lowered.

Finally, the absurdity of an artificially low-degree equation is probably best demonstrated by the fact that it is in no way representative of the original set of proposed equations.

Indeed, imagine that the equations under consideration are, instead, polynomials whose unknowns are unconstrained. Using the same polynomial multipliers, we can apply to these polynomials the same treatment as for the equations mentioned previously. But it is clear that the resulting polynomial would be completely arbitrary, and thus the corresponding final equation would enjoy only an arbitrary relation to the original set of equations. Assuming this final equation to hold would be completely arbitrary. Since this final equation does not reflect the individual conditions expressed by the equations, it could not be impregnated (so to speak) with the original conditions expressed by these equations.

For example, consider the two equations

$$
\begin{aligned}
ax^2 + bxy + cy^2 + dx + ey + f &= 0, \\
a'x^2 + b'xy + c'y^2 + d'x + e'y + f' &= 0.
\end{aligned}
$$

Assume we multiply the first equation by the polynomial

$$Ax^2 + Bxy + Cy^2 + Dx + Ey + F,$$

and the second equation by the polynomial

$$A'x^2 + B'xy + C'y^2 + D'x + E'y + F'.$$

Adding the two products together would yield an equation of the form $(x \ldots 2)^4 = 0$.

Using the second equation, it is always possible to cancel a corresponding term in the polynomial multiplier of the first equation; thus we really only have eleven coefficients that may be used for elimination and that can be used to cancel all ten terms containing y, for example.

But assume instead that we use all twelve coefficients offered by the two polynomial multipliers to cancel all terms containing y and the term x^4; we would then reach a non trivial third-degree equation. However, this equation would not be related in any way to the original problem, since it would not have captured the information contained in the original equations. The values of x arising from this equation would not satisfy the proposed equations: In one word, this "final equation" would in fact be purely arbitrary and not related to the original problem.

Thus many coefficients cannot be used for elimination purposes. We can guarantee that the resulting final equation truly reflects all the conditions set by the original equations only if we do not use them to eliminate terms in the sum equation.

About the number of coefficients in each polynomial multiplier which are useful for the purpose of elimination.

(231.) We have just indicated the importance of not using coefficients that may be cancelled by the proposed equations. We must, however, avoid concluding that we should omit them and that we cannot use them to facilitate the elimination process. On the contrary, we will see shortly that we can advantageously use them to simplify computations, by preserving or generating symmetries inherent to the equations. These would be masked by simply eliminating those coefficients. We must simply remember not to use any of these coefficients to destroy any terms in the sum equation, that is, the equation resulting from the sum of the products of the polynomials specific to each equation by their polynomial multipliers.

(232.) When considering only one product equation, as in the first book, we only needed to know the number of terms of the one polynomial multiplier under consideration then. But since we now use as many polynomial multipliers as there are equations, we must talk about the number of their coefficients that are unfit for elimination purposes. This is easily dealt with, using what we have said up until now.

(233.) Remembering what was said in the first book, we easily see that the number of useful coefficients in the first polynomial multiplier is always equal to the number of coefficients of this polynomial, minus the number of terms that may be cancelled using the $n - 1$ other equations, where n is the total number of equations.

Likewise, the number of useful coefficients of the second polynomial multiplier is the total number of coefficients, minus the number of terms that can be cancelled in this polynomial using the $n - 2$ remaining equations.

The number of useful coefficients in the third polynomial multiplier is the number of terms of that polynomial minus the number of terms that can be cancelled in this polynomial, using the remaining $n - 3$ equations. And so on until the last polynomial, whose number of useful coefficients is precisely the number of its terms.

We have also indicated how to compute the number of terms that may be cancelled in each of these polynomials using the corresponding number of equations. We now introduce a way to determine the form of the polynomials that represent this number of terms. From what was said in (227) about the form of the polynomial multipliers, we will satisfy ourselves with an example.

Assume three equations of the form

$$(u^a \ldots 3)^t = 0,$$
$$(u^{a'} \ldots 3)^{t'} = 0,$$
$$(u^{a''} \ldots 3)^{t''} = 0.$$

The polynomial multiplier of the first equation (227) is

$$(u^{A+a'+a''} \ldots 3)^{T+t'+t''};$$

the polynomial multiplier of the second equation is

$$(u^{A+a+a''} \ldots 3)^{T+t+t''};$$

and that of the third is

$$(u^{A+a+a'} \dots 3)^{T+t+t'}.$$

The number of terms that may be cancelled in the first polynomial, using the second equation, is $N(u^{A+a''} \dots 3)^{T+t''}$. However, the third equation can be used to cancel $N(u^A \dots 3)^T$ terms in the polynomial $(u^{A+a''} \dots 3)^{T+t''}$; thus the true number of terms that can be cancelled in the first multiplier polynomial using the second equation is only

$$N(u^{A+a''} \dots 3)^{T+t''} - N(u^A \dots 3)^T.$$

The number of terms that may be cancelled in that same polynomial using the third equation is $N(u^{A+a'} \dots 3)^{T+t'}$.

Thus, using the second and third equations, the total number of terms that may be cancelled in the first polynomial multiplier is

$$N(u^{A+a''} \dots 3)^{T+t''} - N(u^A \dots 3)^T + N(u^{A+a'} \dots 3)^{T+t'}.$$

Concerning the second polynomial multiplier, we easily observe that only the third equation may be used to cancel terms. Thus the number of terms that may be cancelled is $N(u^{A+a} \dots 3)^{T+t}$.

Thus we see how we can determine the form of the polynomials whose number of terms expresses the number of terms that may be cancelled in each of the polynomial multipliers that are used for elimination purposes.

About the terms that may or must be excluded in each polynomial multiplier

(234.) We have spent enough effort proving that not all coefficients of the polynomial multipliers are admissible for the purpose of elimination, and we have determined the number of these that must be rejected.

But it is not enough to know how many terms must be excluded. We must also know which terms must be excluded, or at least know whether a choice must be made; that is, we must determine whether this exclusion applies to certain terms rather than others.

Although there is considerable freedom in choosing these terms, this freedom is nevertherless not unlimited.

(235.) When, given a polynomial, we cancel as many terms as possible using a given number of equations, we express all conditions of the question stored in these equations into the polynomial.

But the expression of these conditions does not depend upon one particular term of these equations, but rather all these terms; thus it is easy to see that there is no justification for cancelling certain terms rather than others.

For example, we have relied upon this idea when dealing with complete equations in (45): We have assumed we cancel all terms that may be divided by $y^{t'}$, all terms that may be divided by $z^{t''}$, etc. We limited ourselves to this idea because it was sufficient at that time. We also considered that this idea was sufficient and most appropriate to show our concepts. However, it would be wrong to consider that specific powers of single variables or products of variables must be eliminated rather than others. Any terms can be cancelled, provided that the following observations are accounted for.

The number of terms that may be cancelled in the polynomial is limited. Moreover, the number of terms that may be cancelled in each dimension of the polynomial is also limited: This is an important point to remember in order not to exclude more terms than are allowed in any possible dimension. Indeed, it is not enough to remove no more than the appropriately computed number of terms from the total number of terms of a polynomial. We must also be careful not to cancel more terms than explained below in any dimension of the polynomial.

Let us consider the following three equations. We assume they are complete for clarity and simplicity only:

$$(x, y, z)^2 = 0,$$
$$(x, y, z)^2 = 0,$$
$$(x, y, z)^2 = 0.$$

Each of these equations enjoys the expression in (227) as polynomial multiplier; this polynomial is of the form $(x, y, z)^{T+6}$. The total number of terms that may be cancelled in the polynomial multiplier of the first equation is

$$N(x, y, z)^{T+4} - N(x, y, z)^{T+2} + N(x, y, z)^{T+4}.$$

If we want to compute the number of these terms that reach the highest dimension, we get

$$d[N(x, y, z)^{T+4} - N(x, y, z)^{T+2} + N(x, y, z)^{T+4}] \ldots \binom{T}{1},$$

or $N(x, y)^{T+4} - N(x, y)^{T+2} + N(x, y)^{T+4}$.

Thus the number of terms to be eliminated in the first dimension of the polynomial multiplier cannot exceed $N(x, y)^{T+4} - N(x, y)^{T+2} + N(x, y)^{T+4}$.

(236.) Although we cannot eliminate more terms than the number we just computed, we can decide to eliminate fewer such terms and transfer the remaining number of terms to other dimensions, if deemed appropriate.

Consider, for example, the second dimension, where the largest number of terms to be cancelled is no more than $N(x, y)^{T+3} - N(x, y)^{T+1} + N(x, y)^{T+3}$. Assume we have cancelled $N(x, y)^{T+4} - N(x, y)^{T+2} + N(x, y)^{T+4}$ terms in the first dimension. We may increase the number of terms to be cancelled in the second dimension to $N(x, y)^{T+3} - N(x, y)^{T+1} + N(x, y)^{T+3} + q$ if we have only cancelled $N(x, y)^{T+4} - N(x, y)^{T+2} + N(x, y)^{T+4} - q$ in the first dimension.

We now see what is to be said about other dimensions. These are the only limits we must respect. We can choose to cancel any term in each dimension. It does not matter, as long as the number of terms that is cancelled does not exceeed the maximum number of terms we just saw.

(237.) Although we have chosen complete equations as a matter of example, we easily see that this was done for clarity's sake only; it is always easy to determine the number of terms that may be cancelled in each dimension of the polynomial.

About the best use that can be made of the coefficients of the terms that may be cancelled in each polynomial multiplier

(238.) Let us pay attention to all that was said about the number of terms that may be cancelled in each polynomial multiplier; moreover, let us realize that what we have called the *first, second, third* equations are purely arbitrary indices, such that we may permute them arbitrarily; we soon see that the number of terms that may be cancelled in each polynomial multiplier does not really depend upon the sequence that was chosen, unlike what our approach has suggested up until now.

Our real constraint is the total number of useless coefficients in all the polynomial multipliers, as well as the total number of useless coefficients of the same dimension in all polynomial multipliers.

Indeed, depending upon which equation is chosen to be first, we see that the number of terms that may be cancelled in the same polynomial multiplier can change. But we also see that the total number of terms that may be cancelled in all polynomials remains the same, no matter how the equations, and therefore the polynomials, are initially ordered.

Consider, for example, the three equations

$$(x, y, z)^t = 0,$$
$$(x, y, z)^{t'} = 0,$$
$$(x, y, z)^{t''} = 0.$$

The polynomial multiplier of the first equation is $(x, y, z)^{T+t'+t''}$.
The polynomial multiplier of the second equation is $(x, y, z)^{T+t+t''}$.
The polynomial multiplier of the third equation is $(x, y, z)^{T+t+t'}$.

In the first polynomial multiplier, we can cancel $N(x, y, z)^{T+t''}$ $-N(x, y, z)^T + N(x, y, z)^{T+t'}$ terms. In the second polynomial, we can cancel $N(x, y, z)^{T+t}$ terms, and no terms can be cancelled in the third.

Thus, the total number of terms that may be cancelled is

$$N(x, y, z)^{T+t} + N(x, y, z)^{T+t'} + N(x, y, z)^{T+t''} - N(x, y, z)^T.$$

Let us now change the order of the equations. We write them as

$$(x, y, z)^{t'} = 0,$$
$$(x, y, z)^t = 0,$$
$$(x, y, z)^{t''} = 0.$$

The polynomial multiplier of the first equation is $(x, y, z)^{T+t+t''}$.
The polynomial multiplier of the second is $(x, y, z)^{T+t'+t''}$.
The polynomial multiplier of the third is $(x, y, z)^{T+t+t'}$.

In the first polynomial multiplier, we can cancel $N(x, y, z)^{T+t''}$ $-N(x, y, z)^T + N(x, y, z)^{T+t}$ terms. In the second polynomial, we can cancel $N(x, y, z)^{T+t'}$ terms, and we can cancel no terms in the third.

Thus, the total number of terms that may be cancelled is

$$N(x, y, z)^{T+t} + N(x, y, z)^{T+t'} + N(x, y, z)^{T+t''} - N(x, y, z)^T.$$

We therefore see that, indeed, the number of terms that may be cancelled in any one of the polynomial multipliers varies as a function of the order of that polynomial multiplier; however, the total number of terms that may be cancelled in all three polynomials remains constant.

Using the same approach, we can show that the same property holds regarding the total number of terms with highest or first dimension in each polynomial, that the same property holds concerning the second dimension of each polynomial, and so on.

(239.) Since these terms are useless for elimination purposes, until now we have considered that they must always be excluded. However, this exclusion is not required: From (230), it is enough not to count these among the coefficients that must be computed to reach the final equation, or, in general, to attain the stated goal.

Indeed, consider the number of coefficients that we can cancel using a given number of equations; we can use the same equations to assign arbitrary values to the same number of coefficients. We can therefore arbitrarily use those coefficients that do not contribute to the elimination, for as long as none of these are counted among the coefficients that are used for elimination.

(240.) All values of the polynomial multipliers are admissible when proceeding with the elimination. Having formed the sum equation, we can arbitrarily assign values to those coefficients whose number we know does not contribute to solving the question, for as long as these are not used to cancel any new term.

(241.) Considering this single precaution, we can use these coefficients for any purpose, to form arbitrary equations among them and therefore determine the value of these coefficients: We will never be led to an absurd equation.

(242.) It probably goes without saying that we must not include equations that can be used to cancel terms that we would like to keep within those arbitrary equations. Although the resulting final equation would not be absurd, it would not be the one sought: Some terms would remain in front of some of the unknowns that should have been eliminated.

(243.) There is another issue to be avoided when forming these arbitrary equations. This precaution is rarely necessary and is better understood by means of examples. We will deal with this question later.

(244.) These useless coefficients help us make our computations more elegant and thus as simple and as fast as possible. These coefficients are useful to untangle some factors that are important to know, in order to understand all the information implicitly contained in an arbitrary system of equations. If we were not to use these useless coefficients, we would not have been able to recognize the kind of symmetry we feel must be present when several equations of the same form are present. This symmetry would have remained hidden in our computations, and the factors we just mentioned, when combined with nonsymmetrical factors, would be difficult and in practice impossible to recognize as soon as complex or numerous equations arise.

(245.) But since it is very important not to introduce factors that do not relate to the information carried by the system of equations in the result of our computations, we can apply the following general rule: When forming arbitrary equations, behave so as to compute only the coefficients that are useful to elimination, that is, to compute only that number of coefficients. If this is impossible,[5] as may sometimes happen, we must act so as to compute the smallest possible number of coefficients beyond the number of coefficients that are useful for elimination.

The way to reach this goal is to form the same number of arbitrary equations as there are useless coefficients, including as many of these coefficients as possible and no other. We must also remember to bring in only as many of these coefficients as allowed relative to each dimension, from what was said in (235).

Then from (212) and (213), these equations can yield only two things, that is, a condition equation and the requirement that each of these coefficients be zero.

The condition equation, although it is often unimportant regarding our final goal, will always carry some meaning relative to the equations, for example, indicating cases when they can be solved more easily or solved using simpler polynomial multipliers.

The condition that each coefficient be equal to zero will bring considerable simplifications in the computations that follow.

From there we must observe that, if we are to maintain symmetry by taking advantage of these useless coefficients, we must consider all similar coefficients as such, that is, all coefficients entering in similar terms in each polynomial multiplier.

We will not expand further about how to choose, assign and leverage the useless coefficients: Rather, we will clarify our general thoughts by means of examples.

Other applications of the methods presented in this book for the General Theory of Equations

(246.) Reaching the final equation with the simplest expression for an arbitrary number of equations of arbitrary degree is not the only benefit that we can get from our work. We can also use the same methods to find the simplest expression for an arbitrary function containing the unknowns entering into these equations. We can find the value subject to conditions we can choose according to their usefulness.

For example, consider the three arbitrary equations

$$(x, y, z)^t = 0,$$
$$(x, y, z)^{t'} = 0,$$
$$(x, y, z)^{t''} = 0.$$

---

[5] At least when we want to keep the symmetry that facilitates computations.

We ask, given these three equations, for the value of $(x, y, z)^T$ or, in general, the value of any other polynomial in the variables x, y and z, where this value contains as few terms as possible.

In order to keep ideas simple, let us consider the polynomial $(x, y, z)^T$ only. It is clear that

1. If we multiply the first equation by the polynomial $(x, y, z)^{T-t}$, the second equation by the polynomial $(x, y, z)^{T-t'}$, the third equation by the polynomial $(x, y, z)^{T-t''}$, and if we add these three products and the proposed polynomial, the sum will have the same value as that of the proposed polynomial.

2. Using the coefficients introduced by the three polynomial multipliers, and paying attention to the fact that the existence condition of the three equations makes a known number of these coefficients useless, we can always cancel a given number of terms in this sum.

3. The number of terms to which this equation is reduced will be less by one unit than that of the final equation resulting from the proposed equations.

4. The terms that constitute the final polynomial giving the value of the proposed polynomial $(x, y, z)^T$ are, by the way, those we are looking for.

5. Consequently we can get an expression for the polynomial $(x, y, z)^T$ expressed as a function of x alone, or y alone, or z alone.

(247.) These are therefore the means to achieve what we have discussed in (207): To replace x, y and z by their values arising from the proposed equations, or at least replace them by all that can be rationally extracted from the set of equations. The final value of the proposed polynomial can then be obtained by solving the final equation.

(248.) We have assumed in what we just discussed that T is larger than $tt't''$, which by (47) is the expression for the degree of the final equation resulting from the three proposed equations; if instead we have $T < tt't''$, then, defining $T' = tt't''$, we would multiply

the first equation by $\qquad (x, y, z)^{T'-t}$,
the second equation by $\qquad (x, y, z)^{T'-t'}$,
and the third equation by $\quad (x, y, z)^{T'-t''}$,

and we would proceed as previously described.

(249.) It is, however, necessary to use polynomials with high degree only when we want to compute the value of $(x, y, z)^T$ as a function of x only, or y only, or z only. In any other case we can use the polynomial multipliers

$$(x, y, z)^{T-t}, \ (x, y, z)^{T-t'}, \ (x, y, z)^{T-t''}.$$

For example, assume we are looking for the final value of x^3yz given the three equations

$$
\begin{aligned}
(x,y,z)^3 &= 0, \\
(x,y,z)^3 &= 0, \\
(x,y,z)^3 &= 0,
\end{aligned}
$$

expressed as a function of x, y and z, and with the condition that not only x^3, but also y^3 and z^3 not enter in the expression of this value. Since x^3yz has dimension 5, I multiply each of the proposed equations by a polynomial of the form $(x,y,z)^{5-3}$ or $(x,y,z)^2$. Having added the products to x^3yz, I observe that, since the polynomial multipliers have degree less than that of the proposed equations, we cannot cancel any terms using these equations. Consequently, none of the indeterminate coefficients of these equations will be useless. In order to resolve the question, I will therefore have a number of coefficients equal to $3N(x,y,z)^2 = 30$. However, from (59), this number is precisely the number of terms that may be divided by x^3 or y^3 or z^3 in a sum of the form $(x,y,z)^5$; thus it will be possible to get an expression for this sum such that no terms in x, y and z reach a degree higher than 2. Since this sum is the value of x^3yz, we get the value of x^3yz expressed as we asked.

(250.) Considering complete equations, we said in (45) that if we have an arbitrary number of equations

$$
\begin{aligned}
(u\ldots n)^t &= 0, \\
(u\ldots n)^{t'} &= 0, \\
(u\ldots n)^{t''} &= 0, \\
(u\ldots n)^{t'''} &= 0, \\
\text{etc.}
\end{aligned}
$$

we can always find the values of $x^{t'-1}$, $y^{t''-1}$, $z^{t'''-1}$, etc. , using the $n-1$ last equations. We now prove that this assertion is true and how to construct these values to proceed with the elimination.

(251.) We now clarify and prove more rigorously what was said in (56).

We claimed in (56) that we can extract, from a given number of equations, the values of the same number of terms and that if these terms share a common divider, the values of these terms are not the only ones that may be provided by these equations. Consequently, if we used these values only in the process of solving an equation, we would not have answered the question, because we would not have expressed all the information that these equations contain.

Consider, for example, the three equations

$$
\begin{aligned}
(x,y,z)^2 &= 0, \\
(x,y,z)^2 &= 0, \\
(x,y,z)^2 &= 0.
\end{aligned}
$$

We know the degree of the final equation must be eight. Assume we have chosen the polynomial multiplier $(x,y,z)^6$ for the first equation, as must be done to reach the final equation. Assume we extract from the other two equations the values of z^2, yz, and their multiples, in order to substitute them in this polynomial multiplier and into the product equation. Then we would would not have cancelled all terms that may be cancelled, and the resulting final equation would not express the question expressed by the proposed equations. It would still be possible to conclude from the last two equations the value of one and often many other terms. For example, here we could conclude the value of y^3.

Indeed, assume we multiply these two equations, respectively, by

$$A'x + B'y + C'z + D', \text{ and } A''x + B''y + C''z + D'',$$

and that we add the two products together and to y^3; the sum will therefore be the value of y^3. Using the eight indeterminate coefficients,[6] I can cancel all terms that may be divided by z^2 and by yz; I can therefore obtain the value of y^3 resulting from the substitution of the values of z^2 and yz, that is, a value where z^2 and yz do not appear. Thus if I substituted the values of z^2 and yz only, extracted from the last two equations into the product equation that yields the final equation, I would reach a final equation that is unrelated to the question. This does not happen when the terms extracted from the $n-1$ equations do not share a common divider. When these values are substituted in the polynomial multiplier and in the product equation everywhere they can be substituted, they express all the information these $n-1$ equations can express.

Consider the preceding example and assume we have extracted the values of y^2 and z^2 from the last two equations. Assume we could still extract the value of another term, say x^2y. Operating as above, we would only have eight coefficients to cancel the terms that may be divided by y^2 and z^2, and there are eight of these. We would, by (212), be led to a condition equation, without being able to determine a value of x^2y free of y^2 or z^2 or any of their multiples. Thus substituting y^2 and z^2 is enough to express the conditions of the question.

Useful considerations to considerably shorten the computation of the coefficients useful for elimination

(252.) We can add considerably more to the simplifications already brought by the method we have presented in (195 ff), to compute the unknowns in first-order equations, used in the elimination process. We assume in the following that all proposed equations are complete and have the same degree: It will be easy to apply our discussion to incomplete equations as we will show afterwards; but working first with complete equations makes our presentation clearer.

(253.) Let us assume that the known coefficients of the terms in each equation are represented by the same letters and that they differ only by

---

[6]We need eight coefficients to cancel these seven terms, because the seven first-order equations that we will reach are each without any known terms.

their accent, as we have done up until now. Let us assume the same for the undetermined coefficients in the polynomial multipliers of each equation. The coefficient of any term in one of the polynomial multipliers will appear in one of the terms of the sum equation, along with a known coefficient of one of the terms of the equation corresponding to that polynomial multiplier. We feel that the undetermined coefficient of the same term for each other polynomial multiplier will also be found in the term of the sum equation, along with the known coefficient of the same term of the equation corresponding to that polynomial multiplier.

Thus, if there are only two equations, the specific resulting equations will be of the form

$$Aa + A'a' = 0, \ Ab + A'b' + Bc + B'c' = 0, \ Af + A'f' + Bd + B'd' + Ce + C'e' = 0,$$

and so on. If there are three equations, the same specific equations will be of the form

$$Aa + A'a' + A''a'' = 0, \ Ab + A'b' + A''b'' + Bc + B'c' + B''c'' = 0,$$
$$Ad + A'd' + A''d'' + Be + B'e' + B''e'' + Cf + C'f' + C''f'' = 0,$$

and so on.

The number of these equations is $n - 1$, where n is the number of undetermined coefficients.

These equations can be computed much faster than by literally following the rule given in (198).

(254.) We have seen in (198) that the order in which the product of unknowns is written to compute these unknowns is not important, for as long as their order is not changed in the remaining computations. In the equations of interest to our discussion, there is considerable benefit in choosing the order in which the products of unknowns is written, although this choice is not mandatory.

The most convenient order is to group all similar letters together; we are not constrained to follow any particular sequence among these groups.

For example, assume the unknowns are

$$A, B, C, D,$$
$$A', B', C', D'.$$

Considering all possible sequences of these letters, I prefer to use any of the following expressions:

$$AA'BB'CC'DD', \ BB'AA'CC'DD', \ DD'BB'AA'CC', \text{ etc.}$$

Assume the unknowns are

$$
\begin{array}{cccc}
A, & B, & C, & D, \\
A', & B', & C', & D', \\
A'', & B'', & C'', & D''.
\end{array}
$$

Considering all possible sequences made of these twelve unknowns, I prefer to use any of the following expressions:

$$AA'A''BB'B''CC'C''DD'D'',\ BB'B''AA'A''CC'C''DD'D'',$$
$$CC'C''DD'D''AA'A''BB'B'',\ \text{etc.},$$

and so on.

This shows that the order of the groups is completely arbitrary.

(255.) We now examine the consequences that this choice offers when using the rule (198). But we first remark that it is not mandatory to write the product of all unknowns as soon as the computation of lines begins. If some equations are simpler than others, we may prefer to begin with those; when using these equations, we may omit writing down the groups of unknowns that these equations do not contain, and introduce these groups only when we use equations that contain them.

(256.) Assume we have the following three equations

$$Aa + A'a' = 0,$$
$$Ab + A'b' + Bc + B'c' = 0,$$
$$Bd + B'd' = 0.$$

When we compute the value of $AA'BB'$, we obtain

First line $(aA' - a'A)BB'$

Second line $(ab' - a'b)BB' - (aA' - a'A).(cB' - c'B)$

Third line $(ab' - a'b).(dB' - d'B) + (aA' - a'A).(cd' - c'd).$

If we carefully observe the contents of each of these lines, we easily see that each combination such as ab' or cd' or cB' or aA' is always paired with the corresponding combination $a'b$, $c'd$, $c'B$, $a'A$, along with the opposite sign.

We also observe that each combination dB' or $d'B$, aA' or $a'A$ is multiplied by the system $ab' - a'b$ or $cd' - c'd$ in the last line, that is, the line that gives the values of A, A'; B, B'. Elaborating on this observation, we see that we can define the following process to get the values of the coefficients for one of the polynomial multipliers and to find the values of the coefficients of the other polynomial.

(257.) Proceed with the computation of the lines above, performing the swap (198) for only one of the two similar coefficients: Perform this swap always in the same order; that is, for example, swap the coefficient which is written first. Make sure to write the known coefficients, those that are substituted in the swap process, in the same order as the unknown coefficient they substitute.

Then we only get the combinations ab', cd', etc. in the last and following lines, instead of the quantities $ab' - a'b$, $cd' - c'd$. But since we know $a'b$ must come with ab' and $c'd$ must come with cd', we can recover the proper expression quickly in any line, whenever necessary. From now on, we therefore write (ab') instead of $ab' - a'b$; (ac') instead of $ac' - a'c$; (bc') instead of $bc' - b'c.^{7}$

When we have determined the values of the unknown coefficients in the last line of (198), we obtain the values for the analogous quantities by switching the sign of these quantities and the accent of the letter finding itself either alone or outside the parentheses.

Thus, considering the above example, I get

First line $aA'BB'$

Second line $(ab')BB' - aA'cB'$

Third line $(ab')dB' + aA'(cd')$,

from which I conclude $A' = a(cd')$, $B' = d(ab')$; and swapping signs and accents, $A = -a'(cd')$, $B = -d'(ab')$.

(258.) Consider for the second example the following five equations

$$Aa + A'a' = 0,$$
$$Ab + A'b' + Bc + B'c' = 0,$$
$$Ad + A'd' + Be + B'e' + Cf + C'f' = 0,$$
$$Bg + B'g' + Ch + C'h' = 0,$$
$$C' + C'l' = 0.$$

I get the following:

First line $aA'BB'$

Second line $[(ab')BB' - aA'cB']CC'$

Third line $[(ab')eB' - (ad')cB' + aA'(ce')]CC'$
                $+[(ab')BB' - aA'cB']fC'$

Fourth line $[(ab').(eg') - (ad').(cg')CC'$
                $-[(ab')eB' - (ad')cB' + aA'(ce')]hC'$
                $+ [(ab')gB' + aA'(cg')]fC'$,

omitting the terms where BB' and $A'B'$ remain, since they vanish in the next line:

Fifth line $[(ab').(eg') - (ad').(cg')]lC'$
            $+(hl').[(ab')eB' - (ad')cB' + aA'(ce')]$
            $- (fl').[(ab')gB' + aA'(cg')]$;

from which (198) we get

$$A' = -a(cg').(fl') + a(ce').(hl'),$$
$$B' = -g(ab').(fl') + e(ab').(hl') - c(ad').(hl'),$$
$$C' = l[(ab').(eg') - (ad').(cg')];$$

and, by (257), consequently

$$A = a'(cg').(fl') - a'(ce').(hl'),$$
$$B = g'(ab').(fl') - e'(ab').(hl') + c'(ad').(hl'),$$
$$C = -l'[(ab').(eg') - (ad').(cg')].$$

(259.) When there are three polynomial multipliers, we do not only encounter expressions such as ab', or ab'', or $a'b''$, etc. along with the cor-

---

[7]We will assign that meaning to parentheses only when they are applied to monomials; parentheses will retain their ordinary meaning when applied to complete quantities.

responding $a'b$, $a''b$, $a''b'$ with the opposite sign, but also each combination such as $(ab' - a'b)c''$ together with the other two $-(ab'' - a''b)c'$ and $+(a'b'' - a''b')c$; in other terms, the values of the undetermined coefficients are functions of combinations such as

$$(ab' - a'b)c'' - (ab'' - a''b)c' + (a'b'' - a''b')c,$$

and such as $ab' - a'b$, $ac' - a'c$, $ac'' - a''c$, etc.

In addition, consider A, A', A'' three of these coefficients; among the two-dimensional combinations $ab' - a'b$, $ab'' - a''b$, $a'b'' - a''b'$, the combination $a'b'' - a''b'$ enters the expression for A; the combination $ab' - a'b$ enters the value of A''; and the combination $ab'' - a''b$ enters the value of A', with the opposite sign.

For example, consider the five equations

$$Aa + A'a' + A''a'' = 0,$$
$$Ab + A'b' + A''b'' = 0,$$
$$Ac + A'c' + A''c'' + Bd + B'd' + B''d'' = 0,$$
$$Be + B'e' + B''e'' = 0,$$
$$Bf + B'f' + B''f'' = 0.$$

Computing the value of $AA'A''BB'B''$ according to what was said in (198), we get as follows:

First line
$aA'A'' - a'AA'' + a''AA'$

Second line
$[(ab' - a'b)A'' - (ab'' - a''b)A' + (a'b'' - a''b')A]BB'B''$

Third line
$[(ab' - a'b)c'' - (ab'' - a''b)c' + (a'b'' - a''b')c]BB'B''$
$- [(ab' - a'b)A'' - (ab'' - a''b)A' + (a'b'' - a''b')A].(dBB' - d'BB'' + d''BB')$

Fourth line
$[(ab' - a'b)c'' - (ab'' - a''b)c' + (a'b'' - a''b')c]$
$\times(eB'B'' - e'BB'' + e''BB')$
$+ [(ab' - a'b)A'' - (ab'' - a''b)A' + (a'b'' - a''b')A]$
$\times[(de' - d'e)B'' - (de'' - d''e)B' + (d'e'' - d''e')B]$

Fifth line
$[(ab' - a'b)c'' - (ab'' - a''b)c' + (a'b'' - a''b')c]$
$\times[(ef' - e'f)B'' - (ef'' - e''f)B' + (e'f'' - e''f')B]$
$- [(ab' - a'b)A'' - (ab'' - a''b)A' + (a'b'' - a''b')A]$
$\times[(de' - d'e)f'' - (de'' - d''e)f' + (d'e'' - d''e')f].$

Since each quantity A or B, A' or B', etc. takes on the value of its coefficient, these values obviously have the characteristics that we predicted.

From these observations, if we abbreviate a quantity of the form

$$(ab' - a'b)c'' - (ab'' - a''b)c' + (a'b'' - a''b')c,$$

also written $(ab')c'' - (ab'')c' + (a'b'')c$, we can shorten our computations, following what was done in (257), by writing instead $(ab'c''))$ and performing the following steps:

(260.) When computing the *lines*, only swap the first letter present in each group $AA'A''$, $BB'B''$, or $CC'C''$, etc.; swap this letter with the corresponding coefficient in the equation used to compute this line; as we exhaust one particular group, capture the corresponding result between two parentheses. When reaching the last line and computing the values of the unknowns remaining there, capture each combination of dimension 2 between parentheses; from there, proceed as in the following example to conclude the value of similar quantities.

Assume we have found $A'' = (ab').(bc'd'')$. I move from A'' to A' and from A' to A by performing the following operations: I swap the accents $''$ and $'$ and I switch the sign in the quantity of dimension two only; this gives me $A' = -(ab'').(bc'd'')$; in the latter I swap the accent $'$ and zero and and I switch the sign in the quantity of dimension 2 only, which yields $A = (a'b'').(bc'd'')$.

Going back to the five equations

$$Aa + A'a' + A''a'' = 0,$$
$$Ab + A'b' + A''b'' = 0,$$
$$Ac + A'c' + A''c'' + Bd + B'd' + B''d'' = 0,$$
$$Be + B'e' + B''e'' = 0,$$
$$Bf + B'f' + B''f'' = 0,$$

we can proceed with a much faster computation of $AA'A''BB'B''$ as follows:

First line	$aA'A''$
Second line	$ab'A''BB'B''$
Third line	$(ab'c'')BB'B'' - ab'A''dB'B''$
Fourth line	$(ab'c'')eB'B'' + ab'A''de'B''$
Fifth line	$(ab'c'').(ef')B'' - (ab')A''(de'f'').$

We therefore get

$$A'' = -(ab').(de'f''), \quad B'' = (ef').(ab'c'');$$

and consequently

$$A' = +(ab'').(de'f''), \quad B' = -(ef'').(ab'c''),$$
$$\&\ldots \quad A = -(a'b'').(de'f''), \quad B = +(e'f'').(ab'c'').$$

Remembering the meaning of the parentheses, these expressions are the same as those previously found.

(261.) If we had four polynomial multipliers, the groups would consist of four coefficients. We would then swap each letter in each group with its coefficient in the equation used to compute the current line, and would repeat that until the group is exhausted: The result corresponding to each subsequently exhausted group would then be put between parentheses, giving expressions of the form $(ab'c''d''')$. When, reaching the last line, we want to find the value of the remaining unknowns, we put each expression of dimension 3 within parentheses too, and we proceed as in the following example to find the value of similar unknowns.

Assume we have found $A''' = (ab'c'').(de'f''g''')$; we get A'' from A''', A' from A'', and A from A'. We get A'' from A''' by changing $''$ into $'''$ and the sign of the quantity of dimension three; this yields $A'' = -(ab'c''').(de'f''g''')$. We get A' from A'' by changing $'$ into $''$ and the sign of the quantity of dimension 3, yielding $A' = +(ab''c''').(de'f''g''')$. We get A from A' by changing zero into $'$ and the sign of the quantity of dimension 3, yielding $A = -(a'b''c''').(de'f''g''')$.

It is quite easy to extend this rule to a larger number of polynomials.

(262.) It is always easy to obtain the quantities of the form $(ab'c''d''')$ or in general of the form $(ab'c''d'''e^{IV}f^{V}$, etc.), observing that they are the value of the condition equation that must hold for a number n (where n is the number of these quantities) of first-order equations in n unknowns to hold simultaneously.

For example, (ab') is the value of the condition equation that must hold for the two equations

$$
\begin{aligned}
ax + by &= 0, \\
a'x + b'y &= 0
\end{aligned}
$$

to hold simultaneously.

Likewise, $(ab'c'')$ is the value of the condition equation that must hold so that the three equations

$$
\begin{aligned}
ax + by + cz &= 0, \\
a'x + b'y + c'z &= 0, \\
a''x + b''y + c''z &= 0
\end{aligned}
$$

hold simultaneously.

The same is true for $(ab'c''d''')$ and the four equations

$$
\begin{aligned}
ax + by + cz + dt &= 0, \\
a'x + b'y + c'z + d't &= 0, \\
a''x + b''y + c''z + d''t &= 0, \\
a'''x + b'''y + c'''z + d'''t &= 0;
\end{aligned}
$$

and so on.

These quantities are always easy to compute using the rule we have given in (212).

(263.) When computing the value of similar quantities appearing in the last line, we can always avoid putting much care into changes in accents and signs, considering the problem of interest to us; indeed, we can always avoid having to compute the expression specific to each unknown. We seek to compute this value only to substitute it in another quantity where this unknown is present later; however, this substitution occurs as described in (207), by proceding with the computation of a new line, considering the latter quantity as an equation. For example, assume we ask for the value of $Ag + A'g' + A''g'' + Bh + B'h' + B''h''$, subject to the five proposed equations (260). Since the last line is

$$
(ab'c'')ef'B'' - ab'A''(de'f''),
$$

I compute a new line using the quantity

$$
Ag + A'g' + A''g'' + Bh + B'h' + B''h''
$$

as a new equation, and I obtain
$$(ab'c'').(ef'h'') - (ab'g'').(de'f''),$$
resulting from the substitution of the values of A, A', A'', B, B', B'' in the proposed quantity, and I can do that without detailing the expression for the value of each of these quantities.[8]

(264.) The method of computing the unknowns by grouping them is not only applicable to our problem, but also to all first-order equations.

Assume we have the following four equations:
$$\begin{aligned} ax + by + cz + dt + e &= 0, \\ a'x + b'y + c'z + d't + e' &= 0, \\ a''x + b''y + c''z + d''t + e'' &= 0, \\ a'''x + b'''y + c'''z + d'''t + e''' &= 0. \end{aligned}$$

Remembering that by (198) each unknown is equal to its coefficient in the last *line*, divided by the coefficient of the artificial unknown in the same *line*, we see that we can reduce the computations to that of computing the coefficient of only one unknown in the last line; once we have computed one such coefficient, we are able to compute all others the same way. Or, better, when we have computed one coefficient, we are able to extract all the others when the equations enjoy the highest degree of generality.[9] To obtain the value of the coefficient of one of the unknowns in the last line, the question reduces to computing the value of the product of the other unknowns. In order not to make any mistake about signs, we will always keep in mind the order of the unknown of interest in the products of all unknowns. Thus, in the present case, instead of computing the last line in general for $xyztu$, I only compute this last line for $yztu$, and in order to get it as conveniently as possible, I perform the following grouping: $yz.tu$, and I proceed as follows for the computation of lines, keeping in mind that y is assumed to occupy the second place.

First line	$-bz.tu - yz.du$
Second line	$+(bc').tu - bz.d'u + b'z.du + yz.(de')$
Third line	$-(bc').d''u + (bc'').d'u - bz.(d'e'')$
	$-(b'c'').du + b'z.(de'') - b''z.(de')$
Fourth line	$+(bc').(d''e''') - (bc'').(d'e''') + (bc''').(d'e'')$
	$+(b'c'').(de''') - (b'c''').(de'') + (b''c''').(de').$

This is the coefficient of x in the last line.

Likewise, in order to get the coefficient of u, I compute the value of $xyzt$ by grouping it as $xy.zt$. I find the value of the coefficient of u in the last line as follows:
$$\begin{aligned} (ab').(c''d''') - (ab'').(c'd''') + (ab''').(c'd'') \\ +(a'b'').(cd''') - (a'b''').(cd'') + (a''b''').(cd'). \end{aligned}$$

---

[8]This result must naturally have a divider; but since we will have to make these substitutions only in equations where this divider is common to all terms, we will always be able to omit it

[9]See our *Cours de Mathématiques à l'usage des Gardes de la Marine*, Vol. III, page 98.

From this I conclude that

$$x = \frac{\begin{array}{l}+(bc').(d''e''') - (bc'').(d'e''') + (bc''').(d'e'') \\ +(b'c'').(de''') - (b'c''').(de'') + (b''c''').(de')\end{array}}{\begin{array}{l}(ab').(c''d''') - (ab'').(c'd''') + (ab''').(c'd'') \\ +(a'b'').(cd''') - (a'b''').(cd'') + (a''b''').(cd')\end{array}},$$

and so on.

(265.) Consider the following five equations:

$$\begin{aligned}
ax + by + cz + dr + et + f &= 0, \\
a'x + b'y + c'z + d'r + e't + f' &= 0, \\
a''x + b''y + c''z + d''r + e''t + f'' &= 0, \\
a'''x + b'''y + c'''z + d'''r + e'''t + f''' &= 0, \\
a^{IV}x + b^{IV}y + c^{IV}z + d^{IV}r + e^{IV}t + f^{IV} &= 0.
\end{aligned}$$

I would compute, for example, the coefficient of x in the last line by computing $yzr.tu$, or $yz.rtu$, or $yz.rt.u$.

If I had six equations whose unknowns were x, y, z, r, s, and t, I would compute, for example, the coefficient of x by computing $yz.rs.tu$ or $yzrs.tu$ or $yzr.stu$ and so on.

(266.) We now give a striking example illustrating the capacity of our method to take advantage of the simplifications that a few missing terms enable. Assume we have the following twelve equations[10]

$$\begin{aligned}
Aa + A'a' + A''a'' &= 0, \\
Ab + A'b' + A''b'' &= 0, \\
Ac + A'c' + A''c'' + Ba + B'a' + B''a'' &= 0, \\
Bb + B'b' + B''b'' &= 0, \\
Bc + B'c' + B''c'' &= 0, \\
Bd + B'd' + B''d'' + Ca + C'a' + C''a'' &= 0, \\
Cb + C'b' + C''b'' &= 0, \\
Cc + C'c' + C''c'' &= 0, \\
Cd + C'd' + C''d'' + Da + D'a' + D''a'' &= 0, \\
Db + D'b' + D''b'' &= 0, \\
Dc + D'c' + D''c'' &= 0, \\
Ad + A'd' + A''d'' + Da + D'a' + D''a'' &= 0,
\end{aligned}$$

and that we ask for the condition equation that must be satisfied for all these equations to hold simultaneously.

We group the unknowns in groups of three. Introducing each group in the same order as the letters in the equation I use, I compute as follows:

First line	$aA'A''$
Second line	$ab'A''.BB'B''$
Third line	$(ab'c'')BB'B'' - ab'A''.aB'B''$
Fourth line	$(ab'c'')bB'B'' + ab'A''.ab'B''$
Fifth line	$[(ab'c'')bc'B'' - ab'A''.(ab'c'')]CC'C''$
Sixth line	$(ab'c'').(bc'd'')CC'C'' + (ab'c'').ab'A''.aC'C''.$

---

[10]Although we have repeated the same letters in the coefficients in many of these equations, this affects neither our method nor the form of the result: We did this only to avoid having too many different letters.

In these computations I eliminate the term where B' cannot have any influence on the final equation since it is not present in the remaining equations.

Seventh line	$(ab'c'').(bc'd'')bC'C'' - (ab'c'').ab'A''.ab'C''$
Eighth line	$[(ab'c'').(bc'd'')bc'C'' + (ab'c'')ab'A''.(ab'c'')]DD'D''$
Ninth line	$(ab'c'').(bc'd'')^2DD'D'' - (ab'c'')^2.ab'A''.aD'D'',$

by eliminating the term where C'' remains.

Tenth line	$(ab'c'').(bc'd'')^2bD'D'' + (ab'c'')^2.ab'A''.ab'D''$
Eleventh line	$(ab'c'').(bc'd'')^2bc'D'' - (ab'c'')^3.ab'A''$
Twelfth line	$(ab'c'').(bc'd'')^3 - (ab'c'')^3.(ab'd''),$

The condition equation is therefore

$$(ab'c'').[(bc'd'')^3 - (ab'c'')^2(ab'd'')] = 0.$$

(267.) We now reconsider undetermined coefficients in the polynomial multipliers of incomplete equations.

Until now we have assumed that the equations under consideration are complete and have the same degree. The resulting symmetry in the coefficients of these equations and their polynomial multipliers has given us a path towards computing the undetermined coefficients easily. This symmetry is not as perfect when equations have different degrees, or when they are incomplete. However, since we can consider incomplete equations with different degrees as a specific case of complete equations with the same degree (and a number of known coefficients are zero), we guess that we may be able to recover some traces of the advantages we encountered in the computation of complete equations when considering incomplete equations with different degrees.

In order to recover them, consider the question as follows.
(268.) Let

$$A, B, C, D, E, F, \text{etc.}$$
$$A', B', C', D', E', F', \text{etc.}$$
$$A'', B'', C'', D'', E'', F'', \text{etc.}$$
$$\text{etc.}$$

be the coefficients of the polynomial multipliers when they all have the same degree.

If the proposed equations do not all share the same degree, or if they are incomplete, some terms will be missing from the highest dimensions of some of the polynomial multipliers, since these cannot have the same degree.

Assume, for example, that three coefficients are missing in the first multiplier and two in the second. Then the difference between the two cases is that in the first case, we must compute the value of

$$AA'A''BB'B''CC'C''DD'D''EE'E''FF'F'', \text{etc.},$$

whereas in the second case, we must only compute the value of

$$A''B''C'C''DD'D''EE'E''FF'F'', \text{etc.}$$

We therefore mean to keep giving the same letter representations (modulo different accents) to similar terms appearing in both the equations and their polynomial multipliers. If we pay attention and group together similar coefficients, we will reach the result using the same rules for the computation of *lines* as those given until now, and we will take advantage of all the simplifications that can be brought by whatever symmetry remains in the proposed equations and their polynomial multipliers.

We observe that, if the coefficient or unknown does not appear in the equation being employed during the computation of the lines, we should still remember that similar variables are possibly present.

(269.) For example, consider the following three equations:

$$Aa + Bb + Cc + C'c' = 0,$$
$$Bd + Ce = 0,$$
$$Bf + Ch + C'h' = 0.$$

We seek the values of A, B, C, and C'. Since, except for C and C', there are no other groups of similar letters, I group these two only. When proceeding with the calculation of lines, I act implicitly as if $C'e'$ was in fact present in the second equation, except when considering the final result, which is easy to obtain by setting $e' = 0$.

Thus our current task is to compute $AB.CC'$, and we proceed as follows:

First line $(aB - bA).CC' + AB.cC'$,

Second line $ad.CC' - (aB - bA).eC' - dA.cC' + AB.(ce')$,

Third line $ad.hC' - af.eC' + (aB - bA).(eh') + dA.(ch') - fA.(ce')$,

from which we get

$$A = d.(ch') - f(ce') - b(eh'),$$
$$B = a(eh'),$$
$$C' = ad.h - af.e,$$

and consequently (257) $C = -ad.h' + af.e'$.

Since $e' = 0$,

$$A = d.(ch') - fc'e - beh',$$
$$B = aeh',$$
$$C' = adh - aef,$$
$$C = -adh'.$$

(270.) Likewise, consider the following seven equations:

$$Ca + A'd' + A''d'' = 0,$$
$$Cb + A'e' + A''e'' + B'b' + B''d'' = 0,$$
$$Cc + B'e' + B''e'' = 0,$$
$$A'f' + A''f'' = 0,$$
$$Cd + C'd' + C''d'' = 0,$$
$$Ce + C'e' + C''e'' + B'f' + B''f'' = 0,$$
$$Cf + C'f' + C''f'' = 0.$$

We ask for the condition equation.

I group the unknowns as $A'A''B'B''CC'C''$; that is, I build three groups, $A'A''$, $B'B''$, and $CC'C''$. The computations in the group $CC'C''$ are performed as if a, b and c came with similar variables a', b', c' and a'', b'', c''. Introducing these quantities does not lengthen computations and makes them easier by recovering the symmetry of the problem. At the end of the computations, I make sure that

$$a' = 0, \; a'' = 0, \; b' = 0, \; b'' = 0, \; c' = 0, \; c'' = 0.$$

Application of previous considerations to different examples:
Interpretations and usage of various factors that are encountered in the
computation of the coefficients in the final equation

(271.) It is important that the degree of the final equation be no higher than what it is in general, that is, when the coefficients in the given equations are arbitrary. It is also important that the elimination method lead to the final equation with lowest possible degree, even when there exist specific relations among the coefficients that can possibly make the degree of the final equation lower. This method must be able to identify the conditions under which such a lower degree can be achieved and to also give a way to construct it.

(272.) The specific relations that can lead to lowering the degree of the final equation are of two kinds: When there exists a common factor to all terms of this equation, which, by becoming zero, cancels the entire equation; or when the coefficients of a few high-dimensional terms in the final equation vanish. This second kind is the only one that was known up until now. The factor that can likewise lead to a lower degree of the final equation escapes the elimination method for two unknowns, and therefore any known elimination method.

(273.) While not increasing the degree of the final equation beyond the general case is a desirable feature, it is not the only one. Indeed, we may not want to eliminate or avoid some of the factors arising from our analysis: When applied rigorously, our analysis does not give anything that is not related to the original question. If factors arise beyond the equation we try to reach, these factors bear specific information relative to the original question. By neglecting or avoiding them, we may miss important information related to the question; we may even draw illogical conclusions. Thus, if we do not know the factor leading to a final equation with lower degree, we may admit solutions to this equation that are not solutions to the original set of equations.

(274.) The presence of factors in the final equation, following the elimination process, is undesirable only when these factors have no relation to the original set of equations. Conversely, it would be a real concern not to reveal all the properties of the original problem.

(275.) When we propose eliminating variables using several equations, we mean to determine all the possible solutions to these equations. When dealing with this question in general, we usually get two types of factors, one of which indicates the potential to lower the degree of the final equation, and another that indicates some specific ways to solve all proposed equations, in certain cases. For as long as we do not restrict our analysis, it will show these factors. If we narrow down our analysis too much, we will reduce the number of such factors, but I seriously doubt we can always eliminate them all.

(276.) Such is the nature of the method we are about to present. We will give two approaches to reach the final equation. Following the first approach, the final equation will always have the lowest possible degree, but it will require the computation of a large number of coefficients. Indeed, independently of the final equation, the factors of these coefficients will indicate either cases when the degree of the final equation drops, or specific solutions.

The second approach is much shorter, and fewer factors will appear. But these factors will contribute, in certain cases, to complicating the degree of the final equation. Nevertheless, we will see that most of the time these factors can be identified early in the calculations, such that we will be able to remove them before the end of the computations. In the case when the computations get to be too complex to identify these factors, we will show how to use the knowledge acquired in the first book to identify these factors.

(277.) Thus, if our objective is to reach the final equation independent of any specific relationships among coefficients, we will use the second approach.

But if we want to extract all the information about the proposed equations without any spurious statement, we must then use the first approach.

(278.) Let us first compute the equation in x resulting from the two equations

$$
\begin{aligned}
ax^2 &+ bxy + cy^2 = 0, \\
+ \; dx &+ ey \\
+ \; f &
\end{aligned}
$$

and

$$
\begin{aligned}
d'x + e'y &= 0. \\
+ \; f' &
\end{aligned}
$$

The polynomial multiplier of the first equation must be of the form $(x, y)^{T+1}$, and that of the second equation must be of the form $(x, y)^{T+2}$, where T is arbitrary. But since we want to use the simplest polynomial multipliers, and since we see from (47) that the final equation must be second-order only, it is sufficient that the polynomial multiplier of the first equation be degree zero, leading to $T = -1$; and the polynomial multiplier of the second equation is therefore of the form $(x, y)^1$.

We therefore must multiply the first equation by C and the second one by $A'x + B'y + C'$.

However, let us assume $T = 0$ to give an idea of what happens if higher degree polynomial multipliers are used, leading to $(x, y)^1$ and $(x, y)^2$ as polynomial multipliers, that is,

$$
Dx + Ey + F, \text{ and } A'x^2 + B'xy + C'y^2 + D'x + E'y + F'.
$$

The sum equation would be the following equation:

$$
\begin{aligned}
& Da & x^3 & + Db & x^2y & + Dc & xy^2 & + Ec & y^3 = 0.\\
+\ & A'd' & & + Ea & & + Eb & & + C'e'\\
& & & + B'd' & & + C'd'\\
& & & + A'e' & & + B'e'
\end{aligned}
$$

$$
\begin{aligned}
+\ & Dd & x^2 & + De & xy & + Ee & y^2\\
+\ & D'd' & & + D'e' & & + E'e'\\
+\ & Fa & & + Ed & & + Fc\\
+\ & A'f' & & + E'd' & & + C'f'\\
& & & + Fb\\
& & & + B'f'
\end{aligned}
$$

$$
\begin{aligned}
+\ & Df & x & + Ef & y\\
+\ & D'f' & & + E'f'\\
+\ & Fd & & + Fe\\
+\ & F'd' & & + F'e'
\end{aligned}
$$

$$
\begin{aligned}
+\ & Ff\\
+\ & F'f'
\end{aligned}
$$

We first examine how many coefficients are useless for elimination purposes. From what was said up until now, there are $N(x,y)^0 = 1$ such coefficients. Thus one of the coefficients is arbitrary. We put it to best use by assigning the value 0 to it. Since it does not matter which coefficient that is, we will assume $C' = 0$.

Since the final equation must be second order, it must be possible to cancel both y^3 and x^3 terms.

By setting the coefficients corresponding to the terms of highest dimensions to zero, we obtain four first-order equations with no known terms and four unknowns only, since we have set $C' = 0$. Thus, by (213), each of these unknowns is equal to zero, that is,

$$A' = 0,\ B' = 0,\ D = 0,\ E = 0,\ C' = 0.$$

Thus indeed our earlier choice of multipliers was the best.
The sum equation indeed reduces to

$$
\begin{aligned}
& D'd' & x^2 & + D'e' & xy & + E'e' & y^2 = 0.\\
+\ & Fa & & + E'd' & & + Fc\\
& & & + Fb
\end{aligned}
$$

$$
\begin{aligned}
+\ & D'F' & x & + E'f' & y\\
+\ & Fd & & + Fe\\
+\ & F'd' & & + F'e'
\end{aligned}
$$

$$
\begin{aligned}
+\ & Ff\\
+\ & F'f'
\end{aligned}
$$

Thus, computing $D'E'FF'$ according to (198) and (267), we obtain the values of D', E', F, F' as follows:

First line $-D'e'.FF' + D'E'.cF'$, from the term y^2.

Second line $-e'e'.FF' + D'e'.bF' + e'E'.cF' - D'd'.cF' - D'E'.(bc')$, from the term xy.

Third line $-e'e'.eF' - D'e'.(be') + e'f'.cF' - (e'E' - D'd').(ce')$, because $(bc') = bc' - b'c = 0$.

We therefore get $D' = d'(ce') - e'(be')$, $E' = -e'(ce')$, $F' = c.e'f' - e.e'e'$, and consequently $F = +e'.e'e'$. In other terms, since $b' = 0$ and $c' = 0$, $D' = cd'e' - be'e'$, $E' = -ce'e'$, $F' = ce'f' - ee'e'$, $F = e'^3$.

Substituting these values in the sum equation, we get

$$e'[(cd'd' - bd'e' + ae'e')x^2 + [(de' - d'e)e' - f'(be' - 2cd')]x$$
$$+(fe' - f'e)e' + cf'f'] = 0.$$

As for the equation in y, the process is completely similar. It can be obtained by replacing a with c, d with e, and d' with e'.

We must now make a few observations about this result.

(279.) We first notice that the final equation has e' as a common factor. Since there is no superfluous coefficient, we can be assured that the final equation only contains information about the initial set of equations. This being said, what is the meaning of $e' = 0$, which leads to the disappearance of the final equation?

It means that $e' = 0$ satisfies the two proposed equations.

Indeed, the equation $d'x + e'y + f' = 0$ yields $y = \frac{-d'x - f'}{e'}$, which becomes $y = \frac{-d'x - f'}{0}$ when $e' = 0$; since we simultaneously have $d'x + f' = 0$, we get $y = \frac{0}{0}$; but it is clear that, when we substitute this value in the other equation in x and y, we cancel all its terms and therefore this value satisfies the equation.

This value of e' bears another significance. It shows that the equation in x is not second but only of first degree. We will see later in a general statement that when the final equation is computed using the smallest number of coefficients, it will always have two kinds of factors; one of these factors simply indicates that when one of these factors is zero, the equations are satisfied in the sense we just showed; the other kind of factor is the criterion by which we will be able to evaluate whether the degree of the final equation can or cannot be lowered. In our case, there is only one factor e', and it conveys both meanings.

Indeed, if we look for the condition such that the final equation is of first degree only, we see that we must lower the degree of the polynomial multiplier by one unit; this yields a polynomial of the form $(x, y)^{-1}$ for the first equation, whose number of coefficients is zero. The polynomial multiplier of the second equation is of the form $(x, y)^0$, whose number of terms is 1. Thus it is enough to multiply the second equation by the undetermined coefficient A, resulting in

$$Ad'x + Ae'y + Af' = 0.$$

To get the final equation we only have to assume $Ae' = 0$. Since by hypothesis A cannot be zero, we must therefore have $e' = 0$; thus we must have $e' = 0$ so that the final equation be of first degree only.

Conversely, if $e' = 0$, the degree of the final equation can indeed be lowered to one.

(280.) Assume we divide the second-order final equation above by e' and set $e' = 0$. This equation reduces to

$$cd'd'x^2 + 2f'cd'x + cf'f' = 0 \text{ or } c.(d'x + f')^2 = 0,$$

which yields $d'x + f' = 0$, as it should. However, this equation indicates the presence of a double root for x. This conclusion may be considered to be correct, since y has two values corresponding to $x = \frac{-f'}{d'}$. However, it would be a serious mistake to believe these values of y are solutions to the actual final equation, even when its degree is lower.

The following example will show that not all these roots are admissible.

(281.) We propose to obtain the final equation in x resulting from the following two equations

$$\begin{aligned} axy \\ +bx + cy \\ +d \end{aligned} = 0$$

and

$$\begin{aligned} a'xy \\ +b'x + c'y \\ +d' \end{aligned} = 0.$$

From (227), the polynomial multiplier for each equation must be of the form $(x^{A+1}, y^{A+1})^{T+2}$. Since from (62) the final equation must be second-order only, we can simplify our computations by assuming $A = 0$, $\underset{'}{A} = 0$, and $T = 0$.

Multiplying these two equations by the polynomials $(x^1, y^1)^2$ and $(x^1, y^1)^2$, respectively, that is, by

$$Axy + Bx + Cy + D, \quad A'xy + B'x + C'y + D',$$

and adding the two products together, the sum equation has the following form, in which we only write the terms of the first product, since those coming from the second product are similar

$$Aa \quad x^2y^2$$

$$\begin{aligned}+ \quad &Ab \quad x^2y \quad + \quad Ac \quad xy^2\\ + \quad &Ba \qquad\qquad + \quad Ca\end{aligned}$$

$$\begin{aligned}+ \quad &Bb \quad x^2 \quad + \quad Ad \quad xy \quad + \quad Cc \quad y^2\\ &\qquad\qquad\quad + \quad Bc\\ &\qquad\qquad\quad + \quad Cb\\ &\qquad\qquad\quad + \quad Da\end{aligned}$$

$$\begin{aligned}+ \quad &Bd \quad x \quad + \quad Cd \quad y\\ + \quad &Bb \qquad\quad + \quad Dc\end{aligned}$$

$$+ \quad Dd.$$

The number of useless coefficients is $N(x^0, y^0)^0 = 1$. I can therefore assume one arbitrary coefficient is equal to 0; but given the similarity between the two equations, there is no particular reason to pick this coefficient from any particular polynomial multiplier. Thus I will determine this coefficient through an arbitrary equation that is symmetric with respect to both polynomials.

I assume, for example, $Ac + A'c' = 0$; since the equation $Aa + A'a' = 0$ must be satisfied to cancel the x^2y^2 term, I conclude that $A = 0$ and $A' = 0$ from (213), reducing our problem to computing the value of $BB'CC'DD'$.

In order to make sure that the number of coefficients I need to compute is the smallest possible, I first examine whether some equations lead some coefficients to be equal to zero. I observe that this is the case for the equations provided by the terms xy^2 and y^2, where the first is

$$Ac + A'c' + Ca + C'a' = 0,$$

that is, $Ca + C'a' = 0$. The second is

$$Cc + C'c' = 0,$$

which leads to $C = 0$ and $C' = 0$. The question is therefore to compute the value of $BB'.DD'$.

By going through the terms x^2y, xy and y successively, I find

First line $\quad aB'.DD'$
Second line $\quad (ac')DD' - aB'.aD'$
Third line $\quad (ac')cD' + aB'(ac')$,

from which, by (198), we get $D' = c.(ac')$, $B' = a(ac')$, and consequently, by (257), $D = -c'(ac')$ and $B = -a'(ac')$.

Substituting these values in the terms that remain in the sum equation, we get

$$-(ac')[(ab')x^2 + [(ad') - (bc')]x + (cd')] = 0.$$

(282.) There is again a common factor (ac'), and when this factor is equal to zero, the degree of the final equation may be lowered.

Indeed, looking for the condition under which the degree of the final equation can be lowered is equivalent to employing polynomial multipliers whose degree is lower by one unit. That is, we must use polynomial multipliers of the form $(x,y)^0$ since those we have employed so far are of first degree. We must therefore multiply the first equation by A and the second by A'.

The resulting sum equation is

$$\begin{aligned} Aaxy \qquad\quad &= 0, \\ +Abx + Acy& \\ +Ad,& \end{aligned}$$

where we have only written the terms of the first product for the sake of simplicity. The terms of the second product are similar and easy to add.

Since by hypothesis A and A' cannot be zero, the two equations provided by the terms xy and y lead to

First line $\qquad aA'$
Second line $\quad (ac')$;

that is, the condition equation is $(ac') = 0$ and A' and A take the values $A' = a$, $A = -a'$.

When we substitute these values in the terms remaining in the sum equation, it reduces to $(ab')x + (ad') = 0$, leading to the only value x can take in this case.

But if we first divide the original, second-degree final equation by (ac') and then express the condition $(ac') = 0$, that is, $ac' - a'c = 0$, replacing c' by $\frac{a'c}{a}$, the equation becomes

$$(ab')x^2 + \left[(ad') + (ab')\frac{c}{a}\right]x + (ad')\frac{c}{a} = 0,$$

which can be decomposed into two factors

$$(ab')x + (ad')$$
$$\text{and} \qquad x + \frac{c}{a}.$$

However, only the first of these two factors can hold; that is, we can assume $(ab')x + (ad') = 0$, but in no way can we assume $x + \frac{c}{a} = 0$, or $ax + c = 0$. Indeed, the value $x = -\frac{c}{a}$ does not yield any corresponding value for y.

Indeed, if this value of x is substituted in each of the proposed equations, keeping in mind that $(ac') = 0$, y disappears in both equations. Thus no value of y corresponds to $x = -\frac{c}{a}$.

(283.) Despite this irrefutable proof, we could be tempted to formulate one objection.

We could think that the value $x = -\frac{c}{a}$ corresponds to an infinite value for y. Indeed, if we assume y to be infinite, the quantity $axy + by + cy + d$ reduces to $axy + cy$, and the quantity $a'xy + b'x + c'y + d'$ reduces to $a'xy + c'y$; the two equations therefore appear to reduce to

$$axy + cy = 0,$$
$$a'xy + c'y = 0,$$

and assuming $(ac') = 0$, these two equations hold if $ax + c = 0$.

But we must observe that the quantity $axy + bx + cy + d$ reduces to $axy + cy$ for an infinite y only if $axy + cy$ is infinite relative to $bx + d$. However, the converse occurs, since we pretend the equation $axy + cy = 0$ is true. We would therefore have $axy + cy$ simultaneously infinite and zero; this is absurd. Thus we cannot assume that y is infinite and that no value of y, finite or infinite, corresponds to $x = -\frac{c}{a}$.

We could insist and say that when y is infinite, we can neither neglect $bx + d$ relative to $axy + cy$ nor neglect $b'x + d'$ relative to $a'xy + c'y$, and that the two equations

$$\begin{aligned} axy + cy + bx + d &= 0, \\ a'xy + c'y + b'x + d' &= 0 \end{aligned}$$

can hold when $x = -\frac{c}{a}$, by assuming y is infinite. Indeed, the first equation becomes

$$0y + bx + d = 0,$$

and, remembering $c' = \frac{a'c}{a}$, the second equation becomes

$$0.\frac{a'}{a}y + b'x + d' = 0 \text{ or } 0y + \frac{a}{a'}.(b'x + d') = 0.$$

Both equations can hold assuming y to be infinite.

The answer to this would be that it is not sufficient that each equation be satisfied assuming y to be infinite. These infinite values must be the same for each equation. However, the first equation yields $y = -\frac{(bx+d)}{0}$ and the second yields $y = -\frac{(b'x+d')a}{0.a'}$. these values are infinitely far away when we assume $x = -\frac{c}{a}$.

The theory of equations contains many similar cases, where each equation can be satisfied by assuming y to be infinite; however, this infinite value must be the same in each equation, or at least differ by a finite quantity from one equation to another, to be a true solution to our problem.

(284.) We must observe here that the usual elimination method for equations in two unknowns was the only method able to give the true degree of the final equation. However, this method can still yield useless, or even wrong roots. Indeed, by following this method we are immediately led to the equation

$$(ab')x^2 + [(ad') - (bc')]x + (cd') = 0,$$

without any indication of the cases when only one of these roots is admissible.

This elimination method is based on an approach to the problem that is too narrow, because it cannot recognize the symptoms our analysis is able to unveil.

(285.) Our third example is about finding the final equation in x, resulting from the following two equations:

$$\begin{aligned} ax^2 + bxy + cy^2 + dx + ey + f &= 0, \\ a'x^2 + b'xy + c'y^2 + d'x + e'y + f' &= 0. \end{aligned}$$

The polynomial multiplier of each equation has the form $(x, y)^{T+2}$. Since, from (47), the final equation cannot exceed the fourth degree, the simplest polynomial multiplier is $(x, y)^2$, or $Ax^2 + Bxy + Cy^2 + Dx + Ey + F$, for the first equation, and $A'x^2 + B'xy + C'y^2 + D'x + E'y + F'$ for the second equation. If we build the products and add them together, the sum equation takes the following form, where the terms of the second product have been omitted:

$$Aax^4 \quad + \quad Abx^3y \quad + \quad Acx^2y^2 \quad + \quad Bcxy^3 \quad + \quad Ccy^4 = 0.$$
$$+ \quad Ba \qquad + \quad Bb \qquad + \quad Cb$$
$$+ \quad Ca$$

$$+ \quad Adx^3 \quad + \quad Aex^2y \quad + \quad Bexy^2 \quad + \quad Cey^3$$
$$+ \quad Da \qquad + \quad Bd \qquad + \quad Cd \qquad + \quad Ec$$
$$+ \quad Db \qquad + \quad Dc$$
$$+ \quad Ea \qquad + \quad Eb$$

$$+ \quad Afx^2 \quad + \quad Bfxy \quad + \quad Cfy^2$$
$$+ \quad Dd \qquad + \quad De \qquad + \quad Ee$$
$$+ \quad Fa \qquad + \quad Ed \qquad + \quad Fc$$
$$+ \quad Fb$$

$$+ \quad Dfx \quad + \quad Efy$$
$$+ \quad Fd \qquad + \quad Fe$$

$$+ \quad Ff$$

The number of coefficients that are useless for elimination purposes is $N(x, y)^0 = 1$. Thus one of the coefficients can be chosen arbitrarily, but since there is no particular reason to choose that coefficient in one polynomial rather than another, I choose this coefficient by adding a new arbitrary equation that is symmetric with respect to the two equations. Although the choice of this equation is rather arbitrary, I prefer to choose an equation that can help cancel a few coefficients. I would rather assume, for example, $Cb + C'b' = 0$, or $Ca + C'a' = 0$, or $Ce + C'e' =$ or, etc., because making such an assumption, combined with the equation we get to cancel the term y^4 will give $C = 0$ and $C' = 0$.

This being done, we need to compute the value of $AA'BB'DD'EE'FF'$. Following the terms x^3y, x^2y^2, xy^3, x^2y, xy^2, y^3, xy and y, I get the following

First line	$bA'.BB' + AA'.aB'$
Second line	$(bc')BB' - bA'.bB' + cA'.aB' + AA'(ab')$
Third line	$[(bc')cB' + (bc')bA' - (ac')cA']DD'.EE'$
Fourth line	$[(bc').(cd') + (bc').(be') - (ac').(ce')]DD'.EE'$
	$-[(bc')cB' + (bc')bA' - (ac')cA'](bD'.EE' + DD'.aE')$.

For the sake of brevity, we write $(bc').(cd') + (bc').(be') - (ac').(ce') = (1)$, and we get

Fourth line $\quad$ $(1)DD'.EE'$
$\qquad -[(bc')cB' + (bc')bA' - (ac')cA'](bD'.EE' + DD'.aE')$

Fifth line $\quad$ $(1)(cD'.EE' + DD'.bE')$
$\qquad -(bc').(ce').(bD'.EE' + DD'.aE')$
$\qquad +[(bc')cB' + (bc')bA' - (dc')cA'].[(bc')EE'$
$\qquad -bD'.bE' + cD'.aE' + (ab')DD']$

Sixth line $\quad$ $(1)[-cD'.cE' + (bc')DD']$
$\qquad -(bc').(ce').[-bD'.cE' + (ac')DD']$
$\qquad -[(bc')cB' + (bc')bA' - (ac')cA']$
$\qquad \times[(bc')cE' + (bc')bD' - (ac')cD']$

Seventh line
$$\left.\begin{array}{r} (1)[-(ce').cE' + (cd')cD' + (bc')eD'] \\ -(bc').(ce').[-(be')cE' + (cd')bD' + (ac')eD'] \\ -(bc').(cf').[(bc')cE' + (bc')bD' \\ -(ac')cD'] + (1)[(bc')bA' - (ac')cA'] \end{array}\right\} FF'$$
$\qquad -[(1).cD'.cE' - (bc').(ce').bD'.cE'$
$\qquad +[(bc')bA' - (ac')cA'].(bc')cE']bF',$

where we omit the terms where DD', B' and $A'D'$ remain, because they disappear in the lines that follow or yield terms that will disappear later.

Eighth line
$[-(1)(ce')^2 + (bc').(be').(ce')^2] - (bc')^2.(ce').(cf')]FF'$
$$\left.\begin{array}{r} -(1)[-(ce')cE' + (cd')cD' + (bc')eD'] \\ +(bc').(ce').[-(be')cE' + (cd')bD' + (ac')eD' \\ +(bc').(cf').[(bc')cE' + (bc')bD' - (ac')cD'] \\ -(1)[(bc')bA' - (ac')cA'] \end{array}\right\} cF'$$
$+[(1)(ce')cD' - (bc').(ce')^2bD'] + (bc').(ce').[(bc')bA' - (ac')cA']]bF'$
$-(bc').[(1)cD'.cE' - (bc').(ce').bD'.cE' + [(bc')bA' - (ac')cA'].(bc')cE']$

Ninth line
$[-(1)(ce')^2 + (bc').(be').(ce')^2 - (bc')^2.(ce').(cf')]eF' + [(1)(ce').(cf')$
$-(bc').(be').(ce').(cf') + (bc')^2.(cf')^2]cF'$
$+(ce').[(1)(cd')cD' + (bc')eD'] - (bc').(ce').[(cd')bD' + (ac')eD']$
$-(bc').(cf').[(bc')bD' - (ac')cD'] + (1)[(bc')bA' - (ac')cA']]$
$-(be')[(1)(ce')cD' - (bc').(ce')^2bD' + (bc').(ce').[(bc')bA' - (ac')cA']$
$+(bc').(cf')[(1)cD' - (bc').(ce')bD' + (bc').[(bc')bA' - (ac')cA']],$

omitting terms where E' remains but that we do not need.

Before extracting the values of A, A'; D, D'; etc. from the ninth line, we remark that the coefficient of F' and that of A' share the same common factor

$$-(bc').(be').(ce') + (1)(ce') + (bc')^2.(cf').$$

The coefficient of D' has the same factor, although it is not as easy to identify; here is how it may be found.

From the theorems in (219), we have $(bc')e - (be')c + (ce')b = 0$; replacing $(bc')e$ by its value from this equation in the coefficient of eD' the total set of coefficients of D' is

$$\left.\begin{array}{l} -(bc').(ce')^2.(cd') \\ -2(bc')^2.(ce').(cf') \\ +(bc').(be').(ce')^2 \\ -(1).(ce')^2 \\ +(ac').(ce')^3 \end{array}\right\} bD' \qquad \left.\begin{array}{l} +(1).(ce').(cd') \\ +(bc').(ac').(ce').(cf') \\ +(1).(bc').(cf') \\ -(ac').(ce')^2.(be') \end{array}\right\} cD'.$$

Replacing $(ac').(ce')$ by its value $(bc').(be') + (bc').(cd') - (1)$ we get

$$\left.\begin{array}{l} -2(bc')^2.(ce').(cf') \\ +2(bc')^2.(be').(ce')^2 \\ -2(1).(ce')^2 \end{array}\right\} bD' \qquad \left.\begin{array}{l} +(1).(ce').(cd') \\ +(bc')^2.(cf').(cd') \\ -(bc').(ce').(be').(cd') \\ +(1).(ce').(be') \\ +(bc')^2.(cf').(be') \\ -(bc').(ce').(be')^2 \end{array}\right\} cD',$$

that is,

$$-2[(bc')^2.(cf') + (1).(ce') - (bc').(be').(ce')].(ce')bD'$$
$$+[(bc')^2.(cf') + (1).(ce') - (bc').(be').(ce')][(cd') + (be')]cD'.$$

For the sake of brevity, we define

$$(2) = (bc')^2.(cf') + (1)(ce') - (bc').(be').(ce').$$

The ninth line then reduces to the quantity

$$(2)([(cf')c - (ce')e]F' + ([(cd') + (be')]c - 2(ce')b)D' + [(bc')b - (ac')c]A'),$$

from which we get

$$A' = (2).[(bc')b - (ac')c], \; D' = (2).([(cd') + (be')]c - 2(ce')b), \; F' = (2).[(cf')c - (ce')e];$$

and consequently

$$A = -(2).[(bc')b' - (ac')c'], \; D = -(2).([(cd') + (be')]c' - 2(ce')b'),$$
$$F = -(2).[(cf')c' - (ce')e'].$$

Substituting these values in the terms that remain in the sum equation, we obtain the following final equation:

$$(2)\left\{\begin{array}{l} (ac')^2 \\ -(ab').(bc') \end{array}\right\} x^4 \quad \left.\begin{array}{l} +(bc').(bd') \\ -2(ac').(cd') \\ -(be').(ac') \\ +2(ab').(ce') \end{array}\right\} x^3 \quad \left.\begin{array}{l} +(bc').(bf') \\ -2(ac').(cf') \\ +(cd')^2 \\ +(be').(cd') \\ -2(ce').(bd') \\ +(ae').(ce') \end{array}\right\} x^2$$

$$\left.\begin{array}{l} +2(cd').(cf') \\ +(be').(cf') \\ -2(ce').(bf') \\ +(ce').(de') \end{array}\right\} x \quad \left.\begin{array}{l} +(cf')^2 \\ -(ce').(ef') \end{array}\right\} = 0,$$

which, from the equations (220), yields

$$(ab').(ce') - (ac').(be') + (bc').(ae') = 0,$$
$$(bc').(de') - (bd').(ce') + (cd').(be') = 0,$$
$$(bc').(ef') - (be').(cf') + (ce').(bf') = 0$$

and can also be written

$$(2) \left\{ \begin{array}{c} (ac')^2 \\ -(ab').(bc') \end{array} \right\} x^4 + \left\{ \begin{array}{c} +(bc').(bd') \\ -2(ac').(cd') \\ +(ab').(ce') \\ -(bc').(ae') \end{array} \right\} x^3 + \left\{ \begin{array}{c} +(bc').(bf') \\ -2(ac').(cf') \\ +(cd')^2 \\ -(ce').(bd') \\ -(bc').(de') \\ +(ae').(ce') \end{array} \right\} x^2$$

$$+ \left\{ \begin{array}{c} +2(cd').(cf') \\ -(ce').(bf') \\ +(bc').(cf') \\ +(ce').(de') \end{array} \right\} x + \left\{ \begin{array}{c} +(cf')^2 \\ -(ce').(ef') \end{array} \right\} = 0.$$

This equation differs from the formulas resulting from the ordinary elimination method only by the factor (2), which eludes this method, and that we must now discuss.

(286.) The factor (2) is precisely the factor that expresses the cases when the degree of the equation in x can be lowered and reduced to the third degree. The same doe not hold for the equation in y.

If $(ac')^2 - (ab')(bc') = 0$, then the equation in x and the equation in y would be third degree. However, this is the only known case where this happens. But if we had $(2) = 0$, the equation in x would also be third degree, and the ordinary elimination method does not detect this.

Establishing the proof that when $(2) = 0$ the final equation is third degree can only be done by looking for the final equation in x, using only first-degree polynomial multipliers. Following the same process as above, we reach the condition $(2) = 0$. Thus the factor (2) is indeed the indicator of whether the degree of the final equation can be lowered.

(287.) We now make a remark to justify our statement (279) and to clarify what follows later.

We have said in (279) that the final equation, as determined by our method, always offers two kinds of factors, where one indicates when the degree of the equation can be lowered and the other indicates that a solution naturally occurs when certain terms of the proposed equations are missing.

The previous example and its predecessor carried the first kind of factor only. Why is that so? Because we have proceeded so that this factor (which we will call *factor of the second type*) does not appear.

Indeed, considering the current example, we have reduced the question to computing ten coefficients only, although out of the twelve coefficients contained in the two polynomials, only one qualifies as a useless coefficient (230) for elimination purposes. If we had performed our computations using all eleven coefficients instead of assuming $C = 0$ and $C' = 0$, we would have assumed $C = 0$ only and would have computed the value of $AA'BB'C'DD'EE'FF'$. The final equation would have had the factor c', which, if we assume it is zero, does not indicate that the degree of the equation can be lowered; rather, it indicates the presence of a solution whose nature is described in (279). When $c' = 0$, the solution $y^2 = \frac{0}{0}$ satisfies the other equation.

Suppose that, instead of assuming $C = 0$ and keeping C' in our calculation of $AA'BB'$, etc., we had kept the assumption from (285): $Cb + C'b' = 0$; and suppose that, instead of suppressing C and C' right away in all terms where they appear, we had proceeded with the calculation of $AA'BB'CC'DD'EE'FF'$ using the eleven equations then available to us; we then find in the final equation the factor $bc' - b'c$. This factor is more complex than it would have been if we had used only the number of useful coefficients for elimination purposes, but it is a variant of the kind of factors of interest here. Indeed, if we call E the final equation without this factor, the final equation is then $(bc' - b'c)E = 0$.

By choosing $C' = 0$, we get $c'E = 0$. By choosing $C = 0$, we get $cE = 0$; the equation $(bc' - b'c)E = 0$ can thus be seen as the union of these two: $bc'E = 0$ and $-b'cE = 0$. These two equations consist of the six cases $b = 0$, $c' = 0$, $E = 0$, $b' = 0$, $c = 0$, $E = 0$. And each of the four cases $b = 0$, $c' = 0$, $b' = 0$, $c = 0$ is indeed the indicator of a solution of the type described in (279). If $b = 0$, for example, the first of the two equations yields $xy = \frac{-ax^2-cy^2-dx-ey-f}{0} = \frac{0}{0}$, where the latter equality holds because $ax^2 + cy^2 + dx + ey + f = 0$.

(288.) We therefore see that we have not found factors of the second type in the above example because we have avoided them. Since they bring no pertinent information, it is always good to get rid of them whenever possible. I say "whenever possible" because this operation is most often, but not always, possible, as we will soon see.

(289.) We have extended our discussion about these factors in order to prepare the reader to accept our decision to compute a few more coefficients than is absolutely necessary. Indeed, this will allow us to keep symmetry in our computations, which makes it much easier; the converse operation, to maximally reduce the number of factors, can sometimes perturb the symmetry and, although the factor is simpler, it is less easy to find. But when we perform our computations using more coefficients than are necessary for elimination purposes, the factor we will find will only contain, in a more extended way, the same information as that expressed by the factor resulting from using only the useful coefficients for the purpose of elimination.

(290.) We now return to examining the factor (2). If we replace (1) by its value

$$(bc').(be') - (ac').(ce') + (bc').(cd'),$$

we get

$$(2) = (bc').(cd').(ce') - (ac').(ce')^2 + (bc')^2.(cf').$$

Thus each time the equation

$$(bc').(cd').(ce') - (ac').(ce')^2 + (bc')^2.(cf') = 0$$

holds among the coefficients of the two equations, the final equation in x will be third degree. We get this equation by using, as we said, first-degree polynomial multipliers.

Thus, setting

$$A = 0, \ A' = 0, \ B = 0, \ B' = 0, \ C = 0, \ C' = 0$$

the sum equation reduces to the form

$$Dax^3 \quad + \quad Dbx^2y \quad + \quad Dcxy^2 \quad + \quad Ecy^3 = 0.$$
$$\qquad\qquad + \quad Ea \qquad\qquad + \quad Eb$$

$$+ \quad Ddx^2 \quad + \quad Dexy \quad + \quad Eey^2$$
$$+ \quad Fa \qquad + \quad Ed \qquad + \quad Fc$$
$$\qquad\qquad + \quad Fb$$

$$+ \quad Dfx \quad + \quad Efy$$
$$+ \quad Fd \quad + \quad Fe$$

$$+ \quad Ff$$

The terms y^3, xy^2, x^2y, y^2, xy lead to the following last line:

$$-(ce').(bc')bF' - [-(bc').(be') + (ac').(ce') - (bc').(cd')]cF'$$
$$-[-(bc')bD' + (ac')cD' - (bc')cE'].(bc'),$$

used to compute D, D', E, E', F and F'. Combining this equation with the equation given by the y term yields the condition equation

$$(bc').(cd').(ce') - (ac').(ce')^2 + (bc')^2.(cf') = 0,$$

as predicted. At the same time, we get

$$F' = -(ce').(bc')b + [(bc').(be') - (ac').(ce') + (bc').(cd')]c,$$
$$D' = (bc').(bc')b - (ac').(bc')c.$$

Consequently

$$F = (ce').(bc')b' - [(bc').(be') - (ac').(ce') + (bc').(cd')]c',$$
$$D' = (bc')^2b' + (ac').(bc')c'.$$

Replacing the remaining terms in the sum equation, we get

$$\left.\begin{array}{l} -(ab').(bc')^2 \\ +(ac')^2.(bc') \end{array}\right\}x^3 \left.\begin{array}{l} +(bc')^2.(bd') \\ -2(ac').(bc').(cd') \\ +(ab').(bc').(ce') \\ -(bc').(be').(ce') \\ +(ac')^2.(ce') \end{array}\right\}x^2 \left.\begin{array}{l} +(bc')^2.(bf') \\ -(ac').(bc').(cf') \\ -(bc').(ce').(bd') \\ +(bc').(be').(cd') \\ -(ac').(ce').(cd') \\ +(bc').(cd').(cd') \end{array}\right\}x$$

$$\left.\begin{array}{l} -(bc').(ce').(bf') \\ +(bc').(be').(cf') \\ -(ac').(ce').(cf') \\ +(bc').(cd').(cf') \end{array}\right\} = 0,$$

Assume we multiply the latter equation by (ce'), and then replace $(ac').(ce')^2$ by its value

$$(bc').(cd').(ce') + (bc')^2.(cf')$$

given by the condition equation. We then easily see that (bc') is a factor of the equation. This factor indicates the conditions for which the degree of the final equation may be lowered to the second degree.

In other terms, if the two equations

$$(bc').(cd').(ce') - (ac').(ce')^2 + (bc')^2.(cf') = 0$$
$$\text{and } (bc') = 0$$

hold at the same time, or, equivalently if the two equations

$$(ac').(ce')^2 = 0$$
$$\text{and } (bc') = 0$$

hold at the same time, then the final equation reduces to the second degree.

The equation $(ac').(ce')^2 = 0$ can be decomposed in two cases, $(ce') = 0$ and $(ac') = 0$.

In the first case it is obvious that, if we have $(ce') = 0$ and $(bc') = 0$ simultaneously, the final equation will be of degree 2. Indeed, we get the equation $(ac')x^2 - (cd')x - (cf') = 0$ by multiplying the first equation by c' and subtracting the second equation multiplied by c from it.

In the second case, the same operation gives

$$(cd')x + (ce')y + (cf') = 0$$

if $(ac)'$ and (bc') simultaneously vanish. This equation, when combined with any of the two proposed equations, can again only yield an equation of degree 2.

(291.) As can be seen, we have gone into a great deal of detail on the two equations that constitute the latter example; this attention to detail is justified by its consequences. We leave it up to the reader to apply the same techniques to find the equation in y: It is easy to see that the consequences are similar, but not the same.

For example, the equation

$$(bc').(cd').(ce') - (ac').(ce')^2 + (bc')^2.(cf') = 0,$$

which must hold for the equation in x to become third degree, becomes

$$-(ab').(ae').(ad') + (ac').(ad')^2 + (ab')^2.(af') = 0$$

by replacing a by c, a' by c', d by e, and d' by e'. This replacement is necessary to apply what we know about the equation in x to the equation in y. We also see that one equation does not necessarily drop degree when the other does.

(292.) We now take as a matter of example the following three equations:

$$ax^2 \ + \ bxy \ + \ cxz \ + \ dy^2 \ + \ eyz \ + \ fz^2 = 0,$$
$$+ \ gx \ + \ hy \ + \ kz$$
$$+ \ l$$

$$g'x \ + \ h'y \ + \ k'z \ = \ 0,$$
$$+ \ l'$$

$$g''x \ + \ h''y \ + \ k''z \ = \ 0.$$
$$+ \ l''$$

The forms of the polynomial multipliers of the first, second and third equations are

$$(x, y, z)^{T+2}, \ (x, y, z)^{T+3}, \ (x, y, z)^{T+3};$$

since the final equation must not exceed the second degree, from (47), we can assume $T + 2 = 0$ for simplicity.

We will therefore multiply the first equation by L, the second by $G'x + H'y + K'z + L'$, and the third by $G''x + H''y + K''z + L''$. Adding these three products together, the sum equation has the following form

$$
\begin{array}{llllll}
La \ x^2 & +Lb \ xy & +Lc \ xz & +Ld \ y^2 & +Le \ yz & +Lf \ z^2 = 0. \\
+G'g' & +G'h' & +G'k' & +H'h' & +H'k' & +K'k' \\
+G''g'' & +G''h'' & +G''k'' & +H'h'' & +K''k'' & \\
 & +H'g' & +K'g' & & +K'h' & \\
 & +H''g'' & +K''g'' & & +K''h'' &
\end{array}
$$

$$
\begin{array}{lll}
+Lg \ x & +Lh \ y & +Lk \ z \\
+G'l' & +H'l' & +K'l' \\
+G''l'' & +H''l'' & + \quad K''l'' \\
+L'g' & +L'h' & +L'k' \\
+L''g'' & +L''h'' & +L''k''
\end{array}
$$

$$
\begin{array}{l}
+Ll \\
+L'l' \\
+L''l''
\end{array}
$$

In this case, the number of useless coefficients for elimination purposes is

$$N(x, y, z)^{T+1} + N(x, y, z)^{T+1} - N(x, y, z)^{T} + N(x, y, z)^{T+2};$$

that is,

$$2N(x, y, z)^{-1} + N(x, y, z)^{0} - N(x, y, z)^{-2} = 0 + 1 - 0 = 1.$$

Thus one of the coefficients is free. To keep the problem's symmetry, I do not assume one of the coefficients to be zero, but I create an arbitrary, symmetric equation. I assume, for example,

$$K'h' + K''h'' = 0.$$

Then, using this equation and those provided by the terms z^2, yz, y^2, xz, xy, z and y, we compute $G'G''H'H''K'K''LL'L''$, beginning with $K'K''$

for simplicity, then $K'K''L$, and then $K'K''LHH'$, $K'K''LHH'GG'$ and $K'K''LHH'GG'L'L''$. We find from the fourth line that $(k'h'')$ is a common factor, which we can suppress for simplicity.

The last line will reveal the factor $(k'h'')$ again. It can be suppressed again for simplicity. Neglecting from this line the coefficients H, H', K, K', which do not enter the final equation, we get the expression

$$[f(h'l'') - k(k'h'')]h'L'' - [(e(h'l'') - d(k'l'') + k(k'h'')]k'L'' + (k'h'')^2 L$$
$$+ [c(k'h'') + f(h'g'')]h'G'' - [e(h'g'') - d(k'g'') - b(k'h'')]k'G''.$$

Thus we get

$$
\begin{aligned}
L'' &= [f(h'l'') - k(k'h'')]h' - [e(h'l'') - d(k'l'') + k(k'h'')]k', \\
G'' &= [c(k'h'') + f(h'g'')]h' - [e(h'g'') - d(k'g'') - b(k'h'')]k', \\
L &= (k'h'')^2.
\end{aligned}
$$

Consequently

$$
\begin{aligned}
L' &= -[f(h'l'') - k(k'h'')]h'' - [e(h'l'') - d(k'l'') + k(k'h'')]k'', \\
G' &= -[c(k'h'') + f(h'g'')]h'' - [e(h'g'') - d(k'g'') - b(k'h'')]k''.
\end{aligned}
$$

Substituting these in the remaining terms of the sum equation, we easily get the final equation.

(293.) We can doubtlessly reach this final equation much faster by determining y and z as a function of x using the last two equations, and substituting them into the first.

However, our object is not to reach the final equation as fast as possible: Our main interest is to present a method that does not omit any case arising from the original set of equations. By following the shortest route, the factor $(k'h'')^2$ would not be present in the final equation. We now need to examine what it means.

If $(k'h'') = 0$, then the final equation is of first degree only. And indeed, if we multiply the second equation by h'' and we subtract from it the third equation multiplied by h', we get $(g'h'')x - (h'l'') = 0$, which indeed can give only one value for x.

Here is now the reason why we find $(k'h'')$ factored twice.

Assume we had not arbitrarily chosen $K'h' + K''h'' = 0$, but instead $K' = 0$. The final equation would have had the factor $(k'h'')k''$, which decomposes into $(k'h'')$ and k''. The first factor has the significance we just saw. The second factor bears the significance outlined in (279) and (287). By forming the arbitrary equation $K'h' + K''h'' = 0$, we use one more coefficient than we have to, and thus we have one factor whose dimension is higher by one unit: It is the factor $(k'h'')$, which contains by (287) all the information contained in the factor $(k'h'')k''$. Indeed, the factor $(k'h'').(k'h'')$ represents the union of $(k'h'').k'h''$ and of $-(k'h'').k''h'$. These indicate that solutions of the kind described in (279) and (287) exist for $k' = 0$, $h' = 0$, $k'' = 0$, $h'' = 0$. It presents these solutions in a more complete fashion than the factor $(k'h'')k''$. But since we obtain the factor $(k'h'')k''$ by assuming $K' = 0$, we would get the factor $(k'h'')k'$ by assuming $K'' = 0$. Likewise we would get the factor $(k'h'')h''$ by assuming $H' = 0$, and so on.

This confirms all that was said about the nature of the factors in the final equation.

General remarks about the symptoms indicating the possibility of lowering the degree of the final equation, and about the way to determine these symptoms

(294.) From what we said up until now, the indicators of the possibility of a lower degree for the final equations are of two kinds: One occurs when the final equation has a common factor for all its terms. The second occurs when, once the final equation is freed from this common factor, the coefficient of the highest power of the unknown can become zero because of specific relations among the coefficients of the final equations.

(295.) It would therefore seem that computing the final equation is required to identify the conditions under which its degree can drop, so that this final equation can be inspected for a factor that is common to all its terms and for the coefficients of its highest and lower powers of the unknown.

(296.) In order to determine the indicators of the first kind, there is no need to determine the final equation: We can determine the common factor without knowing the equation itself. Concerning the indicators of the second kind, we will discuss them after having taught a general way to determine this factor.

(297.) Assume, therefore, that the proposed equations are represented by the equations

$$(u \ldots n)^t = 0,$$
$$(u \ldots n)^{t'} = 0,$$
$$(u \ldots n)^{t''} = 0.$$

The required polynomial multipliers to reach the final equation are

$$(u \ldots n)^{T+t'+t''+\text{etc.}},$$
$$(u \ldots n)^{T+t+t''+\text{etc.}},$$
$$(u \ldots n)^{T+t+t'+\text{etc.}},$$
etc.

Assume these polynomials have been reduced to the lowest possible number of terms using methods we will soon uncover. Getting the final equation only requires writing the sum equation, as we did in the preceding examples, and solving for the undetermined coefficients.

If the degree of the final equation can be lowered, this may happen in two ways: (i) All the undetermined coefficients entering the highest dimension of the sum equation equal zero; (ii) the known coefficients of terms of the highest dimension of each proposed equation are such that the highest power of the final equation necessarily vanishes.

In the first case, the form of the polynomial multipliers obviously reduces to

$$(u \ldots n)^{T+t'+t''+\text{etc.}-1},$$
$$(u \ldots n)^{T+t+t''+\text{etc.}-1},$$
$$(u \ldots n)^{T+t+t'+\text{etc.}-1},$$

etc.

(298.) Conversely, if we want to know the conditions for which the degree of final equation can be lowered, we write the sum equation using polynomial multipliers of the form

$$(u \ldots n)^{T+t'+t''+\text{etc.}-1},$$
$$(u \ldots n)^{T+t+t''+\text{etc.}-1},$$
$$(u \ldots n)^{T+t+t'+\text{etc.}-1},$$

etc.

Since the number of undetermined coefficients equals the number of necessary coefficients for eliminating all terms that must be eliminated minus 1, we will obtain a condition equation which will be the indicator sought, if this equation only has one factor. If it has several factors, this indicator will point to the different cases in which the degree can be lowered; or it will be such that certain factors indicate cases for which the degree can be lowered, while others indicate particular solutions such as those we have discussed in (279) and (287). This condition equation always contains the factor(s) indicating when the degree can be lowered.

(299.) We have said that the polynomial multipliers

$$(u \ldots n)^{T+t'+t''+\text{etc.}-1}, \text{ etc.}$$

lead to a sum equation where there is only one fewer available coefficient than necessary to cancel all terms that must be cancelled: We now prove this.

Assume for simplicity we only have three unknowns: The number of terms in the highest dimension of the sum equation is $N(u \ldots 2)^{T+t+t'+t''}$ if this is the generic final equation.

The number of terms in the highest dimension of the first polynomial multiplier, minus the number of coefficients that are useless for elimination, is

$$N(u \ldots 2)^{T+t'+t''} - N(u \ldots 2)^{T+t''} - N(u \ldots 2)^{T+t'} + N(u \ldots 2)^{T}.$$

The number of terms in the highest dimension of the second polynomial multiplier, minus the number of coefficients that are useless for elimination, is

$$N(u \ldots 2)^{T+t+t''} - N(u \ldots 2)^{T+t}.$$

The number of terms in the highest dimension of the third polynomial multiplier is $N(u \ldots 2)^{T+t+t'}$.

This being said, I claim that the following equation holds:

$$(A) \quad \ldots N(u\ldots 2)^{T+t+t'+t''} = N(u\ldots 2)^{T+t'+t''} - N(u\ldots 2)^{T+t''}$$
$$-N(u\ldots 2)^{T+t'} + N(u\ldots 2)^{T} + N(u\ldots 2)^{T+t+t''} - N(u\ldots 2)^{T+t}$$
$$+N(u\ldots 2)^{T+t+t'};$$

that is, the total number of terms in the highest dimension of the sum equation is, in general, precisely equal to the number of useful coefficients provided by each of the highest dimensions of the three polynomial multipliers.

Indeed, equation (A) can be written

$$\left. \begin{array}{l} N(u\ldots 2)^{T+t+t'+t''} - N(u\ldots 2)^{T+t'+t''} - N(u\ldots 2)^{T+t+t'} \\ +N(u\ldots 2)^{T+t'} - N(u\ldots 2)^{T+t+t''} + N(u\ldots 2)^{T+t''} + N(u\ldots 2)^{T+t} \\ \hspace{6cm} -N(u\ldots 2)^{T} \end{array} \right\} = 0,$$

that is,

$$ddN(u\ldots 2)^{T+t+t'+t''} \ldots \begin{pmatrix} T+t+t'+t'' \\ t,t'' \end{pmatrix}$$
$$-ddN(u\ldots 2)^{T+t+t''} \ldots \begin{pmatrix} T+t+t'' \\ t,t'' \end{pmatrix} \quad = \quad 0$$

or

$$d^3 N(u\ldots 2)^{T+t+t'+t''} \ldots \begin{pmatrix} T+t+t'+t'' \\ t,t'',t' \end{pmatrix} = 0.$$

However, from (12) the latter identity is indeed true.

Thus, since the highest dimensions of the three polynomial multipliers provide only as many useful coefficients as there are terms in the sum equation, they bring only one more coefficient than there are terms to be cancelled in that dimension to compute a generic expression for the final equation. Thus, if we assume all these coefficients to be zero, the number of remaining coefficients in the final equation will be one coefficient short of what is necessary.

(300.) Thus, finding the condition equation that indicates whether the degree of the final equation can be lowered and which is the common factor of all terms of the equation only requires multiplying all proposed equations by a polynomial whose dimension is one less than that required to compute the generic final equation. By setting to zero each of the terms of the sum equation other than those containing only the final unknown, we will be led to the condition equation, which indicates whether the degree of the final equation can be lowered. The quantity involved in this condition equation will also be the factor that is common to all terms of the generic final equation.

(301.) When the coefficients in the proposed equations take numerical values and we compute the final equation, all the coefficients of the last line, used to compute the final equation, are zero if the degree of the final equation is lower. Conversely, if the last computed line becomes zero when computing the final equation, it proves that the degree of the equation can be lowered by one and that the dimension of the polynomial multipliers must be lowered by one unit.

Indeed, since the last line becomes zero, it shows by (205) that the equations used to compute the lines are not independent; however, there is one arbitrary coefficient among all the undetermined coefficients. We can choose this coefficient to be in the highest dimension and decide it is equal to zero. Then only useful coefficients remain in this highest dimension, and their number equals the number of terms to be cancelled; thus each of these coefficients must become zero by (206). Thus, indeed, we can reach the final equation by using polynomial multipliers whose dimension is less by one unit than the dimension required to compute the generic final equation.

If the degree of the final equation can be lowered by two units, a similar line of thought leads to the conclusion that the final equation can be computed using polynomial multipliers whose dimension is that of the polynomials used to compute the generic final equation, minus two, and so on. Conversely, by using polynomial multipliers whose dimension is smaller by two units, and computing the lines, we will reach two necessary condition equations that must hold so that the final equation equals that of the generic final equation, minus two.

This being said, we could hypothesize that one of these two condition equations should be precisely the same as the necessary condition for lowering the degree of the final equation by one unit only. While the nature of this equation will indeed be the same, it will appear differently: It will be a combination of this equation and the second equation, which will be a common factor for all terms of the final equation when its degree is lowered by only one, and it will indicate whether its degree can be lowered further: This is what we saw in the example in (290).

(302.) The lower the assumed degree of the final equation, the easier it will be to establish whether this degree can be achieved by the final equation. Indeed, the computations involve polynomial multipliers whose degree goes down with the degree of the final equation.

(303.) This conclusion does not hold when determining the second type of indicators that the degree of the final equation can be lowered. Although this task may not be as complex as that of computing the final equation itself, it will at least require computations involving almost all undetermined coefficients. The only exception to this rule is when detecting whether the degree of the final equation can be lowered by one unit only.

We first examine the indicator to determine whether the degree of the final equation can be lowered by one unit.

From (299) the number of useful coefficients in the highest dimension of the polynomial multipliers is precisely equal to the number of terms in the highest dimension of the sum equation. If we set to zero all terms of the highest dimension of the sum equation to determine the conditions when the degree of the final equation drops by one, we are led, by virtue of (206) and (213), to setting each coefficient to zero, or to one condition equation involving coefficients among those in the highest dimension of each proposed equation. We have already examined the first case and we now turn our attention to the second case. We therefore multiply the highest dimension of each equation by the highest dimension of its polynomial multiplier, and add all the resulting products. We then equate to zero each term of this sum, which is also the highest dimension of the sum equation. Proceding with the calculation of *lines*, the last line must be the condition equation sought.

For example, consider the two equations

$$ax^2 \quad + \quad bxy \quad + \quad cy^2 = 0,$$
$$+ \quad dx \quad + \quad ey$$
$$+ \quad f$$

$$a'x^2 \quad + \quad b'xy \quad + \quad c'y^2 = 0.$$
$$+ \quad d'x \quad + \quad e'y$$
$$+ \quad f'$$

We multiply $ax^2 + bxy + cy^2$ by $Ax^2 + Bxy + Cy^2$ on the one hand, and $a'x^2 + b'xy + c'y^2$ by $A'x^2 + B'xy + C'y^2$ on the other hand. The sum of the products is

$$\left.\begin{matrix} Aa \\ A'a' \end{matrix}\right\} x^4 \quad \left.\begin{matrix} +Ab \\ +A'b' \\ +Ba \\ +B'a' \end{matrix}\right\} x^3 y \quad \left.\begin{matrix} +Ac \\ +A'c' \\ +Bb \\ +B'b' \\ +Ca \\ +C'a' \end{matrix}\right\} x^2 y^2 \quad \left.\begin{matrix} +Bc \\ +B'c' \\ +Cb \\ +C'b' \end{matrix}\right\} xy^3 \quad \left.\begin{matrix} +Cc \\ +C'c' \end{matrix}\right\} y^4 .$$

We must remember that there is a useless coefficient; proceding as in (285), we obtain $C = 0$ and $C' = 0$.

We therefore must solve the following equations:

$$Aa + A'a' = 0,$$
$$Ab + A'b' + Ba + B'a' = 0,$$
$$Ac + A'c' + Bb + B'b' = 0,$$
$$Bc + B'c' = 0.$$

We obtain the following:

First line $aA'BB'$

Second line $(ab')BB' - aA'aB'$

Third line $(ab')bB' - (ac')aB' + aA'(ab')$

Fourth line $(ab').(bc') - (ac')^2.$

The fourth line indicates whether the degree of the final equation can be reduced to 3.

Indeed, comparing this line with the final equation found in (285), we see that $(ab').(bc') - (ac')^2$ is the coefficient of x^4 in the final equation, which will thus reduce to degree 3 if $(ab').(bc') - (ac')^2 = 0$. We proceed likewise for an arbitrary number of equations, to find the condition equation that determines whether the final equation drops by one unit by losing its first term only.

(304.) Consider the problem of determining the condition equations that must hold so that the terms immediately following the first term in the final equation vanish. There seems to be no other way than that of computing the entire final equation, as we did in (285). We only observe that, depending upon the term for which we want to obtain the condition equation, only a certain number of coefficients need to be computed, and consequently we will be able to simplify computations from what was given in (201).

For example, assume I want to compute the condition equation that must hold so that the second term of the final equation found in (285) vanishes, without computing the final equation. I notice that the term in x^3 in the sum equation is $(Ad + A'd' + Da + D'a')x^3$. Thus I only have to compute A, A', D, D', that is, A' and D' only; I then simplify this computation using what was given in (201).

(305.) It takes as much effort to compute the condition equations leading to the cancellation of the highest-order terms in the final equation as to compute the final equation itself. However, these condition equations are far less important to compute than those leading to the indicators of the first kind.

Indeed, if the degree of the equation can be lowered by cancelling the coefficients of higher order terms in the generic final equation, we do not have to fear that the resulting equation will be of unjustifiably high degree. Computations will automatically identify and cancel these terms.

When the degree of the final equation can be lowered because a factor common to all its terms becomes zero, the situation is different. A final equation where this factor is cancelled is not able to provide any clue as to whether its degree can be lowered: Any method that would neglect this factor would be erroneous regarding the true number of roots that are relevant to the question. It is important to know this factor or, at least, to account for its presence in any method that claims to be reliable.

To be exact, we do not need to know in advance whether this factor is zero or not, because calculations will indicate this, as we have observed (301). However, polynomial multipliers are simpler than those determining the generic final equation when the factor is zero. Thus, it is useful to have methods to determine this before proceeding with the computation of the final equation.

Conversely, when the degree of the final equation drops because the coefficients of the highest-order terms in the final equation are destroyed, the polynomial multipliers keep the same degree as for the generic final equation, such that there is no real advantage to knowing this in advance.

About means to considerably reduce the number of coefficients used for elimination. Resulting simplifications in the polynomial multipliers

(306.) We have previously shown how to compute the number of useless coefficients for elimination purposes, and we have named the others *useful coefficients.* Using these coefficients only, we can be certain to know all the information contained in the proposed equations, either considering them in general or relative to specific relations that may bind some of the known coefficients together.

The previous examples, however, indicate that all of what we named useful coefficients are not always required to gather all the important knowledge about the proposed equations. Indeed, the more undetermined coefficients we admit in our computations, the more factors will appear in the final equation. But these new factors only replicate, by their presence and their meaning, those that appear when we use the smallest possible number of coefficients, or they only express the existence of solutions of the type described in (279) and (287). Thus, they do not express anything that was not known before. Therefore it is worth improving our method by giving ways to exclude, whenever possible, the coefficients that can give rise to such factors, and indeed this is possible in many cases, one of which was shown in (292).

(307.) We first illustrate our point by means of an example.

Assume we must find the final equation in x resulting from three equations of the form $(x, y, z)^2 = 0$. We can, by (224), use a polynomial of the form $(x, y, z)^{T+4}$ for each equation. But since, by (47), the degree of the final equation must not exceed 8, we cannot assume T to be less than 2 in general. Thus the polynomial multiplier with lowest degree that may be used is $(x, y, z)^6$.

However, what would happen if we admitted the polynomial $(x, y, z)^{T+4}$ without assigning a degree higher than 8 to the final equation?

All coefficients in dimensions higher than 6, in each polynomial multiplier, would be equal to zero, or at least we could always assume they are equal to zero.

Indeed, the highest dimension $T + 4$ of the three polynomial multipliers would provide a number of coefficients equal to $3N(x, y)^{T+4}$.

But among these, a number $3N(x, y)^{T+2} - N(x, y)^T$ of them would not be admissible for elimination purposes. Thus we only have

$$3N(x, y)^{T+4} - 3N(x, y)^{T+2} + N(x, y)^T$$

coefficients available to cancel all terms in the highest dimension of the sum equation, that is, all terms in the dimension $T + 6$. On the other hand, the number of terms in this dimension of the sum equation is $N(x, y)^{T+6}$; the difference between these two numbers is therefore

$$N(x, y)^{T+6} - 3N(x, y)^{T+4} + 3N(x, y)^{T+2} - N(x, y)^T,$$

that is,

$$[N(x, y)^{T+6} - N(x, y)^{T+4} - N(x, y)^{T+4} + N(x, y)^{T+2}]$$
$$-[N(x, y)^{T+4} - N(x, y)^{T+2} - N(x, y)^{T+2} + N(x, y)^T],$$

or $d^3 N(x, y)^{T+6} \dots \begin{pmatrix} T \\ 2, 2, 2 \end{pmatrix}$, that is, zero.

Thus, in order to cancel all terms in the sum equation relative to dimensions higher than 8, we only have as many undetermined coefficients as there are terms to be cancelled. Thus, by (213), each of these coefficients is zero, or at least can be assumed to be such.

By allowing these coefficients to be present in polynomial multipliers of higher order, we only multiply the sum equation by a factor expressing the condition resulting from cancelling all terms with dimension higher than 8.

We therefore can, and should, limit each polynomial multiplier to the form $(x, y, z)^6$. A higher dimensional form would only indicate solutions whose nature is described in (286) and (287), or repetitions of what the simplest form can already point out.

The form $(x, y, z)^6$ is not the simplest possible. Beginning with this form, we find from (231 ff) that the number of coefficients that are useful for elimination is

$$3N(x, y, z)^6 - 3N(x, y, z)^4 + N(x, y, z)^2,$$

that is, 157. But these coefficients, deemed to be useful in the sense we explained previously, are not all required; a relatively large number of them would not have any other effect on the final equation than to give factors indicating solutions of the type described in (279) and (287), or would only repeat the factors found in the final equations derived from the minimum possible number of coefficients.

Let us identify the new terms that may be excluded. First, I consider the terms in the sum equation where y and z reach the dimension 8, either together or separately. The number of such terms is $N(y)^8$. The number of coefficients introduced by the three polynomial multipliers in these eight-dimensional terms is $3N(y)^6$. Among these, $3N(y)^4 - N(y)^2$ are useless. Thus, we really have $3N(y)^6 - 3N(y)^4 + N(y)^2$ coefficients to cancel $N(y)^8$ terms where y and z reach the dimension 8 in the sum equation. Thus, to eliminate nine terms, we have $21 - 15 + 3 = 9$ coefficients only; thus, by (213), each of these coefficients will be zero. So we can skip all terms of dimension 6 in y and z in the polynomial multipliers. Thus the form of the multipliers can be reduced to $[x, (y, z)^5]^6$.

Likewise we can now analyze the terms of the sum equation where y and z can reach the dimension 7, given the new form of the polynomial multipliers; their number is $2N(y)^7$. The polynomials introduce $6N(y)^5 - 6N(y)^3 + 2N(y)^1$ useful coefficients. We therefore have $36 - 24 + 4 = 16$ coefficients to cancel the 16 terms where y and z reach, together, the dimension 7 in the sum equation. Thus, by (213), each of these coefficients will be zero.

The form of the multiplier polynomials therefore reduces to $[x\ (y, z)^4]^6$.

Likewise, there are

$$9N(y)^4 - 9N(y)^2 + 3N(y)^0 = 45 - 27 + 3 = 21$$

useful coefficients to cancel the twenty-one terms where y and z reach, together or separately, the dimension 6 in the sum equation resulting from this new form. Thus each of these coefficients will be zero, and the form of the polynomial multipliers can be reduced to $[x, (y, z)^3]^6$.

Likewise, there are

$$12N(y)^3 - 12N(y)^1 + 4N(y)^{-1} = 48 - 23 + 0 = 24$$

useful coefficients to cancel the twenty-four terms where y and z reach, together or separately, the dimension 5 in the sum equation resulting from this new form. Thus each of these coefficients will be zero and the form of the polynomial multipliers can be reduced to $[x, (y, z)^2]^6$.

This is the simplest form relative to the total dimension of x, y and z and the total dimension of y and z. We will see later that it can be reduced further, but, using this form, we have already reduced the number of coefficients from 157 in the form $(x, y, z)^6$ to only 87.

(308.) In general, let t, t', t'', etc. be the exponents of the degree of each equation. We will assume these equations to be complete for simplicity. Let D be the degree of the final equation, which we know from (47) is $tt't''$etc.

Let $(u \ldots n)^{T-t}$ be the multiplier of the first equation, $(u \ldots n)^{T-t'}$ the multiplier of the second equation, and $(u \ldots n)^{T-t''}$ the multiplier of the third equation, and so on.

The number of useful coefficients in the first polynomial multiplier is

$$d^{n-1}[N(u\ldots n)^{T-t}] \ldots \left(\begin{array}{c} T-t \\ t', t'', t''', \text{etc.} \end{array} \right),$$

keeping in mind the number of terms that can be cancelled in this polynomial, using the $n-1$ last equations.

For the same reason, the number of useful coefficients in the second polynomial multiplier is

$$d^{n-2}[N(u\ldots n)^{T-t'}] \ldots \left(\begin{array}{c} T-t' \\ t'', t''', \text{etc.} \end{array} \right),$$

and

$$d^{n-3}[N(u\ldots n)^{T-t''}] \ldots \left(\begin{array}{c} T-t'' \\ t''', \text{etc.} \end{array} \right)$$

is the number of useful coefficients in the third polynomial multiplier, and so on.

Hence, the total number of coefficients to obtain the final equation is

$$\begin{aligned} & d^{n-1}[N(u\ldots n)^{T-t}] \ldots \left(\begin{array}{c} T-t \\ t', t'', t''', \text{etc.} \end{array} \right) \\ & + d^{n-2}[N(u\ldots n)^{T-t'}] \ldots \left(\begin{array}{c} T-t' \\ t'', t''', \text{etc.} \end{array} \right) \\ & + d^{n-3}[N(u\ldots n)^{T-t''}] \ldots \left(\begin{array}{c} T-t'' \\ t''', \text{etc.} \end{array} \right). \end{aligned}$$

On the other hand, the number of terms to be cancelled in the sum equation to get the final equation is $N(u\ldots n)^T - D - 1$; thus we must have

$$\begin{aligned} N(u\ldots n)^T - D - 1 = {} & d^{n-1}[N(u\ldots n)^{T-t}] \ldots \left(\begin{array}{c} T-t \\ t', t'', t''', \text{etc.} \end{array} \right) \\ & + d^{n-2}[N(u\ldots n)^{T-t'}] \ldots \left(\begin{array}{c} T-t' \\ t'', t''', \text{etc.} \end{array} \right) \\ & + d^{n-3}[N(u\ldots n)^{T-t''}] \ldots \left(\begin{array}{c} T-t'' \\ t''', \text{etc.} \end{array} \right) + \text{ etc.} - 1, \end{aligned}$$

that is,

$$(A)\ldots N(u\ldots n)^T - d^{n-1}[N(u\ldots n)^{T-t}]\ldots\begin{pmatrix} T-t \\ t',t'',t''',\text{etc.} \end{pmatrix}$$

$$-d^{n-2}[N(u\ldots n)^{T-t'}]\ldots\begin{pmatrix} T-t' \\ t'',t''',\text{etc.} \end{pmatrix}$$

$$-d^{n-3}[N(u\ldots n)^{T-t''}]\ldots\begin{pmatrix} T-t'' \\ t''',\text{etc.} \end{pmatrix} - \text{etc.} = D = tt't'', \text{etc.}$$

for any T.

In other terms, the difference between the number of terms in the sum equation and the number of useful coefficients of all polynomial multipliers is equal to the exponent of the final equation, $D = tt't''$etc.

We must note, however, that when we say this equality must hold for any T, we mean for any value of T greater than $tt't''$, etc.

Let us now assume that T is greater than $tt't''$, etc., and that the difference between these quantities is an arbitrary number q.

Then the number of terms in the sum equation is greater by an amount $d[N(u\ldots n)^T]\ldots\begin{pmatrix} T \\ q \end{pmatrix}$. The total number of useful coefficients of the polynomial multipliers will increase by the quantity

$$d\left(d^{n-1}[N(u\ldots n)^{T-t}]\ldots\begin{pmatrix} T-t \\ t',t'',t''',\text{etc.} \end{pmatrix}\right.$$

$$+d^{n-2}[N(u\ldots n)^{T-t'}]\ldots\begin{pmatrix} T-t' \\ t'',t''',\text{etc.} \end{pmatrix}$$

$$\left.+d^{n-3}[N(u\ldots n)^{T-t''}]\ldots\begin{pmatrix} T-t'' \\ t''',\text{etc.} \end{pmatrix} + \text{etc.}\right)\ldots\begin{pmatrix} T \\ q \end{pmatrix}.$$

We therefore get from (12) and because of equation (A)

$$\left.\begin{array}{l} dN(u\ldots n)^T \\ -d\left(d^{n-1}[N(u\ldots n)^{T-t}]\ldots\begin{pmatrix} T-t \\ t',t'',t''',\text{etc.} \end{pmatrix}\right) \\ -d\left(d^{n-2}[N(u\ldots n)^{T-t'}]\ldots\begin{pmatrix} T-t' \\ t'',t''',\text{etc.} \end{pmatrix}\right) \\ -d\left(d^{n-3}[N(u\ldots n)^{T-t''}]\ldots\begin{pmatrix} T-t'' \\ t''',\text{etc.} \end{pmatrix} + \text{etc.}\right) \end{array}\right\}\ldots\begin{pmatrix} T \\ q \end{pmatrix} = 0.$$

Thus, the number of coefficients available to cancel the terms with dimension greater than $tt't''$etc. is precisely equal to the number of these terms; thus from (213) we can assume that each of these terms is equal to zero.

It would therefore be useless to consider polynomial multipliers with order higher than $N(u\ldots n)^{D-t}$, $N(u\ldots n)^{D-t'}$, $N(u\ldots n)^{D-t''}$, etc.

(309.) Before inspecting the other terms that may be rejected, let us pause for a moment to see that the expression of D given in equation (A) does not differ from the expression found in (46); that is, it is not different from $d^n N(u\ldots n)^T\ldots\begin{pmatrix} T \\ t,t',t'',\text{etc.} \end{pmatrix}$. This is not obvious. An example will suffice.

Assume there are three equations only; then equation (A) becomes

$$N(u\ldots3)^T - dd[N(u\ldots3)^{T-t}]\ldots\left(\begin{array}{c}T-t\\t',t''\end{array}\right)$$

$$-d[N(u\ldots3)^{T-t'}]\ldots\left(\begin{array}{c}T-t'\\t''\end{array}\right) - N(u\ldots3)^{T-t''} = D.$$

But $N(u\ldots3)^T - N(u\ldots3)^{T-t''} = d[N(u\ldots3)^T]\ldots\left(\begin{array}{c}T\\t''\end{array}\right)$. Thus we have

$$d[N(u\ldots3)^T]\ldots\left(\begin{array}{c}T\\t''\end{array}\right) - dd[N(u\ldots3)^{T-t}]\ldots\left(\begin{array}{c}T-t\\t',t''\end{array}\right)$$

$$-d[N(u\ldots3)^{T-t'}]\ldots\left(\begin{array}{c}T-t'\\t''\end{array}\right) = D.$$

However,

$$d[N(u\ldots3)^T]\ldots\left(\begin{array}{c}T\\t''\end{array}\right)$$

$$-d[N(u\ldots3)^{T-t'}]\ldots\left(\begin{array}{c}T-t'\\t''\end{array}\right)$$

$$= dd[N(u\ldots3)^{T-t'}]\ldots\left(\begin{array}{c}T\\t',t''\end{array}\right).$$

We therefore have

$$dd[N(u\ldots3)^{T-t'}]\ldots\left(\begin{array}{c}T\\t',t''\end{array}\right) - dd[N(u\ldots3)^{T-t}]\ldots\left(\begin{array}{c}T-t\\t',t''\end{array}\right) = D,$$

that is, $d^3[N(u\ldots3)^T]\ldots\left(\begin{array}{c}T\\t',t'',t'''\end{array}\right) = D.$

That argument applies for any other value of n, and it therefore follows that equation (A) is in fact equivalent to the equation

$$d^n[N(u\ldots n)^T]\ldots\left(\begin{array}{c}T\\t,t',t'',\text{etc.}\end{array}\right) = D.$$

(310.) We now turn our attention to other terms that may be omitted from the polynomial multipliers.

Assume again for simplicity that the proposed equations are complete. Assume their degrees are t, t', t'', t''', etc.

I want to identify the terms in the sum equation whose number of useful coefficients is precisely the same as the number of equations to determine them. I remark that, if there indeed exist such terms, the sum equation must be of the form $[u,(x,\ldots n-1)^B\ldots n]^T$ after these are suppressed. That is, assuming u is the unknown to be present in the final equation, the other unknowns cannot exceed, together or separately, the dimension B.

Instead of giving B its lowest possible value, assume it takes an arbitrary value between this lowest possible value and T. From all that was said in the first book, we easily find that

1. The first polynomial multiplier is $[u, (x \ldots n-1)^{B-t} \ldots n]^{T-t}$.

   The second polynomial mutliplier is $[u, (x \ldots n-1)^{B-t'} \ldots n]^{T-t'}$.

   The third polynomial multiplier is $[u, (x \ldots n-1)^{B-t''} \ldots n]^{T-t''}$, and so on.

2. The number of useful coefficients in the first polynomial multiplier is

$$d^{n-1}(N[u, (x \ldots n-1)^{B-t} \ldots n]^{T-t}) \ldots \left(\begin{array}{c} T-t \\ t', t'', t''', \text{ etc.} \end{array} : \begin{array}{c} B-t \\ t', t'', t''', \text{ etc.} \end{array} \right).$$

The number of useful coefficients in the second polynomial multiplier is

$$d^{n-2}(N[u, (x \ldots n-1)^{B-t'} \ldots n]^{T-t'}) \ldots \left(\begin{array}{c} T-t' \\ t'', t''', \text{ etc.} \end{array} : \begin{array}{c} B-t' \\ t'', t''', \text{ etc.} \end{array} \right).$$

The number of useful coefficients in the third polynomial multiplier is

$$d^{n-3}(N[u, (x \ldots n-1)^{B-t''} \ldots n]^{T-t''}) \ldots \left(\begin{array}{c} T-t'' \\ t''', \text{ etc.} \end{array} : \begin{array}{c} B-t'' \\ t''', \text{ etc.} \end{array} \right).$$

and so on.

Thus, reasoning as we have done in (308), we conclude that

$$(A) \ldots N[u, (x \ldots n-1)^{B} \ldots n]^{T} - d^{n-1}(N[u, (x \ldots n-1)^{B-t} \ldots n]^{T-t}) \ldots$$
$$\ldots \left(\begin{array}{c} T-t \\ t', t'', t''', \text{ etc.} \end{array} : \begin{array}{c} B-t \\ t', t'', t''', \text{ etc.} \end{array} \right.$$
$$-d^{n-2}(N[u, (x \ldots n-1)^{B-t'} \ldots n]^{T-t'}) \ldots \left(\begin{array}{c} T-t' \\ t'', t''', \text{ etc.} \end{array} : \begin{array}{c} B-t' \\ t'', t''', \text{ etc.} \end{array} \right.$$
$$-d^{n-3}(N[u, (x \ldots n-1)^{B-t''} \ldots n]^{T-t''}) \ldots \left(\begin{array}{c} T-t'' \\ t''', \text{ etc.} \end{array} : \begin{array}{c} B-t'' \\ t''', \text{ etc.} \end{array} \right) - \text{ etc.}$$
$$= D.$$

Assuming T remains the same, we vary B by an arbitrary amount q; then we get

$$(C) \ldots d \left\{ \begin{array}{c} N[u, (x \ldots n-1)^{B} \ldots n]^{T} \\ -d^{n-1}(N[u, (x \ldots n-1)^{B-t} \ldots n]^{T-t}) \ldots \\ \ldots \left(\begin{array}{c} T-t \\ t', t'', t''', \text{ etc.} \end{array} : \begin{array}{c} B-t \\ t', t'', t''', \text{ etc.} \end{array} \right) \\ -d^{n-2}(N[u, (x \ldots n-1)^{B-t'} \ldots n]^{T-t'}) \ldots \\ \ldots \left(\begin{array}{c} T-t' \\ t'', t''', \text{ etc.} \end{array} : \begin{array}{c} B-t' \\ t'', t''', \text{ etc.} \end{array} \right) \\ -d^{n-3}(N[u, (x \ldots n-1)^{B-t''} \ldots n]^{T-t''}) \ldots \\ \ldots \left(\begin{array}{c} T-t'' \\ t'', \text{ etc.} \end{array} : \begin{array}{c} B-t'' \\ t''', \text{ etc.} \end{array} \right) \end{array} \right\} \ldots \left(\begin{array}{c} B \\ q \end{array} \right) = 0.$$

Let us observe that the value of B is unconstrained, whereas that of T cannot be below $tt't''$, etc. Thus, the only condition B must satisfy is that equation (C) holds, and this condition will hold until $B = t + t' + t'' + t''' +$ etc. $- n + 2$.

Indeed, we need the expressions $N(x \ldots n - 1)^{B-q}$, $N(x \ldots n - 1)^{B-t-q}$, $N(x, \ldots n-1)^{B-t'-q}$, etc., the expressions $N(x \ldots n-1)^{B-t-t'-q}$, $N(x \ldots n-1)^{B-t-t''-q}$, etc., those of $N(x \ldots n - 1)^{B-t-t'-t''-q}$, etc. to all be positive integers. If we remember that, in general,

$$N(x \ldots n - 1)^{B-r} = \frac{(B - r + 1).(B - r + 2) \ldots (B - r + n - 1)}{1.2.3 \ldots n - 1},$$

we see that this expression is a positive integer until $B - r + n - 1 = 0$, that is, until $r = B + n - 1$.

The largest actual value for r is $r = t + t' + t'' + t''' +$ etc. $+ q$; thus $B = t + t' + t'' + t''' +$ etc. $+ q - n + 1$.

However, the smallest value we can assign to q is $q = 1$; thus the smallest possible value for B is $B = t + t' + t'' + t''' +$ etc. $- n + 2$, and B can be further reduced.

Thus B cannot be further reduced if $B = t + t' + t'' + t''' +$ etc. $- n + 1$. Assuming B to be larger than this value would introduce superfluous coefficients.

(311.) Thus, considering the complete equations $(u \ldots n)^t = 0$, $(u \ldots n)^{t'} = 0$, $(u \ldots n)^{t''} = 0$, etc., it is enough to use the polynomial multipliers

$$[u, (x \ldots n - 1)^{t'+t''+\text{etc.}-n+1} \ldots n]^{D-t},$$
$$[u, (x \ldots n - 1)^{t+t''+\text{etc.}-n+1} \ldots n]^{D-t'},$$
$$[u, (x \ldots n - 1)^{t+t'+\text{etc.}-n+1} \ldots n]^{D-t''}, \text{ etc.}$$

(312.) Consider, for example, complete equations in two unknowns. The two simplest polynomial multipliers are in general

$$(x^{D-t}, y^{t'-1})^{D-t} \text{ and } (x^{D-t'}, y^{t-1})^{D-t'}.$$

In complete equations in three unknowns, the simplest polynomial multipliers, relative to the total dimension of the three unknowns and the total dimension of the two unknowns to be eliminated, are

$$[x^{D-t}, (y, z)^{t'+t''-2}]^{D-t}, \ [x^{D-t'}, (y, z)^{t+t''-2}]^{D-t'}, \ [x^{D-t''}, (y, z)^{t+t'-2}]^{D-t''}.$$

(313.) We will almost always be able to reject other terms in each polynomial multiplier. We will proceed as we will explain later to determine these terms.

(314.) We remark that if all proposed equations are of first degree, then all polynomial multipliers would be of the form $[u^0, (x \ldots n - 1)^0]^0$, that is, it would be enough to multiply each equation by a single undetermined coefficient, and it is obvious that this must indeed be the case.

(315.) Assume now that the proposed equations are not complete, but are incomplete and of the form $[u^a, (x \ldots n - 1)^b \ldots n]^t = 0$, where b is the highest dimension that the $n - 1$ unknowns to be eliminated can reach, either together or separately. We conclude that, if D is the degree of the final equation, the polynomial multiplier of each equation cannot be more complicated than

$$[u^{D-a}, (x \ldots n-1)^{b'+b''+b'''+\text{etc.}-n+1} \ldots n]^{T-t},$$
$$[u^{D-a'}, (x \ldots n-1)^{b+b''+b'''+\text{etc.}-n+1} \ldots n]^{T-t'},$$
$$[u^{D-a''}, (x \ldots n-1)^{b+b'+b'''+\text{etc.}-n+1} \ldots n]^{T-t''},$$
$$[u^{D-a'''}, (x \ldots n-1)^{b+b'+b''+\text{etc.}-n+1} \ldots n]^{T-t'''}, \text{ etc.},$$

respectively.

As for the value of T, it is determined by observing that it must satisfy the following inequalities:

$$D - a + b' + b'' + b''' + \text{ etc.} - n + 1 > T - t,$$
$$D - a' + b + b'' + b''' + \text{ etc.} - n + 1 > T - t',$$
$$D - a'' + b + b' + b''' + \text{ etc.} - n + 1 > T - t'',$$
$$D - a''' + b + b' + b'' + \text{ etc.} - n + 1 > T - t''';$$

and so on.

In other words, we can pick T as the smallest of the quantities

$$D + t - a + b' + b'' + b''' + \text{ etc.} - n + 1,$$
$$D + t' - a' + b + b'' + b''' + \text{ etc.} - n + 1,$$
$$D + t'' - a'' + b + b' + b''' + \text{ etc.} - n + 1,$$
$$D + t''' - a''' + b + b' + b'' + \text{ etc.} - n + 1, \text{etc.}$$

If T were larger, the polynomial would only apparently be of degree T. If T were smaller, it would sometimes not be general enough, and the resulting polynomial multipliers would not be able to answer the question.

(316.) In other incomplete polynomials it will also be possible to considerably reduce the number of coefficients, and we will, in general, be able to apply what we have just said.

However, we must follow what we will soon learn about equations of the form $[x, (y, z)^1]^2 = 0$, to avoid making mistakes about the true number of useful coefficients for the purpose of elimination.

Indeed, it becomes difficult to evaluate the largest number of terms that may be cancelled in each polynomial multiplier after they are altered from their most general form. Consequently, it also becomes difficult to evaluate the number of arbitrary coefficients and equations, either in the final equation taken globally or relative to each of its dimensions. This could lead us to count more coefficients than needed. To be sure, knowing the degree of the final equation would allow us to determine how many superfluous coefficients are available and therefore the number of arbitrary equations that may be formed. But we must also determine how many such equations must be formed for each dimension of the sum equation; not doing so could lead to a wrong or trivial final equation. Our forthcoming remark will allow us to use the simplifications we pointed out, by giving the means to compute how many arbitrary coefficients remain in each polynomial multiplier and what dimension they belong to.

More applications, etc.

(317.) We propose to compute, in general, the simplest polynomial multipliers for first-order incomplete equations in two unknowns, represented by

$$(x^a, y^{\underset{\prime}{a}})t = 0 \text{ and } (x^{a'}, y^{\underset{\prime}{a'}})t' = 0.$$

From what was said in (233), the most general form of the polynomial multiplier of the first equation is $(x^{A+a'}, y^{A+\underset{\prime}{a'}})T+t'$, and that of the polynomial multiplier of the second equation is $(x^{A+a}, y^{A+\underset{\prime}{a}})T+t$.

Let D be the degree of the final equation, whose value is known to be

$$tt' - (t-a).(t'-a') - (t-\underset{\prime}{a}).(t'-\underset{\prime}{a'});$$

we thus have $A + a + a' = D$ and consequently $A = D - a - a'$; this is the lowest possible value for A.

Now consider $\underset{\prime}{A}$; since $\underset{\prime}{a}$ is the highest dimension reached by y in the first equation and $\underset{\prime}{a'}$ is the highest dimension of y in the second, we can assume $\underset{\prime}{A} + \underset{\prime}{a'} = \underset{\prime}{a'} - 1$ or $\underset{\prime}{A} + \underset{\prime}{a} = \underset{\prime}{a} - 1$, that is $\underset{\prime}{A} = -1$, from the observation (315).

The two polynomial multipliers thus become $(x^{D-a}, y^{\underset{\prime}{a'}-1})T+t'$ and $(x^{D-a'}, y^{\underset{\prime}{a}-1})T+t$. We must now compute T. Following what was said in (315), T must be chosen as the smallest of the two values allowed by the following inequalities:

$$D - a + \underset{\prime}{a'} - 1 > T + t',$$
$$D - a' + \underset{\prime}{a} - 1 > T + t,$$

or $T < D - a - t' + \underset{\prime}{a'} - 1$ and $T < D - a' - t - \underset{\prime}{a} - 1$. We therefore pick

$$T = D - a - t' + \underset{\prime}{a'} - 1 \text{ or } T = D - a' - t + \underset{\prime}{a} - 1,$$

depending upon whether

$$D - a - t' - \underset{\prime}{a'} - 1 \left\{ \begin{matrix} < \\ \text{or} > \end{matrix} \right\} D - a' - t + \underset{\prime}{a} - 1,$$

that is, depending upon whether

$$\underset{\prime}{a'} + a' - t' \left\{ \begin{matrix} < \\ \text{or} > \end{matrix} \right\} \underset{\prime}{a} + a - t,$$

and we get the simplest polynomial multipliers that may be obtained in general.

(318.) Consider the example we gave in (281). We obtain $T = -1$ by applying what we just said; thus each appropriate polynomial multiplier would be of the form $(x^1, y^0)^1$, that is, $Ax + B$. This is indeed the simplest polynomial we could reach (281).

(319.) Assume the two proposed equations are of the form $(x^2, y^2)^3 = 0$; the form of the two polynomial multipliers is $(x^5, y)^6$, and $(x^7, y^3)^9$ is the form of the sum equation. Since two terms must be cancelled in the highest dimension, and since each polynomial multiplier can provide only one coefficient, each of these two coefficients must be zero. Thus the form of each polynomial multiplier can be reduced to $(x^5, y)^5$. However, this reduction is specific to this example, and it depends on a close examination of the sum equation. The reduction could not be performed blindly, as the following example shows.

Assume the two proposed equations are of the form $(x^3, y^3)^6 = 0$. The degree of the final equation is 18 and the polynomial multipliers are of the form $(x^{15}, y^2)^{17}$, from what was said in (315). This is the simplest form that may be used. The sum equation only has one term in its highest dimension, and each polynomial multiplier provides one coefficient. Therefore these coefficients do not have to be zero. If we used the form $(x^{15}, y^2)^{16}$, we would find fewer undetermined coefficients than are necessary to perform the elimination.

Thus we see that, although it is sometimes possible to reduce the total dimension of the polynomial multipliers, this cannot always be done. It happens by accident and can be identified only through the examination of the sum equation.

(320.) As a new example to illustrate what we said until now, we consider three equations with the general form

$$
\begin{aligned}
ax^2 &+ bxy + cxz = 0. \\
+ \; dx &+ ey + fz \\
+ \; g &
\end{aligned}
$$

We propose to get the equation in x.

The general form of the polynomial multipliers is $[x, (y, z)^{T+2}]^{T+4}$ (see (231 ff)). From what was said in (311), we must use the much simpler form $[x, (y, z)^0]^{T+4}$ or $(x)^{T+4}$. Since the final equation (131) must only be fourth degree, the simplest form for each polynomial multiplier is $(x)^2$.

Assume that we multiply each equation by a polynomial of the form $Ax^2 + Bx + C$ and that we add the three products together; the sum equation will be of the form

$$
\begin{array}{llllll}
Aa & x^4 & + \; Ab & x^3y & + \; Ac & x^3z = 0. \\
+ \; Ad & x^3 & + \; Ae & x^2y & + \; Af & x^2z \\
+ \; Ba & & + \; Bb & & + \; Bc & \\
+ \; Ag & x^2 & + \; Be & xy & + \; Bf & xz \\
+ \; Bd & & + \; Cb & & + \; Cc & \\
+ \; Ca & & & & & \\
+ \; Bg & x & + \; Ce & y & + \; Cf & z \\
+ \; Cd & & & & & \\
+ \; Cg & & & & &
\end{array}
$$

In this case no coefficient is useless; although we can cancel two terms in the polynomial $Ax^2 + Bx + C$ using the last two equations, this would require introducing new terms, which would invalidate the form that we have shown must be assumed by the polynomial multiplier.

We now only have to compute the value of $AA'A''BB'B''CC'C''$. We get it as follows, going sucessively through x^3z, x^3y, x^2z, x^2y, xz, xy, z and y.

First line
$cA'A''$.
Second line
$-(bc')A''BB'B''$ [because $(cb') = -(bc')$].
Third line
$-(bc'f'')BB'B'' + (bc')A''cB'B''$.
Fourth line
$[-(bc'f'')bB'B'' + (bc'e'')cB'B'' + (bc')A''(bc')B'']CC'C''$.
Fifth line
$[-(bc'f'').(bf')B'' + (bc'e'').(cf')B'' - (bc'f'').(bc')A'']CC'C''$
$+[-(bc'f'')bB'B'' + (bc'e'')cB'B'' + (bc')A''(bc')B'']cC'C''$.
Sixth line
$[(bc'f'').(be'f'') - (bc'e'').(ce'f'')]CC'C''$
$-[-(bc'f'').(bf')B'' + (bc'e'').(cf')B'' - (bc'f'').(bc')A'']bC'C''$
$+[-(bc'f'').(be')B'' + (bc'e'').(ce')B'' - (bc'e'').(bc')A'']cC'C''$
$-[-(bc'f'').bB'B'' + (bc'e'').cB'B'' + (bc')A''(bc')B''].(bc')C''$.
Seventh line
$[(bc'f'').(be'f'') - (bc'e'').(ce'f'')]fC'C''$
$+[-(bc'f'').(bf')B'' + (bc'e'').(cf')B'' - (bc'f'').(bc')A''](bf')C''$
$-[-(bc'f'').(be')B'' + (bc'e'').(ce')B'' - (bc'e'').(bc')A''].(ce'f'')$.

We have omitted terms where $B'B''$ and $A''B''$ remain, since these would disappear at the end anyway.

Eighth line
$-[(bc'f'').(be'f'') - (bc'e'').(ce'f'')].(ef')C''$
$+[-(bc'f'').(bf')B'' + (bc'e'').(cf')B'' - (bc'f'').(bc')A''].(be'f'')$
$-[-(bc'f'').(be')B'' + (bc'e'').(ce')B'' - (bc'e'').(bc')A''].(ce'f'')$.

From these we get

$$
\begin{aligned}
A'' &= (b'ce'').(ce'f'').(bc') - (bc'f'').(be'f'').(bc'), \\
B'' &= -(bc'e'').(ce'f'').(ce') + (bc'e'').(be'f'').(cf') \\
&\quad - (bc'f'').(be'f'').(bf') + (bc'f'').(ce'f'').(be'), \\
C'' &= [(bc'e'').(ce'f'') - (bc'f'').(be'f'')].(ef').
\end{aligned}
$$

But from what was said in (218), we have

$$(bc'e'')f - (bc'f'')e + (be'f'')c - (ce'f'')b = 0$$
$$\text{and} \quad (bc'e'')f' - (bc'f'')e' + (be'f'')c' - (ce'f'')b' = 0.$$

Thus we get

$$(be'f'').(cf') = (ce'f'').(bf') + (bc'f'').(ef')$$

by multiplying the first of these two equations by f', the second by f, and subtracting one from the other.

Likewise we get

$$(ce'f'').(be') = -(bc'e'').(ef') + (be'f'').(ce')$$

by multipliying the first equation by e', the second by e, and subtracting one from the other.

Substituting in the value of B'', this value becomes

$$B'' = [(bc'e'').(ce'f'') - (bc'f'').(be'f'')][(bf') - (ce')].$$

Introducing the shorthand $(1) = (bc'e'').(ce'f'') - (bc'f'').(be'f'')$, we get

$$\left. \begin{array}{l} A'' = (1)(bc') \\ B'' = (1)[(bf') - (ce')] \\ C'' = (1)(ef') \end{array} \right\}$$

and consequently

$$\left\{ \begin{array}{l} A' = -(1)(bc'') \\ B' = -(1)[(bf'') - (ce'')] \\ C'' = -(1)(ef'') \end{array} \right\} \text{ and } \left\{ \begin{array}{l} A = (1)(b'c'') \\ B = (1)[(b'f'') - (c'e'')] \\ C = (1)(e'f''). \end{array} \right.$$

Substituting into the remaining terms of the sum equation yields

$$(1) \times \left[(ab'c'')x^4 \begin{array}{c} +(bc'd'') \\ +(ab'f'') \\ -(ac'e'') \end{array} \right\} x^3 \begin{array}{c} +(bc'g'') \\ +(bf'g'') \\ -(ce'g'') \\ +(ae'f'') \end{array} \right\} x^2 \begin{array}{c} +(bf'g'') \\ -(ce'g'') \\ +(de'f'') \end{array} \right\} x + (ef'g'') \right] = 0.$$

If the factor (1) had been ignored, this equation would have been very easy to find, by simply substituting into one equation the values of y and z found in the others; we will talk later about how to indeed expedite the computation of the final equation with as few such factors as possible. However, our current interest is precisely the factor (1).

This factor is the indicator of whether the degree of the final equation can be lowered to degree 3. The degree can be lowered if

$$(1) \text{ or } (bc'e'').(ce'f'') - (bc'f'').(be'f'') = 0.$$

This is easy to confirm by picking polynomial multipliers of the form $Bx + C$, leading to the condition equation

$$(bc'e'').(ce'f'') - (bc'f'').(be'f'') = 0.$$

(321.) The equation in y or z, although it is also fourth degree, is not as simple, by far. Since our goal is not so much to perform computations as to present methods for performing them, we will be all the more willing to skip the details, and we will give later a much faster method to reach either one of these two equations.

About the care to be exercised when using simpler polynomial multipliers
than their general form (231 ff), when dealing with incomplete equations

(322.) In the previous example, we have reduced the form of each polynomial multiplier, first to $[x, (y, z)^0]^{T+4}$, and then to $(x)^2$.

Assume, however, that we reduce the general form $[x, (y, z)^{T+2}_{,}]^{T+4}$ to
$[x, (y, z)^{T+2}_{,}]^2$ and then to $[x, (y, z)^2]^2$ or $(x, y, z)^2$, since we know that the
final equation can only be fourth degree and $T_{,}$ cannot be larger than zero.

The three polynomial multipliers then provide thirty coefficients, three of
which are arbitrary, such that twenty-seven coefficients are available for the
purpose of elimination. Since the sum equation contains only twenty-five
terms in y and z, there is one more coefficient than necessary. We could
be inclined to think this coefficient could be used to lower the degree of the
equation by one degree.

This inclination would be all the more justified since no more than three
coefficients in the polynomial multipliers can indeed be cancelled using the
proposed equations; among these there are two coefficients in the first polynomial and one in the second. Multiplying the second and third equation by
A and A', respectively, and adding them to the first polynomial multiplier,
it becomes apparent that only two terms can be eliminated. Likewise, multiplying the third equation by A'' and adding it to the second polynomial
multiplier, we find that only one term can be eliminated.

It would be in vain to try and cancel more terms in the first polynomial,
even by adding products of each of the last two equations with higher degree
polynomials. No more than two terms can be eliminated.

Thus it seems that twenty-seven coefficients are indeed available for elimination purposes and that the degree of the final equation can be reduced to
3.

This paradox can be resolved by remembering that the polynomial multiplier form $(x, y, z)^2$ can be used only if two conditions are satisfied. First,
each coefficient of the highest dimensions of each polynomial multiplier must
be zero; second, cancelling each of these coefficients does not imply cancelling
a few terms in the remaining polynomial $(x, y, z)^2$.

Let us pick a higher degree polynomial, for example, the polynomial
$[x, (y, z)^2]^3$ as the first polynomial multiplier.

Eight terms can be cancelled in this polynomial, using the last two equations: Six in the third dimension, and two in all the other dimensions. But
if we cancelled six terms in the third dimension, we would contradict the
assumption that the polynomial is degree 3; thus we must accept that only
five coefficients can be cancelled in dimension 3, and the three others must
be cancelled among all lower dimensions, that is, in the polynomial $(x, y, z)^2$.

In the second polynomial, we can cancel only three terms in the dimension
3 and one in all the other dimensions, using the last equation. Thus three
terms will remain in the highest dimension.

Since there is no equation available to cancel any term in the third polynomial multiplier, we obtain a total of ten coefficients in the highest dimensions
of the three polynomial multipliers; that is precisely as many terms as will
have to be cancelled in the dimension 5 of the product equation, so each of
these coefficients must equal zero.

But conserving a term in dimension 3 of the first polynomial multiplier
results in one's ability to cancel one more term in the lower dimensions of the
polynomial $(x, y, z)^2$. Thus cancelling terms in the highest dimensions of the
polynomial multipliers implies that one of the lower dimensional terms in one
of these polynomials must be cancelled too. Thus, using polynomials of the
form $(x, y, z)^2$ to address our problem really gives us twenty-six coefficients
rather than twenty-seven. Using the twenty-seventh coefficient to cancel the
term x^4 would lead to a trivial or erroneous equation. See (230).

(323.) We can see here the confirmation and the proof of what we said
in (236): We cannot arbitrarily dispose of more coefficients in the higher
dimensions than what we then said, but we can at the same time arbitrarily
dispose of fewer coefficients and redirect other arbitrary conditions on lower
dimensional coefficients.

Indeed, consider, for example, the task of cancelling as many terms as
possible in the first dimension of the first polynomial multiplier, using the
remaining equations. This task would often be impossible and there would
often be not enough undetermined coefficients left to cancel the terms in the
product equation.

Assume, for example, that we take polynomial multipliers of the form
$[x, (y, z)^3]^4$ in the example we just discussed; eleven terms would need to
be cancelled in the highest dimension of the first polynomial multiplier, but
only ten are available.

If we cancelled them all, only six coefficients would be available from
the highest dimensions of the two other polynomial multipliers, since six
coefficients could be cancelled in the second polynomial multiplier. But ten
terms would still need cancelling in the product equation.

(324.) Through this remark, we make an important observation concerning the use of polynomial multipliers whose form is simpler than the general
form given in (224), when applied to incomplete equations.

The polynomial multipiers for this kind of equation can without a doubt
be much simpler than their general form, as in the case of complete equations. However, it is possible to run into difficulties similar to those we
encountered in the previous example when we use the simplest form based
on the knowledge of the degree of the final equation. We may end up finding too many coefficients, as in the present example, or too few of them, as
seen in example (319). In the first case, we can be misled as to how many
coefficients are redundant. In the second case we missed our goal.

(325.) Here is the process we must follow in order to properly use simpler
polynomial multipliers when such an opportunity arises.

We begin by determining the most general form these polynomials may
take, based on what was said in (224). We then determine the highest

degree reached in each polynomial multiplier by the unknown used in the final equation, based on the known degree of that equation.

The highest exponents of the other unknowns and the combinations of any two or more unknowns are much more arbitrary. But they are subject to the conditions (83 ff) for the existence of the polynomial multipliers and the existence of all their by-products used to compute the expression for the number of arbitrary coefficients. They are also subject to the conditions specifying the form of all these polynomials (see (120 ff)). We therefore assign to each undetermined exponent the smallest value that may satisfy these conditions.

That being done, two cases may arise when beginning with the highest dimension of the sum equation. This dimension can be higher than D, where D is the degree of the final equation, or it can equal D.

If the second case arises, the total dimension of the polynomial multipliers cannot be lowered.

In the first case, we will very often be able to pick polynomial multipliers with lower dimension. Here is how to recognize such a situation.

We first compute the number of terms in the highest dimension of the sum equation. This can be easily obtained by varying the general expression for the number of terms in this equation by -1.

Likewise, we determine the number of coefficients that are useful for elimination purposes and that are contributed by the highest dimension of each polynomial multiplier. If the total number of these useful coefficients exceeds the number of terms in the highest dimension of the sum equation, the total dimension of each polynomial multiplier cannot be lowered. But if the total number of useful coefficients in every highest dimension of the polynomial multipliers is lower than or equal to the number of terms in the highest dimension of the sum equation, then we can bring down the total dimension of each of the polynomial multipliers.

If these numbers are equal, we can repeat the same process for the next dimension.

But assume instead that the total number of useful coefficients contributed by the highest dimension of each polynomial multiplier is lower than the number of terms in the highest dimension of the sum equation. It proves there are many useless coefficients for the purpose of elimination. These coefficients may not all be used in this dimension by (236) and (323). We can then use a number of these coefficients as if they were in fact useful. Their number would then equal that necessary to match the number of terms in the highest dimension of the sum equation.

For example, let N be the number of terms in the highest dimension of the sum equation; N' the total number of useful coefficients in each of the highest dimension of the polynomial multipliers; and N'' the sum of the number of useless coefficients in the highest dimensions of the polynomial multipliers. Assume $N' < N$. Consider the possibility that, in fact, we do not have N'' useless coefficients in the highest dimension of the sum equation; rather we only have $N'' - (N - N') = N'' + N' - N$ such coefficients, while assuming the $N - N'$ others are useful for elimination. Since there are now as many coefficients as there are terms to be cancelled, each coefficient can be made zero (213), and the highest dimension of each polynomial multiplier can be lowered by one unit.

Among the N'' potential arbitrary equations, we are supposed to have created only $N'' + N' - N$ of them. Thus there will remain $N - N'$ arbitrary equations to create for the lower dimensions of the sum equation.

We will proceed the same way regarding the next dimension in the sum equation, accounting for the arbitrary equations relative to the first dimension.

If, past this investigation, the highest dimension of the polynomial multiplier can still be lowered by one unit, we will have to account for the total number of arbitrary equations that had to be created while dealing with the two highest dimensions.

We will keep this process going until the total number of useful coefficients in the highest current dimension of each polynomial multiplier becomes higher than the number of terms in the highest current dimension of the sum equation; we do this unless this highest dimension becomes equal to the degree of the final equation. In both cases, we will have reached polynomial multipliers whose degree is as low as possible. I say polynomials of lowest degree and not simplest polynomials, because many terms can still be cancelled in these polynomials, as we will soon see.

(326.) We will therefore be able to limit ourselves to polynomials whose dimension is determined as described earlier to proceed with the elimination. However, we will also have to keep in mind the number of unused arbitrary equations in order to account for the true number of useless coefficients or, equivalently, the true number of arbitrary equations to be formed.

(327.) Assume we now have determined the simplest total dimension of the polynomial multipliers; we now want to determine the other terms that may be eliminated from these polynomial multipliers. For that purpose, we must perform, relative to the $n - 1$ unknowns to be eliminated, the same investigation as what we have just done, relative to the total dimension of the n unknowns. When we have determined the lowest total dimension that these $n - 1$ unknowns may assume, we can further our investigation to the total dimension of the $n-2$ unknowns reaching the highest dimensions arising from combinations of $n - 2$ of these unknowns. We then perform a similar examination of the total dimension among $n - 3$ of these $n - 1$ unknowns, which reach the highest possible dimension formed by combinations of $n - 3$ of these unknowns. Doing this, we will be able to determine, prior to any elimination operation, the simplest polynomials that may be used; keeping a proper accounting of the unused arbitrary equations will keep us from being fooled by the new aspect of the final equations and to know the true number of arbitrary equations that remain to be formed. We now clarify our statements with a few examples.

More applications, etc.

(328.) We propose to find the simplest form of the polynomial multipliers that can be used for elimination purposes in three equations of the form

$$
\begin{aligned}
axy &+ bxz + cyz = 0. \\
+\ dx &+ ey + fz \\
+\ g &
\end{aligned}
$$

These three equations are of the form $(x^1, y^1, z^1)^2 = 0$.

We know how to compute the degree of the final equation for these equations (130); and this degree is $8 - 1 - 1 - 1 = 5$.

According to what was said in (224), I first use polynomial multipliers of the form $(x^{A+2}, y_{\prime}^{A+2}, z_{\prime\prime}^{A+2})^{T+4}$.

The form of the polynomials whose number of terms contribute to the number of arbitrary coefficients is $(x^{A+1}, y_{\prime}^{A+1}, z_{\prime\prime}^{A+1})^{T+2}$ and $(x^A, y_{\prime}^A, z_{\prime\prime}^A)^T$.

Since all these polynomials must have the same form as that of the proposed equations (105), we must have

$$
A + A_{\prime} > T, \ A + A_{\prime\prime} > T, \ A_{\prime} + A_{\prime\prime} > T;
$$
$$
A + 1 + A_{\prime} + 1 > T + 2, \ A + 1 + A_{\prime\prime} + 1 > T + 2, \ A_{\prime} + 1 + A_{\prime\prime} + 1 > T + 2,
$$

and so on: All these inequalities must hold or be, at least, equalities.

Since the degree of the final equation is 5, I see I cannot assume $A < 2$; thus I assume $A = 2$; from there I see T cannot be less than 2. I therefore assume $T = 2$.

We now turn our attention to $A_{\prime}$ and $A_{\prime\prime}$. Since we must have $A_{\prime} + A_{\prime\prime} > T$ or $A_{\prime} + A_{\prime\prime} = T = 2$, I can assign no value smaller than 1 to either $A_{\prime}$ or $A_{\prime\prime}$; I therefore write $A_{\prime} = 1$ and $A_{\prime\prime} = 1$.

Thus the simplest general expression for each polynomial multiplier is $(x^4, y^3, z^3)^6$, without considering the simplifications that may be brought about by the arbitrary coefficients.

For the present case, I proceed according to what was said in (325 ff) to know whether this form can be simplified not only regarding the dimension of the polynomial multipliers (it is 6 so far), but also regarding the total dimension of the two unknowns y and z (it is 6 so far), and also regarding their specific dimensions (they are 3 and 3 so far).

Since the form of the sum equation is $(x^5, y^4, z^4)^8 = 0$, the eighth dimension contains nineteen terms. But the number of useful coefficients in the dimension 6 of the three polynomial multipliers is nineteen. Thus we have as many coefficients as we have terms to be cancelled, and each of these coefficients is zero.

The form of the polynomial multipliers can therefore be reduced to $(x^4, y^3, z^3)^5$.

We now examine dimension 7 of the sum equation and find that it has twenty-one terms; dimension 5 of the three polynomial multipliers contains twenty-one useful coefficients; thus each of these coefficients is zero and the form of each polynomial multiplier reduces to $(x^4, y^3, z^3)^4$.

Examining dimension 6 of the sum equation, we find twenty-one terms. Dimension 4 of the three polynomial multipliers yields twenty-two coefficients. Thus each coefficient is not necessarily zero; the additional coefficient can be used to cancel terms in the lower dimensions; thus the dimension 4 cannot be lowered. Thus the simplest form, relative to the total dimension of the polynomial multipliers, is $(x^4, y^3, z^3)^4$.

We must now examine the form $(x^4, y^3, z^3)^4$ of the polynomial multipliers, relative to the highest dimension 4 that the two unknowns y and z can reach.

The form of the product equation is now $(x^5, y^4, z^4)^6$. It can give only three terms involving y and z only, which reach the dimension 6. But the three polynomial multipliers have only one useful coefficient, among those terms of dimension 4 in y or z only; the other 8 coefficients are arbitrary. Thus, from (325), we can cancel the terms where y and z reach the dimension 4 together in each polynomial multiplier. To do so, we must assume that only six of the possible eight arbitrary equations are formed; thus we only have to account for the remaining two arbitrary equations in the sum equation, and we will perform this as follows.

The form of the polynomial multipliers is therefore now $[x^4, (y^3, z^3)^3]^4$, and the form of the sum equation is therefore $[x^5, (y^4, z^4)^5]^6$.

The number of terms in the final equation where y and z reach the dimension 5 together is eight, and the number of useful coefficients coming from those terms in the polynomial multipliers where y and z reach the dimension 3 is twelve. But since two arbitrary equations have not been employed, we can reduce the number of useful coefficients by two, which thus reduces to ten: Since this number is larger than the number of terms (eight) that we must cancel, we must conclude by (325) that we cannot lower the total dimension of y and z any further unless the specific dimension of each variable can be lowered, wich remains to be examined.

We can then take polynomial multipliers of the form $[x^4, (y^3, z^3)^3]^4$ remembering that two more coefficients can be disposed of than would naturally occur.

The form of the sum equation is now $[x^5, (y^4, z^4)^5]^6$, and there are five terms containing z^4. In order to cancel them, the polynomial multipliers provide six useful coefficients; since two arbitrary equations have not been used, only four of these coefficients can be used. Then, from (325), we will conclude, as we have done above, that we can exclude the terms in z^3 in each of the polynomial multipliers, and one unused arbitrary equation will remain.

We now follow the same process for y^4, and we see that, if we include the remaining arbitrary equation, we have as many useful coefficients as there are terms in y^4 to be cancelled. Thus the form of the polynomial multipliers can be reduced to $[x^4, (y^2, z^2)^3]^4$.

Given the current form of the polynomial multipliers, we can still lower the total dimension of y and z.

Indeed, the sum equation is now of the form $[x^5, (y^3, z^3)^5]^6$, and only four terms remain where y and z can reach the dimension 5 together. But the number of useful coefficients for the terms that can yield those is zero, with twelve useless coefficients; thus if we assume by (325) that we form only eight out of twelve arbitrary equations, we can reduce the form $[x^4, (y^2, z^2)^3]^4$ to the form $[x^4, (y^2, z^2)^2]^4$, keeping in mind that there are four more arbitrary coefficients than what the new form of the polynomial multiplier naturally provides.

We now examine whether it is possible to lower the specific dimension of z. The sum equation, which is of the form $[x^5, (y^3, z^3)^4]^6$, has seven terms in z^3. We also see that the polynomial multipliers give nine matching useful coefficients; since four arbitrary coefficients or equations remain, we must count only five useful coefficients; thus z^2 can be cancelled. In addition, two arbitrary coefficients will appear beyond those naturally present in the new form $[x^4, (y^2, z^1)^2]^4$.

A similar investigation for y^2 reveals seven terms in y^3 in the sum equation, with nine useful coefficients from the polynomial multipliers, two of which must be removed because of the two arbitrary equations that remain to be used. Thus we see that we can eliminate y^2 as well. Consequently the polynomial multipliers can be reduced to $[x^4, (y^1, z^1)^2]^4$.

We can reach an even simpler form: Indeed, the sum equation is, at this point, of the form $[x^5, (y^2, z^2)^4]^6 = 0$, and it contains only three terms where y and z reach the dimension 4 together. But the terms produced by the polynomial multipliers contain no useful coefficient for the purpose of elimination. The nine appropriate coefficients are all arbitrary; thus assuming six of these coefficients are determined via arbitrary equations and three are used to cancel the three terms of interest, the form of the polynomial multipliers reduces to $[x^4, (y^1, z^1)^1]^4$, with three arbitrary coefficients or three arbitrary equations in the sum equation.

Finally, we find that the simplest form for the polynomial multipliers is $[x^4, y^1, z^0]^4$, or simply $(x^4, y^1)^4$.

Indeed, considering the form $[x^4, (y^1, z^1)^1]^4$, the sum equation is of the form $[x^5, (y^2, z^2)^3]^6 = 0$ and has nine terms in z^2. These nine terms can be matched by twelve useful coefficients provided by the polynomial multipliers. But three arbitrary coefficients remain; if we determine three among the twelve coefficients using three arbitrary equations, there remain nine coefficients to cancel the terms in z^2, and thus each of these twelve coefficients can be assumed to be zero. So we can suppress the terms in z in each of the three polynomial multipliers; their form now reduces to $(x^4, y^1)^4$ and this is the simplest possible expression: Indeed, there are nine terms in y^2 in the sum equation, and the polynonial multipliers provide twelve useful coefficients. We cannot assume that any of these coefficients will be equal to zero.

(329.) We have seen in (320) that the simplest polynomial multipliers were of the form $(x)^2$ for three equations of the form $[x, (y, z)^1]^2 = 0$. But we have seen from (322) that we could use polynomial multipliers of the form $(x, y, z)^2$, remarking that, in this case, we would be able to form one more arbitrary equation in the sum equation than the three equations naturally provided by the form $(x, y, z)^2$.

Since one arbitrary coefficient remains, we suspect that this form can be further reduced, and this is indeed the case.

Indeed, the sum equation is of the form $[x, (y, z)^3]^4$ and has eight terms where y and z reach the dimension 3 either together or separately. However, the terms in the three polynomial multipliers that provide these eight terms give nine useful coefficients, but this number reduces to eight because of the arbitrary equation we just pointed out. Thus there are as many coefficients as there are terms to cancel, and each of these coefficients is zero. Thus the form $(x, y, z)^2$ reduces to $[x, (y, z)^1]^2$. Since we have seen from (315) that this one could reduce to $(x)^2$, we have reduced all the apparently different forms of polynomial multipliers down to one.

(330.) In (307), we have reduced the form of the polynomial multipliers of three equations of the form $(x, y, z)^2 = 0$ to $[x, (y, z)^2]^6$. This form can be further reduced, as we have already said.

Indeed, the sum equation is of the form $[x, (y, z)^4]^8$ and has five terms in z^4; but the terms of the three polynomial multipliers leading to these terms in z^4 provide no useful coefficient; instead, they provide only fifteen arbitrary coefficients; we can decide that only ten of these are arbitrary and use the five others to destroy terms in z^4. Then each of these coefficients will be zero; consequently, the form $[x, (y, z)^2]^6$ can be reduced to $[x, (y^2, z^1)^2]^6$, with five arbitrary coefficients available from the three polynomial multipliers.

We must not lose sight of the fact that the five arbitrary coefficients corresponding to five arbitrary equations in the sum equation are not really arbitrary to the point that these five equations could be chosen anywhere in the sum equation. Remembering what we said (234), we see that we can form only one arbitrary equation in the highest dimension of the sum equation; likewise we can form only one equation in the following dimension, if we have formed one such equation in the highest dimension, or two only if that was not done. We can form only one equation in the third dimension (counting from the highest dimension down), if one equation were used in each of the two highest dimensions, or three if no such equation was built before, and so on.

Assume we had decided to simplify the calculations all at once, using the polynomials of the form $[x, (y^2, z^1)^2]^6$ right away for equations of the form $(x, y, z)^2 = 0$, without having followed the process described above. We would have found five more coefficients than are really needed. From the observation made in (322), we would not have been led to use any of these coefficients to destroy the highest-order terms in the final equation. They would therefore have been determined via any other arbitrary equation. However, there are limits to these arbitrary equations: Assume we

have formed more arbitrary equations in the highest dimensions of the sum
equation than the number we discussed above, although fewer than we have
arbitrary coefficients available; we would then miss the final equation and
would end up either with a trivial equation or with a wrong equation.

(331.) These observations allow us to evaluate whether our theory brings
knowledge of any importance to perfecting and making algebraic analysis
more rigorous, and what can be thought of solutions where, using the method
of undetermined coefficients, we end up with more coefficients than are available for the solution of interest to us.

(332.) So far we have given examples of our theory, regarding the way
to compute the value of the coefficients of the polynopmial multipliers and
regarding how to reduce them to their lowest possible number. It now remains to give an example of how to determine these polynomial multipliers
in cases where the general expression for the number of their terms can have
several different forms, as we have seen in (120 ff).

(333.) Pick, for example, the following three equations:

$$
\begin{aligned}
fyz \qquad &= 0, \\
+hx + ky + lz \\
+m
\end{aligned}
$$

$$
\begin{aligned}
e'xz \qquad &= 0, \\
+h'x + k'y + l'z \\
+m'
\end{aligned}
$$

$$
\begin{aligned}
+h''x + k''y + l''z \quad &= 0. \\
+m''
\end{aligned}
$$

When bringing these equations back to the proposed form (82), their form
is

$$
\begin{aligned}
[(x^1, y^1)^1, (x^1, z^1)^1, (y^1, z^1)^2]^2 &= 0, \\
[(x^1, y^1)^1, (x^1, z^1)^2, (y^1, z^1)^1]^2 &= 0, \\
[(x^1, y^1)^1, (x^1, z^1)^1, (y^1, z^1)^1]^1 &= 0.
\end{aligned}
$$

From what we said in (224) and (233), the form of the sum equation is

$$
[(x^{A+3}, y'^{A+3})'^{B+3}, (x^{A+3}, z''^{A+3})'^{B+4}, (y'^{A+3}, z''^{A+3})''^{B+4}]^{T+5} = 0.
$$

The form of the polynomial multiplier for the first equation is

$$
[(x^{A+2}, y'^{A+2})^{B+2}, (x^{A+2}, z''^{A+2})'^{B+3}, (y'^{A+2}, z''^{A+2})''^{B+2}]^{T+3}.
$$

The form of the polynomial multiplier for the second equation is

$$
[(x^{A+2}, y'^{A+2})^{B+2}, (x^{A+2}, z''^{A+2})'^{B+2}, (y'^{A+2}, z''^{A+2})''^{B+3}]^{T+3}.
$$

The form of the polynomial multiplier for the third equation is

$$
[(x^{A+2}, y'^{A+2})^{B+2}, (x^{A+2}, z''^{A+2})'^{B+3}, (y'^{A+2}, z''^{A+2})''^{B+3}]^{T+4}.
$$

The expressions for the three polynomials whose number of terms participate in the expression of the number of terms that may be cancelled in the first of the three polynomial multipliers, using the second and third equations, are as follows:

$$[(x^{A+1}, y'^{A+1})^{B+1}, (x^{A+1}, z''^{A+1})'^{B+1}, (y'^{A+1}, z''^{A+1})''^{B+1}]^{T+1},$$

$$[(x^{A+1}, y'^{A+1})^{B+1}, (x^{A+1}, z''^{A+1})'^{B+2}, (y'^{A+1}, z''^{A+1})''^{B+1}]^{T+2},$$

$$\text{and } [(x^A, y'^A)^B, (x^A, z''^A)'^B, (y'^A, z''^A)''^B]^T.$$

Finally, the expression of the polynomial whose number of terms expresses the number of terms that may be cancelled in the second equation, using the third equation, is

$$[(x^{A+1}, y'^{A+1})^{B+1}, (x^{A+1}, z''^{A+1})'^{B+1}, (y'^{A+1}, z''^{A+1})''^{B+2}]^{T+2}.$$

This being said, the three proposed equations belong to the general forms presented in (120 ff). They also belong to the specific case (129). Thus we know (1) that the degree of the final equation is 3; and (2) that the polynomials we have just introduced, and which by (105) must all have the same form within those presented in (120 ff), can belong to any of these forms. Let us therefore pick the first form from (120) as if they could belong to that form only.

The conditions that lead to this form (replacing C by T) are

$$T - B < B_{/} - A;\ T - B < B_{//} - A_{/};\ T - B_{/} < B_{//} - A_{//}.$$

Since the product equation and all other polynomials above must have the same form, we get

$$T + 5 - B - 3 < B + 4 - A - B - 3;\ T + 5 - B - 3 < B_{//} + 4 - A_{/} - 3;$$

$$T + 5 - B_{/} - 4 < B_{//} + 4 - A_{//} - 3;$$

that is,

$$T - B + 1 < B_{/} - A;\ T - B + 1 < B_{//} - A_{/};\ T - B_{/} < B_{//} - A_{//}.$$

Likewise

$$T - B + 1 < B_{/} - A + 1;\ T - B + 1 < B_{//} - A_{/};\ T - B_{/} < B_{//} - A_{//};$$

$$T - B + 1 < B_{/} - A;\ T - B + 1 < B_{//} - A_{/} + 1;\ T - B_{/} + 1 < B_{//} - A_{//} + 1;$$

$$T - B + 2 < B_{/} - A + 1;\ T - B + 2 < B_{//} - A_{/} + 1;\ T - B_{/} + 1 < B_{//} - A_{//} + 1;$$

$$T - B < B_{/} - A;\ T - B < B_{//} - A_{/};\ T - B_{/} < B_{//} - A_{//};$$

$$T - B + 1 < B_{/} - A + 1;\ T - B + 1 < B_{//} - A_{/};\ T - B_{/} < B_{//} - A_{//};$$

$$T - B < B_{/} - A;\ T - B < B_{//} - A_{/};\ T - B_{/} < B_{//} - A_{//};$$

$$T - B + 1 < B_{/} - A;\ T - B + 1 < B_{//} - A_{/} + 1;\ T - B_{/} < B_{//} - A_{//}.$$

These inequalities always reduce to three for equations in three unknowns and are

$$T - B < B' - A - 1;\quad T - B < B'' - A' - 1;\quad T - B' < B'' - A''.$$

Thus, as long as the quantities T, B, B', B'', A, A', A'' satisfy these three inequalities, the product equation, and the seven other polynomials all have the same form, as they should.

We can therefore choose arbitrary values for these quantities as long as (1) they satisfy these conditions; (2) they also satisfy the general conditions for the existence of the polynomials mentioned in (83); and (3) $A+3$ cannot be smaller than 3. We are seeking the equation in x and the final equation must be third degree.

In order to satisfy all the existence conditions for all these polynomials, it is sufficient that they be satisfied by the polynomial

$$[(x^A, y^{A'})^B, (x^A, z^{A''})^{B'}, (y^{A'}, z^{A''})^{B''}]^T.$$

This being said, I set

$$T - B = B' - A - 1;\quad T - B = B'' - A' - 1;\quad T - B' = B'' - A'';$$

right away, since the inequalities above can be equalities. And I conclude that

$$T = 2B + A'' - A' - A - 2;\quad B' = B + A' - A - 1;\quad B'' = B + A'' - A - 1.$$

I assume arbitrarily that $A = A' = A''$ and I get

$$T = 2B - A - 2,\quad B' = B - 1,\quad B'' = B - 1.$$

We now note that the smallest possible value for A such that the polynomial above exists is $A = 2$; I therefore assume $A = A' = A'' = 2$; then the smallest possible value of B that meets the existence conditions for this polynomial is $B = 4$. I therefore get

$$T = 4,\ B = 4,\ B' = 3,\ B'' = 3,\ A = 2,\ A' = 2,\ A'' = 2,$$

and the generating polynomial becomes

$$[(x^2, y^2)^4, (x^2, z^2)^3, (y^2, z^2)^3]^4.$$

This being said, the product equation and the seven polynomials above will have the following forms:

$$[(x^5, y^5)^7, (x^5, z^5)^7, (y^5, z^5)^7]^9 = 0,$$
$$[(x^4, y^4)^6, (x^4, z^4)^6, (y^4, z^4)^5]^7$$
$$[(x^4, y^4)^6, (x^4, z^4)^5, (y^4, z^4)^6]^7$$
$$[(x^4, y^4)^6, (x^4, z^4)^6, (y^4, z^4)^6]^8$$
$$[(x^3, y^3)^5, (x^3, z^3)^4, (y^3, z^3)^4]^5$$
$$[(x^3, y^3)^5, (x^3, z^3)^5, (y^3, z^3)^4]^6$$
$$[(x^2, y^2)^4, (x^2, z^2)^3, (y^2, z^2)^3]^4$$
$$[(x^3, y^3)^5, (x^3, z^3)^4, (y^3, z^3)^5]^6.$$

From these and from what was said in (325 ff), it is now easy to determine surely the simplest forms these three polynomial multipliers can have.

For example, we find that ten terms must be cancelled in the highest dimension of the sum equation and that the total number of useful coefficients in the highest dimension in each polynomial multiplier is only ten out of a total of twenty-four coefficients. Thus since there are only as many coefficients as there are terms to cancel, and assuming that each of the fourteen arbitrary coefficients is zero, then each of the ten useful coefficients will also be zero. Thus the total dimension of each polynomial multiplier can be lowered by 1. Examining the next dimension in the sum equation, we find it has eighteen terms, whose destruction can be performed by nineteen useful coefficients; thus the total dimension cannot be lowered anymore, unless it can be done after lowering the total dimension of y and z or after lowering each of them. The same investigation can be done relative to the total dimension of y and z, then relative to the degree of y and relative to the degree of z as we have seen above.

(334.) We now see that, when the proposed equations do not have the same form, the form of the polynomial multipliers is in general more complex: In the example given in (328), the final equation must be fifth degree, and in the current example it must be third degree. In the first case, we have reached the general form and the simplified form of the polynomial multipliers much faster because the proposed equations all had the same form. However, the equations of interest to us now are only particular cases of the former.

Although the polynomial multipliers are currently much more complicated than in (328), they may still be reduced to a simpler form. But, to reach this simpler form, we must necessarily begin with a form that can be surely determined only by following the process we have just illustrated. Only by starting from this general expression will we be sure that each step uses the true number of arbitrary coefficients successively entering in the simplifying forms leading to the simplest possible multipliers.

Assume we begin up front with a simpler polynomial multiplier form, for example, a simpler form than that determined in (328). It seems there is no risk of getting lost, since the current equations are particular cases of those in (328), and the resulting polynomial multipliers must therefore be simpler, or, at worst, as complicated.

The truth is, however, that if we were to proceed this way, we would not be assured that the form of the polynomials that express the number of arbitrary coefficients is the one that expresses their largest number. Devoid of any guidance, we could be led to either a wrong or trivial final equation.

(335.) Thus we see that, if we want to reach the final equation without any errors starting from the equations as they are, we must follow the method we have described. We must absolutely know the degree of the final equation and the general form of the polynomial multipliers for each equation, as well as that of the polynomials whose number of terms expresses that of the arbitrary equations we can use.

(336.) By the way, we can, if we want to, avoid the more complex forms by computing the final equation resulting from a similar number of equations of the same form, and which includes the proposed equations. For example, in the current cases, we could seek the final equation resulting from three equations of the form

$$(x^1, y^1)^2, (x^1, z^1)^2, (y^1, z^1)^2 = 0,$$
$$(x^1, y^1)^2, (x^1, z^1)^2, (y^1, z^1)^2 = 0,$$
$$[(x^1, y^1)^1, (x^1, z^1)^1, (y^1, z^1)^1]^1 = 0.$$

This final equation surely contains the final equation sought as a particular case and provides it by equating the coefficients of these equations with those in the proposed equations. But it would almost always happen that this final equation is of a higher degree than what it must be. However, we know from what was said in (294 ff) which indicators can show whether the degree can be lowered, and how to do so; thus, this method can be used to find the final equation with lowest degree for the three equations of interest.

But this method is not the simplest possible.

Indeed, the lowest final equation would be reached only after having completed the entire elimination process, starting from more complex equations than those initially given: It is easy to get an idea of the amount of time involved and to realize that this work should be undertaken only if no other option exists.

Finding the true form of the simplest polynomial multipliers for the proposed equations only requires a methodical enumeration of the number of terms of the sum equation, of the polynomial multipliers and of the polynomials whose number of terms express the number of terms that may be cancelled. This enumeration leads to excluding several terms from these polynomials, without ever having to compute the sum equation.

(337.) Our observations lead us to the importance of looking for the degree of the final equation. This is, first, because of all the applications where it is more important to determine the number of roots than their value, and these applications are not infrequent (see (48), for example). We also see that knowing the degree of the final equation is extremely important in determining the form that the polynomial multipliers must have to get the final equation securely. As long as we ignore this, we will indeed have arbitrary coefficients, but we will still face the possibility of making mistakes regarding the conclusions these can lead to.

About equations where the number of unknowns is lower by one unit than the number of these equations. A fast process to find the final equation resulting from an arbitrary number of equations with the same number of unknowns

(338.) When the number of equations exceeds the number of unknowns by one unit, then the final equation is a condition equation involving the coefficients of the proposed equations. This condition can be more or less simple, depending on the process we follow to get it. The process we are about to present follows what we have said up until now, and we believe it is the simplest. At the same time, it is the fastest process to reach the final equation resulting from an arbitrary number of equations in the same number of unknowns.

When the number of unknowns is the same as the number of equations, we can always group under the same label all terms in x that stand in front of the same power or product of the other unknowns; we can always turn the question into a question where the number of unknowns is less than the number of equations by one unit.

Consider, for example, the equation

$$ax^2 + bxy + cy^2 + dx + ey + f = 0.$$

Writing $c = A$, $bx + e = B$, $ax^2 + dx + f = C$, we can write the equation as

$$Ay^2 + By + C = 0,$$

which is an equation in one unknown.

Consider the equation in three unknowns:

$$ax^2 + bxy + cxz + dy^2 + eyz + fz^2 = 0,$$
$$+gx + hy + kz$$
$$+l$$

Writing $d = A$, $e = B$, $f = C$, $bx + h = D$, $cx + k = E$, $ax^2 + gx + l = F$, we can write this equation as

$$Ay^2 + Byz + Cz^2 = 0,$$
$$+Dy + Ez$$
$$+F$$

that is, as one equation in two unknowns.

By taking this viewpoint, it is possible to considerably shorten the computations required by our first method, because many fewer coefficients need computing. Before presenting all the advantages of the second method, we begin with a comparison of both.

(339.) By letting the equations take all possible forms within a particular degree, the first method never leads to exceeding the degree reached by the final equation, even when specific relations exist among the coefficients that lead to possibly lowering the degree of the final equation. However, this advantage can be gained only through the computation of a very large number of coefficients. When we use the processes previously described to reduce these coefficients to their smallest possible number, we are sure to find not only the final equation, but also all the indicators of when the degree of the final equation can be lowered. No existing method was able to do so previously. In one word, the result provides us with everything there is to know about the proposed equations, and we avoid irrelevant roots as we have seen in (282). This drawback is present in the usual elimination method for two unknowns, and it would be even more considerable if this method were applied to a larger number of unknowns, even if this method were not to exaggerate the general degree of the final equation. In short, our first method is, in my opinion, as perfect as possible from a theoretical standpoint.

On the practical side of things, that is, when dealing with computational and complexity issues, the second method presents many advantages. By using a much smaller number of coefficients, the results are simpler, and so are the means to obtain them. Assuming the proposed equations have no relations linking their coefficients in a way that the degree of the final equation can be lowered, it yields the final equation as quickly as seems to be possible.

We say as quickly as seems to be possible, but it does not always yield the simplest possible equation. Indeed, the results of this second method are immensely less complex than those we would try to obtain using successive elimination, and are devoid of any excessively complicated and unrelated factors. It will, however, sometimes bring one or more factors to the final equation. These factors are not unrelated to the question, but they almost always indicate solutions of the form described in (279) and (287); since they only provide trivial insight to the question, it would probably be better that they do not appear. Although these factors can be avoided in many cases, in particular when there are only two equations, there seems to be no general method to reach the final equation from an arbitrary number of equations without these parasitic factors. As soon as we talk about general methods, the nature of our analysis indifferently considers general solutions and particular solutions. This is the reason why we doubt we will ever avoid these factors in the second method.

But if, on the one hand, it seems impossible to avoid these factors in general, on the other hand these factors will often appear before the end of the computations, as we have already seen and will see further. Then we will be able to extract them and thus simplify the remaining computations. In the few cases where the factor appears only with the final equation, it will be harder to extract; we will, however, provide the means to do so.

Here is, I think, all that we may expect from this second elimination method: Either it avoids the unimportant factors or, if it cannot, it reveals them so that they can be extracted from the final equation.

About polynomial multipliers that are appropriate for elimination using this second method

(340.) We can apply what was said about the general form of the polynomial multipliers in equations with the same number of unknowns to the case where the number of unknowns is smaller than the number of equations by one unit.

This form must always be such that the degree of the final equation is an exact differential, whose order equals that of the equations. In the case when there is one more equation than there are unknowns, the result of the

elimination must be a condition equation; that is, it must contain no unknowns, and the degree of the final equation must be zero. This is also what
will always happen, when we pick the form of the polynomial multipliers
appropriately. Consider, for example, three unknowns and four equations
represented by

$$(u \ldots 3)^t = 0,$$
$$(u \ldots 3)^{t'} = 0,$$
$$(u \ldots 3)^{t''} = 0,$$
$$(u \ldots 3)^{t'''} = 0.$$

Their respective polynomial multipliers are

$$(u \ldots 3)^{T+t'+t''+t'''},$$
$$(u \ldots 3)^{T+t+t''+t'''},$$
$$(u \ldots 3)^{T+t+t'+t'''},$$
$$(u \ldots 3)^{T+t+t'+t''}.$$

The number of useful coefficients in the first polynomial multiplier is

$$d^3 N(u \ldots 3)^{T+t'+t''+t'''} \ldots \begin{pmatrix} T+t'+t''+t''' \\ t',t'',t''' \end{pmatrix}.$$

The number of useful coefficients in the second polynomial is

$$d^2 N(u \ldots 3)^{T+t+t''+t'''} \ldots \begin{pmatrix} T+t+t''+t''' \\ t'',t''' \end{pmatrix}.$$

The number of useful coefficients in the third polynomial is

$$d N(u \ldots 3)^{T+t+t'+t'''} \ldots \begin{pmatrix} T+t+t'+t''' \\ t''' \end{pmatrix}.$$

Finally, the number of useful coefficients in the fourth polynomial is

$$N(u \ldots 3)^{T+t+t'+t''}.$$

Since the number of terms to be cancelled is the total number of terms in
the sum equation, minus one, we must have

$$N(u \ldots 3)^{T+t+t'+t''+t'''} = d^3 N(u \ldots 3)^{T+t'+t''+t'''} \ldots \begin{pmatrix} T+t'+t''+t''' \\ t',t'',t''' \end{pmatrix}$$
$$+ d^2 N(u \ldots 3)^{T+t+t''+t'''} \ldots \begin{pmatrix} T+t+t''+t''' \\ t'',t''' \end{pmatrix}$$
$$+ d N(u \ldots 3)^{T+t+t'+t'''} \ldots \begin{pmatrix} T+t+t'+t''' \\ t''' \end{pmatrix} + N(u \ldots 3)^{T+t+t'+t''}.$$

This equation, which we have already seen in examples in (309), reduces
to the following:

$$d^4 N(u \ldots 3)^{T+t'+t''+t'''} \ldots \begin{pmatrix} T+t+t'+t''+t''' \\ t,t',t'',t''' \end{pmatrix} = 0.$$

This equation obviously holds, since $N(u \ldots 3)^{T+t'+t''+t'''}$ is only a function of three dimensions (see (12) and (39)).

(341.) Regarding incomplete equations, the general form that we have taught when the number of unknowns equals that of the equations is again appropriate, when the number of unknowns is less than the number of equations by one unit. But we must add a few observations.

(342.) Remember from (84 ff) that the form of the polynomial multipliers is not unique. We could thus believe that we need to perform the same verifications as those indicated in (120 ff) to identify, among all possible forms, those which are admissible. In the current case, we want to show that all different forms presented in (120 ff) and all those which could occur in all other equations are admissible. No conditions will have to be satisfied, except to make sure that all polynomial multipliers, the sum equation, and all polynomials whose number of terms express the number of terms that can be cancelled in each polynomial multiplier belong to the same form, whichever it is.

(343.) Indeed, assume for simplicity that we consider only one polynomial multiplier, as was done in the first book; the expression of the number of remaining terms after cancelling all those that can be cancelled, using all equations except that considered for the polynomial multiplier, is an exact differential of order n, where $n + 1$ is the total number of equations.

For the same reason, the expression of the number of remaining terms, introducing the fictitious terms described in (110), is also an exact differential of order n. Thus the difference between the number of remaining terms without fictitious terms and the number of remaining terms with fictitious terms is an exact differential of order $n + 1$. Since the number of unknowns is n, the total dimension of the variables entering in the expression of the number of these terms can also only be n; thus the latter differential is zero. Thus introducing fictitious terms does not cancel more terms than could be cancelled without. Since this argument applies to each of the forms that can be taken by the expression for the number of terms, we can take any form for the polynomial multiplier.

Thus the form of the polynomial multiplier is subject to none of the above mentioned conditions (120 ff).

(344.) We must therefore only make sure that all polynomial multipliers have the same form among those mentioned in (120 ff), and to assign the same form to the sum equation and all polynomials whose number of terms contibute to expressing the number of terms that may be cancelled in each polynomial multiplier.

Details of the method

(345.) We will imitate what was done in (224) in the case when the number of unknowns equals the number of equations, to determine the general form of the polynomial multipliers. Moreover, we will follow the process described in (306 ff) in the same case, to reduce the polynomials to their simplest possible form, that is, to the smallest possible number of terms.

When the chosen expression for the number of terms can have several different forms, we will arbitrarily pick one of these forms. All polynomials used as polynomial multipliers or used to express the number of terms that can be cancelled in those polynomial mutlipliers will then be of the same form.

Having chosen the polynomial multipliers, and having reduced them to their simplest possible form, the computation of the final equation will follow the same process as in the first case; however, we will not be looking for the values of the undetermined coefficients, since these values are not useful; in fact, they are not useful in the first method either. We will sequentially compute the different *lines*, successively going through all different terms in the sum equation, following the most appropriate sequence. Setting the last line to zero will yield the condition equation, which is also the final equation sought.

We now clarify these considerations by means of examples.

First General Example

(346.) We propose our first general example to be equations in two unknowns and with arbitrary degrees.

When written as equations in one unknown, they are represented by $(x\ldots1)^t = 0$ and $(x\ldots1)^{t'} = 0$.

The polynomial multiplier of the first equation is, in general, of the form $(x\ldots1)^{T+t'}$ (see (224)), and the degree of the second polynomial multiplier is $(x\ldots1)^{T+t}$.

Under this format, the degree of the final equation must be zero. Thus nothing constrains the value of T except that $T+t'$ must be no smaller than t'; otherwise we would exclude terms that the equation $(x\ldots1)^{t'}$ does not allow.

I therefore assume that $T = 0$; then the polynomial multipliers become $(x\ldots1)^{t'}$ and $(x\ldots1)^t$.

To determine whether this form can be further reduced, I note that the number of useless coefficients is 1 and in the highest dimension. The sum equation is of the form $(x\ldots1)^{t+t'} = 0$; thus I have only one useful coefficient to cancel the term in $x^{t+t'}$ in the highest dimension of that equation. This coefficient is therefore zero if I assume its counterpart in the other polynomial multiplier to be zero as well, something I can always do. The form of the two polynomial multipliers therefore reduces to $(x\ldots1)^{t'-1}$ and $(x\ldots1)^{t-1}$; this is in complete agreement with what we said in the *Mémoires de l'Académie des Sciences, Year 1764* and that we had then found using a very different method.

(347.) Consider two equations of the form

$$ax^2 + bx + c = 0.$$

I multiply each equation by a polynomial of the form $Ax + B$, and the sum equation is then

$$Aa\ x^3\ +\ Ab\ x^2\ +\ Ac\ x\ +\ Bc = 0.$$
$$+\ Ba\quad\ +\ Bb$$

By setting the total coefficients of x^3, x^2, etc. to zero, I proceed with the computation of $AA'BB'$ as follows:

First line	$aA'BB'$
Second line	$(ab')BB' - aA'aB'$
Third line	$(ab')bB' - (ac')aB'$

The term where A' remains is rejected since it is not in the last equation and thus cannot change the final equation.

Fourth line $\quad (ab')(bc') - (ac')^2$

Thus the final equation is $(ab').(bc') - (ac')^2 = 0$.

(348.) Assume the two proposed equations are of the form

$$ax^3 + bx^2 + cx + d = 0.$$

Each polynomial multiplier is (346) of the form

$$Ax^2 + Bx + C.$$

Thus the sum equation is of the form

$$\begin{array}{lllllll}
Aa & x^5 & +Ab & x^4 & +Ac & x^3 & +Ad & x^2 & +Bd & x & +Cd = 0. \\
 & & +Ba & & +Bb & & +Bc & & +Cc & & \\
 & & & & +Ca & & +Cb & & & & \\
\end{array}$$

We therefore write

First line	$aA'BB'$
Second line	$[(ab')BB' - aA'aB']CC'$
Third line	$[(ab')bB' - (ac')aB' + aA'(ab')]CC' + [(ab')BB' - aA'aB']aC'$
Fourth line	$[(ab').(bc') - (ac').(ac') + (ad').(ab')]CC'$
	$\quad -[(ab')bB' - (ac')aB']bC'$
	$\quad + [(ab')cB' - (ad')aB']aC',$

rejecting the terms where A' and BB' remain but cannot influence the degree of the final equation.

Fifth line $\quad \ldots \quad [(ab').(bc') - (ac').(ac') + (ad').(ab')]cC'$
$\quad\quad\quad\quad\quad -[(ab').(bd') - (ac').(ad')]bC'$
$\quad\quad\quad\quad\quad + [(ab').(cd') - (ad').(ad')]aC',$

rejecting terms where B' remains.

Sixth line $\quad \ldots \quad [(ab').(bc') - (ac').(ac') + (ad').(ab')](cd')$
$\quad\quad\quad\quad\quad -[(ab').(bd') - (ac').(ad')](bd')$
$\quad\quad\quad\quad\quad + [(ab').(cd') - (ad')^2](ad').$

Thus the final equation is

$$\left.\begin{array}{l}
[(ab').(bc') - (ac')^2 + (ad').(ab')](cd') - [(ab').(bd') - (ac').(ad')](bd') \\
\quad\quad\quad\quad\quad\quad\quad\quad\quad\quad\quad + [(ab').(cd') - (ad')^2](ad')
\end{array}\right\} = 0.$$

This process is easy to apply to higher degrees, and we will not pursue these computations.

Second General Example

(349.) The second general example shows how to perform eliminations for complete equations in three unknowns.

These equations can be written as equations in two unknowns and be represented by three equations of the form

$$(x \ldots 2)^t = 0,$$
$$(x \ldots 2)^{t'} = 0,$$
$$(x \ldots 2)^{t''} = 0.$$

The polynomial multiplier of the first equation is $(x \ldots 2)^{T+t'+t''}$. The second polynomial multiplier is $(x \ldots 2)^{T+t+t''}$. The third one is $(x \ldots 2)^{T+t+t'}$. Since the degree of the final equation must be zero, T does not have to satisfy any condition other than

$$T + t' + t'' > t' + t'', \; T + t + t'' > t + t'', \; T + t + t' > t + t',$$

where equalities are allowed. These conditions result from the requirement that the number of terms that can be cancelled in any of the three polynomial multipliers using the other two equations must be a positive integer; for the first polynomial multiplier, this expression is

$$N(x \ldots 2)^{T+t''} + N(x \ldots 2)^{T+t'} - N(x \ldots 2)^{T};$$

thus if we had $T + t' + t'' < t' + t''$, that is, $T < 0$, $N(x \ldots 2)^T$ would be negative and we would not be able to use the expressions found in (39) for $(x \ldots n)^T$.

Let us assume $T = 0$ right away and pick the following polynomial multipliers:

For the first: ... $(x \ldots 2)^{t'+t''}$

For the second: ... $(x \ldots 2)^{t+t''}$

For the third: ... $(x \ldots 2)^{t+t'}$.

To determine whether this form is the simplest possible, I compute the sum equation as $(x \ldots 2)^{t+t'+t''} = 0$, and I observe that this equation has $t + t' + t'' + 1$ terms to be cancelled in the highest dimension.

The highest dimension of the first polynomial multiplier provides $t + t'' + 1 - t'' - 1 - t' - 1 + 1 = 0$ useful coefficients.

The highest dimension of the second polynomial multiplier provides $t + t'' + 1 - t - 1 = t''$ useful coefficients.

The highest dimension of the third polynomial multiplier provides $t + t' + 1$ useful coefficients.

In other terms, the highest dimensions of the three polynomial multipliers provides $t + t' + t'' + 1$ useful coefficients.

Thus we have as many coefficients as there are terms to be cancelled in the highest dimension, and each of these coefficients is zero. Thus the polynomial multipliers of the three proposed equations can be chosen as follows

First polynomial: ... $(x \ldots 2)^{t'+t''-1}$

Second polynomial: ... $(x \ldots 2)^{t+t''-1}$

Third polynomial: ... $(x \ldots 2)^{t+t'-1}$.

(350.) Likewise we examine the highest dimension of the sum equation resulting from these polynomial multipliers, and we see that $t + t' + t''$ of its terms must be cancelled.

We also see that the highest dimension of the first polynomial multiplier gives $t' + t'' - t' - t'' = 0$ useful coefficients;

that the highest dimension of the second polynomial multiplier gives $t + t'' - t = t''$ useful coefficients;

and that the highest dimension of the third polynomial multiplier provides $t + t'$ useful coefficients.

Thus the highest dimensions of the three polynomial multipliers provide $t + t' + t''$ useful coefficients, a number equal to the number of terms to be cancelled. Thus each of these coefficients is zero, and the three polynomial multipliers can be taken as follows:

First equation	...	$(x \ldots 2)^{t'+t''-2}$
Second equation	...	$(x \ldots 2)^{t+t''-2}$
Third equation	...	$(x \ldots 2)^{t+t'-2}$.

Performing a similar investigation on the highest dimension of the sum equation resulting from these new polynomial multipliers, we see that the number of terms in that dimension is $t + t' + t'' - 1$;

that the highest dimension of the first polynomial multiplier provides $t' + t'' - 1 - t' + 1 - t'' + 1 = 1$ useful coefficients;

that the highest dimension of the second polynomial multiplier provides $t + t'' - 1 - t + 1 = t''$ useful coefficients;

and that the highest dimension of the third polynomial multiplier provides $t + t' - 1$ useful coefficients.

Thus the highest dimensions of the three polynomial multipliers provide $t + t' + t''$ useful coefficients, which is one more coefficient than there are terms to be cancelled. Thus we cannot assume that each coefficient is zero.

And the degrees of the three polynomial multipliers $(x \ldots 2)^{t'+t''-2}$, $(x \ldots 2)^{t+t''-2}$, $(x \ldots 2)^{t+t'-2}$ cannot be lowered to a smaller dimension.

(351.) We now need to examine whether x and y each reach the total dimension of the polynomial (x and y are the two unknowns to be eliminated).

I first remark that there is only one term of the sum equation in y that reaches the dimension $t + t' + t'' - 2$ and that the number of useful coefficients in the three polynomial multipliers to cancel this term is zero. Thus there are fewer useful coefficients than there are terms to be cancelled. Thus, according to (325), we can form two arbitrary equations instead of the three that can be formed here, and we can use the third equation to cancel the term of interest. Each one of the three arbitrary coefficients is zero, and there remains one arbitrary equation for the set of polynomial multipliers, which are now

First equation   ... $(x^{t'+t''-2}, y^{t'+t''-3})t'+t''-2$
Second equation   ... $(x^{t+t''-2}, y^{t+t''-3})t+t''-2$
Third equation   ... $(x^{t+t'-2}, y^{t+t'-3})t+t'-2.$

(352.) Given this new form of the polynomial multipliers, the sum equation is of the form

$$(x^{t+t'+t''-2}, y^{t+t'+t''-3})t+t'+t''-2 = 0.$$

The unknown y reaches the degree $t + t' + t'' - 3$ in two terms. These are $xy^{t+t'+t''-3}$ and $y^{t+t'+t''-3}$.

The three polynomial multipliers provide $6 - 6 = 0$ useful terms to cancel these two terms. Assume that only four out of six arbitrary equations are actually formed and that the other two are used to cancel these two terms. Then the six coefficients of the three polynomial multipliers that led to the terms $y^{t+t'+t''-3}$ in the sum equation are zero. The polynomial multipliers then reduce to the following forms:

First equation   ... $(x^{t'+t''-2}, y^{t'+t''-4})t'+t''-2$
Second equation   ... $(x^{t+t''-2}, y^{t+t''-4})t+t''-2$
Third equation   ... $(x^{t+t'-2}, y^{t+t'-4})t+t'-2,$

with three arbitrary equations to be formed: One comes from the first reduction and two come from the second reduction. But we must observe that only two of these three arbitrary equations can be used for the highest dimension.

(353.) Following a similar argument, the polynomial multipliers can reduce to the form:

First equation   ... $(x^{t'+t''-2}, y^{t'+t''-5})t'+t''-2.$
Second equation   ... $(x^{t+t''-2}, y^{t+t''-5})t+t''-2.$
Third equation   ... $(x^{t+t'-2}, y^{t+t'-5})t+t'-2,$

along with six arbitrary equations to be formed in the sum equation: One comes from the first reduction, two from the second, and three from the third. Out of these six arbitrary equations, no more than three can be used in the highest dimension of the sum equation; no more than two in the second highest dimension, if three have been attributed to the first; and no more than one can be attributed to the third highest dimension, if five have been attributed to the first two dimensions.

(354.) In general, and following the same argument, the polynomial multipliers can be reduced to the forms

First equation   ... $(x^{t'+t''-2}, y^{t'+t''-2-q})t'+t''-2$
Second equation   ... $(x^{t+t''-2}, y^{t+t''-2-q})t+t''-2$
Third equation   ... $(x^{t+t'-2}, y^{t+t'-2-q})t+t'-2,$

with $\frac{(q+1)\cdot(q)}{2}$ arbitrary equations to form in the sum equation: These consist of q equations in the first or highest dimension, $q - 1$ in the second, $q - 2$ in the third, and so on.

(355.) We assume $t > t' > t''$ to find the highest value of q; this is always possible, possibly by permuting the equations.

Then what we have said previously will occur until $q = t'' - 1$ included. Thus the form of the three polynomial multipliers can be reduced as follows:

First equation	...	$(x^{t'+t''-2}, y^{t'-1})^{t'+t''-2}$
Second equation	...	$(x^{t+t''-2}, y^{t-1})^{t+t''-2}$
Third equation	...	$(x^{t+t'-2}, y^{t+t'-t''-1})^{t+t'-2},$

with the same number of arbitrary equations as we have just outlined.

(356.) This is not the simplest general form relative to y.

Indeed, it is now possible to cancel terms in the first polynomial multiplier not by using the second equation, but by using the third equation only:

To cancel the t'' terms in $y^{t+t'-1}$ appearing in the sum equation, the first polynomial multiplier provides zero useful coefficients; the second polynomial provides zero useful coefficients; the third polynomial provides t'' such useful coefficients.

Thus, there are precisely as many useful coefficients as there are terms to be cancelled. Hence, each one of these terms is zero.

Thus, the form of the polynomial multipliers can be reduced as follows:

First equation	...	$(x^{t'+t''-2}, y^{t'-2})^{t'+t''-2}$
Second equation	...	$(x^{t+t''-2}, y^{t-2})^{t+t''-2}$
Third equation	...	$(x^{t+t'-2}, y^{t+t'-t''-2})^{t+t'-2},$

and in general,

First equation	...	$(x^{t'+t''-2}, y^{t'-q'})^{t'+t''-2}$
Second equation	...	$(x^{t+t''-2}, y^{t-q'})^{t+t''-2}$
Third equation	...	$(x^{t+t'-2}, y^{t+t'-t''-q'})^{t+t'-2},$

until $t' - q' = t'' - 1$, that is, until $q' = t' - t'' + 1$. Indeed, our argument is valid up until then. As soon as $q' = t' - t'' + 1$, we are not able to further lower the form relative to y.

Indeed, the form of the polynomial multipliers is then

First equation	...	$(x^{t'+t''-2}, y^{t''-1})^{t'+t''-2}$
Second equation	...	$(x^{t+t''-2}, y^{t-t'+t''-1})^{t+t''-2}$
Third equation	...	$(x^{t+t'-2}, y^{t-1})^{t+t'-2}.$

It is then impossible to cancel any term in the first polynomial multiplier using the first or the second equation. We thus have t' useful coefficients from the first polynomial mutliplier to cancel all t' terms in $y^{t+t''-1}$. Likewise, we have zero useful coefficients from the second multiplier, and t' coefficients from the third.

We therefore have $2t'$ useful coefficients to cancel t' terms: We cannot assume these coefficients are all zero: The last form of the polynomial multipliers is as simple as possible relative to y.

(357.) We now examine this form relative to x. In the sum equation, only one term in $x^{t+t'+t''-2}$ exists; the three polynomial multipliers provide three coefficients to cancel this term. Out of these, only two can be considered to be useful, because we can cancel one term in the second polynomial. We therefore have more useful coefficients than terms to be cancelled and we cannot assume they are all zero in each polynomial multiplier. But we must remember (354) that we must form $\frac{(q+1)q}{2} = \frac{t''(t''-1)}{2}$ arbitrary equations. Thus, assuming we use one such equation in the present case, we recast our problem to a problem where there are as many useful coefficients as there are terms to be cancelled. Then the form of the polynomial multipliers can be reduced as follows:

First polynomial multiplier   ... $(x^{t'+t''-3}, y^{t''-1})^{t'+t''-2}$
Second polynomial multiplier   ... $(x^{t+t''-3}, y^{t-t'+t''-1})^{t+t''-2}$
Third polynomial multiplier   ... $(x^{t+t'-3}, y^{t-1})^{t+t'-2}$.

We now pursue the same argument with terms containing $x^{t+t'+t''-3}$ in the sum equation. We see that (1) there are two such terms; and (2) the three polynomial multipliers provide six coefficients, where four can be considered to be useful. Indeed, two of them can be cancelled in the second polynomial; since $\frac{t''(t''-1)}{2} - 1$ arbitrary equations remain to be formed, two of them can be used here. The form of the polynomial multipliers therefore reduces to

First polynomial   ... $(x^{t'+t''-4}, y^{t''-1})^{t'+t''-2}$
Second polynomial   ... $(x^{t+t''-4}, y^{t-t'+t''-1})^{t+t''-2}$
Third polynomial   ... $(x^{t+t'-4}, y^{t-1})^{t+t'-2}$.

Following the same argument, this form reduces to

First equation   ... $(x^{t'+t''-2-q''}, y^{t''-1})^{t'+t''-2}$
Second equation   ... $(x^{t+t''-2-q''}, y^{t-t'+t''-1})^{t+t''-2}$
Third equation   ... $(x^{t+t'-2-q''}, y^{t-1})^{t+t'-2}$.

until $q'' = t'' - 1$. Thus, the most reduced general form is

First equation   ... $(x^{t'-1}, y^{t''-1})^{t'+t''-2}$
Second equation   ... $(x^{t-1}, y^{t-t'+t''-1})^{t+t''-2}$
Third equation   ... $(x^{t+t'-t''-1}, y^{t-1})^{t+t'-2}$.

By most reduced general form, we do not mean that no arbitrary coefficient remains. Rather, $N(x^{t-t''-1}, y^{t-t'-1})^{t-2}$ such coefficients remain in the second polynomial. But this means it is not possible to cancel any new terms in all three polynomial multipliers altogether.

(358.) There are a few exceptions to our presentation; we now present them.

1. We must consider the case $t = t'$ separately. In this case, we cannot cancel any terms in the first polynomial, at least without using arbitrary equations; and the same holds for the second polynomial, as soon as we have reached the general form that is most reduced relative to y. Thus the most reduced form we have described, relative both to y and x, does not hold anymore when $t = t'$. In this case, the following form for the three polynomial multipliers:

$$(x^{t'+t''-2}, y^{t''-1})^{t'+t''-2} \dots (x^{t+t''-2}, y^{t-t'+t''-1})^{t+t''-2} \dots (x^{t+t'-2}, y^{t-1})^{t+t'-2}$$

cannot lose terms, unless we use some of the available arbitrary equations, whose number is $\frac{t''(t''-1)}{2}$; $t'' - 1$ of them are in the highest dimension, $t'' - 2$ are in the second, and so on. Only by examining the specific values of t'' can we decide whether a few terms can be cancelled in the polynomial multipliers.

Assume, for example, $t'' = 2$; the number of available arbitrary equations is only one; thus the term all in x cannot be cancelled in all three polynomial multipliers. Thus only one arbitrary equation can be formed in the highest dimension of the sum equation.

Assume now $t'' = 3$; the number of available arbitrary equations is three, two for the highest dimension of the sum equation, and one for the second. Thus, the term all in x can be cancelled in each polynomial multiplier, and one arbitrary equation remains to be formed in the second dimension of the sum equation.

2. We must also distinguish the most reduced, general form relative to x and y, when $t'' > t - t'$ or $t < t'' + t'$; we must then limit ourselves to the general form

$$(x^{t'+t''-2-q''}, y^{t''-1})^{t'+t''-2} \dots (x^{t+t''-2-q''}, y^{t-t'+t''-1})^{t+t''-2}$$
$$\dots (x^{t+t'-2-q''}, y^{t-1})^{t+t'-2},$$

where the value of q'' is $t - t' - 1$ if $t - t' - 1 < t'' - 1$, that is, if $t < t'' + t'$.

Indeed, the argument leading us to the most reduced general form relative to x and y assumes that the polynomial $(x^{t-2-q''}, y^{t-t'-1})^{t-2}$ (that expresses the number of terms that may still be cancelled in the second polynomial multiplier) is a true polynomial with degree $t - 2$; for this to be true, we must have $t - 2 - q'' + t - t' - 1 > t - 2$, or at least equality holds between the terms. Equivalently, we must have $q'' < t - t' - 1$, or at least equality holds; thus if $t'' - 1$ were greater than or equal to $t - t' - 1$, we would have to stop when $q'' = t - t' - 1$; otherwise the form we get would be wrong.

Thus, if $t < t' + t''$, the most reduced general form relative to x and y is as follows:

First equation   ... $(x^{2t'+t''-t-1}, y^{t''-1})t'+t''-2$
Second equation   ... $(x^{t'+t''-1}, y^{t-t'+t''-1})t+t''-2$
Third equation   ... $(x^{2t'-1}, y^{t-1})t+t'-2$.

There are still $\frac{t''(t''-1)}{2} - \frac{(t-t')\cdot(t-t'-1)}{2}$ coefficients left and a number of arbitrary equations to form, according to the number of terms that can still be cancelled in the second polynomial.

THIRD GENERAL EXAMPLE

(359.) The third general example considers the elimination in first-order incomplete equations in three unknowns.

Recasting these equations to equations in two unknowns, they can be represented in general by

$$(x^a, y^{\overset{a}{\prime}})t = 0,$$

$$(x^{a'}, y^{\overset{d}{\prime}})t' = 0,$$

$$(x^{a''}, y^{\overset{d''}{\prime}})t'' = 0.$$

The general form of the polynomial multipliers is therefore

First equation   ... $(x^{A+a'+a''}, y^{\overset{A+d+d''}{\prime}})T+t'+t''$

Second equation   ... $(x^{A+a+a''}, y^{\overset{A+a+d''}{\prime}})T+t+t''$

Third equation   ... $(x^{A+a+a'}, y^{\overset{A+a+d}{\prime}})T+t+t'$.

The apparent degree of the final equation must be zero, and T, A, $\underset{\prime}{A}$ are subject to no constraint, except that the expression of the degree of the final equation must be zero. This condition is satisfied by setting $T = 0$, $A = 0$, $\underset{\prime}{A} = 0$; thus, I make this assumption right away, and the form of the polynomial multipliers becomes

First polynomial   ... $(x^{a'+a''}, y^{\overset{d+d''}{\prime}})t'+t''$

Second polynomial ... $(x^{a+a''}, y^{\overset{a+d''}{\prime}})t+t''$

Third polynomial   ... $(x^{a+a'}, y^{\overset{a+d}{\prime}})t+t'$.

From (351), the highest dimension of the sum equation has

$$a + a' + a'' + \underset{\prime}{a} + \underset{\prime}{a'} + \underset{\prime}{a''} - t - t' - t'' + 1$$

terms. The first dimension of the first polynomial multiplier provides no useful coefficient. The first dimension of the second polynomial multiplier provides

$$a + a'' + \underset{\prime}{a} + \underset{\prime}{a''} - t - t'' + 1 - a - \underset{\prime}{a} + t - 1 = a'' + \underset{\prime}{a''} - t''$$

such coefficients. The first dimension of the third polynomial multiplier provides

$$a + a' + a + a' - t - t' - 1$$
\qquad{\tiny /}\qquad{\tiny /}

such coefficients. Thus we have as many useful coefficients as there are terms to be cancelled, and each coefficient of the highest-dimensional terms in each polynomial multiplier is zero. Thus, we can reduce the dimension of each polynomial multiplier.

We can apply a similar argument to the highest dimension of each new polynomial multiplier, to show that the total dimension of each polynomial can be lowered by one unit but no further. Thus, the simplest general form relative to the total dimension of each polynomial multiplier is the following:

First polynomial $\quad \dots \quad (x^{a'+a''}, y'^{\,d+d''})t'+t''-2$

Second polynomial $\quad \dots \quad (x^{a+a''}, y'^{\,a+d''})t+t''-2$

Third polynomial $\quad \dots \quad (x^{a+a'}, y'^{\,a+d})t+t'-2.$

(360.) Assuming this form can be reduced relative to y, we now investigate the largest value that q can reach in the following form:

First polynomial $\quad \dots \quad (x^{a'+a''}, y'^{\,d+d''-q})t'+t''-2$

Second polynomial $\quad \dots \quad (x^{a+a''}, y'^{\,a+d''-q})t+t''-2$

Third polynomial $\quad \dots \quad (x^{a+a'}, y'^{\,a+d-q})t+t'-2.$

The sum equation then has

$$t + t' + t'' - 2 - a - a' - a'' + q + 1$$
\qquad{\tiny /}\quad{\tiny /}\quad{\tiny /}

terms containing $y'^{\,a+d+d''-q}$.

The first polynomial multiplier provides no useful coefficients to cancel these terms, but there are $q - 1$ available arbitrary equations.

The second polynomial provides $t'' - a''$ useful coefficients.

The third provides $t + t' - 2 - a - a' + 1 + q$ coefficients.

Thus we have

$$t + t' + t'' - 2 - a - a' - a'' + 1 + q - q + 1$$
$$= t + t' + t'' - a - a' - a''$$

useful coefficients to cancel

$$t + t' + t'' - a + a' - a'' + q - 1$$

terms.

Thus, assuming we use $q - 1$ arbitrary equations to cancel the terms in the sum equation, we get as many equations as there are coefficients, and each coefficient can be assumed to be zero. We can therefore reduce the polynomial multipliers if the argument above holds. Then, $q - 1$ arbitrary equations remain to be formed, that is, we have $q - 1$ spare arbitrary equations, without accounting for those provided by the opportunity to cancel more terms in the polynomial multipliers.

(361.) We now look at the requirements for our argument to hold and what determines the highest value of q.

This argument assumes that the value of q does not prevent the existence of any of the three polynomial multipliers, or any of those whose number of terms contribute to expressing the number of terms that may be cancelled in the first and in the second polynomial multipliers. For that purpose, we must have $q < \underset{/}{a}; q < \underset{/}{a}'; q < \underset{/}{a}''$. Moreover, we must have

$$a'' + \underset{/}{a}'' - q > t'' - 2; \; a' + \underset{/}{a}' - q > t' - 2; \; a + \underset{/}{a} - q > t - 2;$$

thus, any q bigger than the smallest of the following six values:

$$q < \underset{/}{a}; \; q < \underset{/}{a}'; \; q < \underset{/}{a}''; \; q < a + \underset{/}{a} - t + 2; \; q < a' + \underset{/}{a}' - t' + 2;$$
$$q < a'' + \underset{/}{a}'' + t'' + 2$$

is admissible; equivalently, any q bigger than the smallest of the last three quantities is admissible.

Thus we can pick q to be the smallest of the latter three quantities plus one; we then get the most reduced possible form, according to the computations above. However, this is not necessarily the most reduced form in general.

Giving this value to q and then increasing it towards higher values, we reach a situation where it will not be possible to cancel any terms in the first or second polynomial multiplier, using the last two equations. Reasoning along the same argument as we did in (356 ff), we still are able to cancel some terms in the polynomial multipliers relative to y, until q' equals the smallest of the five largest quantities among the six above.

But this new reduction does not change the number of available equations, which is $q - 1$ for each cancelled power of y in the sum equation. From our first argument, this gives a total of $\frac{q \cdot (q-1)}{2}$ available arbitrary equations from $q = 0$ to the largest value of q, or until $q' = 0$. Since each value of q' yields precisely as many useful coefficients as there are terms to be cancelled, the same number of arbitrary equations remains available when reaching the largest value of q', that is, the smallest power of y.

(362.) Examining x now, we will perform as in (357) to determine whether its degree can also be lowered, using the available arbitrary equations.

(363.) In what we said, we have assumed that

$$a' + a'' < t' + t'' - 2; \; \underset{/}{a}' + \underset{/}{a}'' < t' + t'' - 2; \; a + a'' < t + t'' - 2,$$

and so on; if the converse were to occur, we would, for example, immediately reduce the form $(x^{a'+a''}, y^{\frac{\underset{/}{a}+\underset{/}{a}'}{}})t'+t''-2$ to $(x^{t'+t''-2}, y^{\frac{\underset{/}{a}+\underset{/}{a}'}{}})t'+t''-2$ if we only have $a' + a'' > t' + t'' - 2$. We would reduce this form to $(x^{t'+t''-2}, y^{t'+t''-2})^{t'+t''-2}$ if we also had $a' + a'' > t' + t'' - 2$, and we would then proceed as above to examine further possible reductions.

FOURTH GENERAL EXAMPLE

(364.) We limit the development of our investigations to equations in four unknowns. In addition, we limit ourselves to complete equations, and we pay attention to the total dimension of their polynomial multipliers only: We will only expand briefly on further possible reductions, because all the previous material we have developped justifies our not spending more effort on explaining ways to simplify forms.

(365.) Once recast as equations in three unknowns, complete equations in four unknowns are represented by

$$
\begin{aligned}
(x \ldots 3)^t &= 0, \\
(x \ldots 3)^{t'} &= 0, \\
(x \ldots 3)^{t''} &= 0, \\
(x \ldots 3)^{t'''} &= 0.
\end{aligned}
$$

From what was said about the general examples 1, 2 and 3, the form of the polynomial multipliers (224) reduces to

$$
\begin{aligned}
(x \ldots 3)^{t'+t''+t'''} &= 0, \\
(x \ldots 3)^{t+t''+t'''} &= 0, \\
(x \ldots 3)^{t+t'+t'''} &= 0, \\
(x \ldots 3)^{t+t'+t''} &= 0.
\end{aligned}
$$

However, the total dimension of the polynomials can be reduced further.

Indeed, the highest dimension in the sum equation has $N(x \ldots 2)^{t+t'+t''+t'''}$ terms.

The highest dimension of the first polynomial multiplier provides

$$
\begin{aligned}
N(x \ldots 2)^{t'+t''+t'''} &- N(x \ldots 2)^{t''+t'''} - N(x \ldots 2)^{t'+t'''} - N(x \ldots 2)^{t'+t''} \\
+N(x \ldots 2)^{t'''} &+ N(x \ldots 2)^{t''} - N(x \ldots 2)^{0} + N(x \ldots 2)^{t'} \\
= d^3 N(x \ldots 2)^{t'+t''+t'''} & \ldots \begin{pmatrix} t'+t''+t''' \\ t',t'',t''' \end{pmatrix} = 0
\end{aligned}
$$

useful coefficients.

The highest dimension of the second polynomial multiplier provides

$$
\begin{aligned}
N(x \ldots 2)^{t+t''+t'''} &- N(x \ldots 2)^{t+t'''} - N(x \ldots 2)^{t+t''} + N(x \ldots 2)^{t} \\
= d^2 N(x \ldots 2)^{t+t''+t'''} & \ldots \begin{pmatrix} t+t''+t''' \\ t'',t''' \end{pmatrix}
\end{aligned}
$$

useful coefficients.

The highest dimension of the third polynomial multiplier provides

$$
dN(x \ldots 2)^{t+t'+t'''} \ldots \begin{pmatrix} t+t'+t''' \\ t''' \end{pmatrix}
$$

useful coefficients.

Finally, the highest dimension of the fourth polynomial multiplier provides $N(x\dots2)^{t+t'+t''}$ useful coefficients.

Thus the difference between the number of terms to cancel and the number of useful coefficients is

$$N(x\dots2)^{t+t'+t''+t'''} - d^2 N(x\dots2)^{t+t''+t'''} \dots \left(\begin{array}{c} t+t''+t''' \\ t'',t''' \end{array} \right)$$

$$-dN(x\dots2)^{t+t'+t'''} \dots \left(\begin{array}{c} t+t'+t''' \\ t''' \end{array} \right) - N(x\dots2)^{t+t'+t''}$$

$$-d^3 N(x\dots2)^{t+t'+t''+t'''} \dots \left(\begin{array}{c} t+t'+t''+t''' \\ t',t'',t''' \end{array} \right) = 0.$$

Thus each coefficient in the highest dimension of each polynomial multiplier is zero: We can lower the degree of each polynomial multiplier by one unit.

If we perform the same investigation for the next two dimensions, we see they also vanish: The form of the polynomial multipliers reduces to

$$(x\dots3)^{t'+t''+t'''-3},$$
$$(x\dots3)^{t+t''+t'''-3},$$
$$(x\dots3)^{t+t'+t'''-3},$$
$$(x\dots3)^{t+t'+t''-3}.$$

(366.) Thus, in general, the dimension of the simplest polynomial multipliers is always such that the total dimension of the sum equation equals the sum of the dimensions of all given equations, minus as many units as there are unknowns.

Indeed, the difference between the number of terms in the highest dimension in the sum equation and the number of useful coefficients in the highest dimension of all polynomial multipliers will always be

$$d^n N(x\dots n-1)^{t+t'+t''+t'''+ \text{ etc.}} \dots \left(\begin{array}{c} t+t'+t''+t''' + \text{ etc.} \\ t',t'',t''' + \text{ etc.} \end{array} \right) = 0,$$

where n is the number of unknowns.

The difference between the number of terms in the highest dimension in the new sum equation and the number of useful coefficients coming from the highest dimension of the new polynomial multipliers will always be

$$d^n N(x\dots n-1)^{t+t'+t''+t'''+ \text{ etc. } -1} \dots \left(\begin{array}{c} t+t'+t''+t''' + \text{ etc.} -1 \\ t',t'',t''' + \text{ etc.} \end{array} \right) = 0.$$

The difference between the number of terms in the highest dimension of the second new sum equation and the number of useful coefficients coming from the highest dimension of the new polynomial multipliers will always be

$$d^n N(x\dots n-1)^{t+t'+t''+t'''+ \text{ etc. } -2} \dots \left(\begin{array}{c} t+t'+t''+t''' + \text{ etc.} -2 \\ t',t'',t''' + \text{ etc.} \end{array} \right) = 0,$$

and so on until $t + t' + t'' + t''' + $ etc. $- n$.

In general, we may convince ourselves about this by remembering, from (39),

$N(x \dots n - 1)^{t+t'+t''+t'''+ \text{ etc. } -q}$
$= \frac{(t+t'+t''+t'''+ \text{ etc. } -q+1).(t+t'+t''+t'''+ \text{ etc. } -q+2)\dots(t+t'+t''+t'''+ \text{ etc. } -q+n-1)}{1.2.3\dots(n-1)}.$

Assume we cancel combinations of one, two, three, etc. quantities t', t'', t''', etc. in this expression. This yields the various expressions implicitly contained in

$$d^n N(x \dots n - 1)^{t+t'+t''+t'''+ \text{ etc.}} \quad \dots \left(\begin{array}{c} t + t' + t'' + t''' + \text{ etc.} \\ t', t'', t''', \text{ etc.} \end{array} \right).$$

We therefore easily see that all these expressions will occur as long as they are not negative, that is, as long as $q < n - 1$, including $q = n - 1$. Thus the equation

$$d^n N(x \dots n - 1)^{t+t'+t''+t'''+ \text{ etc. } -n+1} \quad \dots \left(\begin{array}{c} t + t' + t'' + t''' \text{ etc. } - n + 1 \\ t', t'', t''', \text{ etc.} \end{array} \right) = 0$$

still holds. Thus the form of the sum equation generally reduces to

$$(x \dots n)^{t+t'+t''+t'''+ \text{ etc. } -n} = 0,$$

from which we can easily extract the form of the polynomial multipliers.

(367.) Having determined, in general, the total dimension of each polynomial multiplier, the shortest path is now to also determine the highest power each unknown must reach in each polynomial multiplier; we will not get into this detail because of the large number of subcases that arise when dealing with the highest possible level of generality. But what we said in (351) and elsewhere is enough to give a roadmap for any case.

(368.) We now consider specific examples for the purpose of discussing what we have already said more extensively and enlightening the reader about the factors that may arise during the computations leading to the condition equation, that is, the final equation.

(369.) Assume we seek the final equation resulting from the following three equations

$$ax^2 + bxy + cy^2 + dx + ey + f = 0,$$
$$d'x + e'y + f' = 0,$$
$$d''x + e''y + f'' = 0.$$

The general form of the polynomial multipliers would be, following (224 ff), $(x,y)^{T+2}$, $(x,y)^{T+1}$, $(x,y)^{T+1}$. The number of arbitrary coefficients would be $2N(x,y)^{T+1} - N(x,y)^T$ in the first and $N(x,y)^T$ in the second, that is, $2N(x,y)^{T+1}$ arbitrary equations in the sum equation. From (349 ff) this general form of the polynomial multipliers reduces to the form $(x,y)^0$, $(x,y)^1$, $(x,y)^1$; the number of arbitrary coefficients is one in the second multiplier, that is, one arbitrary equation in the sum equation.

We therefore multiply the first equation by C, the second equation by $A'x + B'y + C'$, and the third equation by $A''x + B''y + C''$, leading to the following sum equation:

$$
\begin{array}{lll}
\quad Ca \quad x^2 & + \quad Cb \quad xy & + \quad Cc \quad y^2 = 0. \\
+ \quad A'd' & + \quad A'e' & + \quad B'e' \\
+ \quad A''d'' & + \quad A''e'' & + \quad B''e'' \\
& + \quad B'd' & \\
& + \quad B''d'' &
\end{array}
$$

$$
\begin{array}{lll}
+ \quad Cd \quad x & + \quad Ce \quad y & \\
+ \quad A'f' & + \quad B'f' & \\
+ \quad A''f'' & + \quad B''f'' & \\
+ \quad C'd' & + \quad C'e' & \\
+ \quad C''d'' & + \quad C''e'' &
\end{array}
$$

$$
\begin{array}{l}
+ \quad Cf \\
+ \quad C'f' \\
+ \quad C''f''
\end{array}
$$

I use the arbitrary equation $B'd' + B''d'' = 0$, and I compute the value of $A'A''B'B''CC'C''$ as follows, by going through the terms x^2, xy, the arbitrary equation, and the terms y^2, x, y and the term without x or y, successively. I first take $A'A''CC'C''$.

First line $d'A''CC'C'' + A'A''aC'C''$

Second line $[(d'e'')CC'C'' - d'A''bC'C'' + e'A''aC'C'' + A'A''(ab')C'']B'B''$

Third line $-[(d'e'')CC'C'' - d'A''bC'C'' + e'A''aC'C'' + A'A''(ab')C'']B'B''$

Fourth line $-[(d'e'')cC'C'' + d'A''(bc')C'' - e'A''(ac')C'' + A'A''(ab'c')]d'B''$

               $+ [(d'e'')CC'C'' - d'A''bC'C'' + e'A''aC'C'' + A'A''(ab')C''](d'e'')$.

I now remark that $(ab')C'' = 0$, $(bc')C'' = 0$, $(ac')C'' = 0$, and $(ab'c'') = 0$, remembering that $(bc')C''$ is only the abbreviated representation of

$$
(bc' - b'c)C'' - (bc'' - b''c)C' + (b'c'' - b''c')C.
$$

Indeed, the latter expression turns out to be zero, since $b' = 0$, $b'' = 0$, $c' = 0$, $c'' = 0$. Likewise we see that $(ac')C'' = 0$, $(ab')C'' = 0$, and $(ab'c'') = 0$.

     Hence, the fourth line reduces to

$$
-(d'e'')cC'C''d'B'' + (d'e'')[(d'e'')CC'C'' - d'A''bC'C'' + e'A''aC'C''].
$$

Extracting the common factor $(d'e'')$ (we will examine this factor later),

$$
-cC'C''d'B'' + [(d'e'')CC'C'' - d'A''bC'C'' + e'A''aC'C'']
$$

Fifth line $-(cd')C''d'B'' + [(d'e'').dC'C'' - (d'f'')bC'C'' + (e'f'')aC'C''].$

Omitting the terms where A'' remains,

Sixth line $+(cd')C''(d'f'')$

              $+[(d'e'').(de')C'' - (d'f'').(be')C'' + (e'f'').(ae')C'']$

Seventh line $(cd'f'').(d'f'') + (d'e'').(de'f'')$

              $-(d'f'').(be'f'') + (e'f'').(ae'f'') = 0.$

That is the final equation where the terms with a', b', c'; a'', b'', c'' have been omitted. Thus the true final equation is

$$c(d'f'')^2 + (d'e'').(de'f'') - b(e'f'').(d'f'') + a(e'f'')^2 = 0.$$

Observation

(370.) We can reach the last equation much faster by computing the values of x and y using the last two equations and substituting them in the first. In general, when $n - 1$ out of n proposed equations are of first degree, we can obtain the final equation much faster using simple substitutions. But the cases where this easier computation occurs are rare, and we miss the factor $(d'e'')$ that we have encountered above. This factor turns out to be quite useful.

Indeed, we generally observe that each time we encounter a factor before the end of the computation of the lines, it proves that the final equation can be simplified when this factor is zero, and it can be reached with fewer coefficients.

Thus, assume $(d'e'') = 0$ in the current example. I claim the final equation is then much simpler than the one we have just found. Indeed, multiplying the second equation by e'', the third equation by e', and subtracting the second product from the first, we get

$$(d'e'')x - (e'f'') = 0.$$

Since $(d'e'') = 0$, this expression reduces to $(e'f'') = 0$, which is the final equation when $(d'e'') = 0$.

We now turn our attention to our next claim, which says we can reach the final equation using fewer coefficients. Here is a factual proof.

Assume we form the arbitrary equation $B'e' + B''e'' = 0$ in addition to the equation $B'd' + B''d'' = 0$, or equivalently we assume $B' = 0$, $B'' = 0$. Then the sum equation shows that $C = 0$ and thus reduces to

$$
\begin{array}{llll}
A'd' & x^2 & + & A'e' & xy = 0. \\
+\ A''d'' & & & +\ A''e'' \\
\\
+\ A'f' & x & + & C'e' & y \\
+\ A''f'' & & & +\ C''e'' \\
+\ C'd' & & & \\
+\ C''d'' & & & \\
\\
+\ C'f' & & & \\
+\ C''f'' & & &
\end{array}
$$

It is easy to see that this equation yields $A' = 0$ and $A'' = 0$; thus, the sum equation reduces to

$$
\begin{array}{llll}
C'd' & x & + & C'e' & y = 0, \\
+\ C''d'' & & & +\ C''e'' \\
\\
+\ C'f' & & & \\
+\ C''f'' & & &
\end{array}
$$

with only two coefficients C and C' to be determined.

But the two equations

$$C'd' + C''d'' = 0, \text{ and } C'e' + C''e'' = 0$$

lead to the condition equation $(d'e'') = 0$. Since this condition holds by hypothesis, only one of these two equations, combined with the equation $C'f' + C''f'' = 0$, is enough to answer the question.

One gives the final equation $(d'f'') = 0$ and the other gives $(e'f'') = 0$, and it is easy to see that they are equivalent since $(d'e'') = 0$.

(371.) We now prove the postulate that each time we encounter a factor before reaching the last line, it proves that fewer coefficients are needed when this factor is zero. The argument is that as soon as we reach the line which yields this factor, the equation used to compute this line is superfluous when the factor is zero. It can therefore be omitted and there is one more unknown than necessary. Thus, we can form a new arbitrary equation, in such a way that a larger number of unknowns or coefficients can be made zero. We will see examples of this.

(372.) The same does not hold when this factor occurs in the last line only, that is, in the final equation. Since it arises only with the final equation, it means it is not a common factor of the values of the unknowns. Consequently, assuming this factor is zero does not cancel any unknowns, unlike what happens when the factor occurs before the last line.

The following three equations:

$$ax^2 + bxy + cy^2 + dx + ey + f = 0,$$
$$a'x^2 + b'xy + c'y^2 + d'x + e'y + f' = 0,$$
$$d''x + e''y + f'' = 0$$

are an example when the factor occurs within the final equation only. We will see the three polynomial multipliers of these equations reduce to

$$Dx + F, \ D'x + F', \ A''x^2 + B''xy + D''x + E''y + F''.$$

If we proceed with the computation of the final equation, we find no common factor in any of the *lines*, except in the last one, where the factor e'' appears.

Assume we use instead the polynomial multipliers

$$Dx + Ey + F, \ D'x + E'y + F', \ B''xy + D''x + E''y + F'',$$

and we form the arbitrary equation due to the presence of one useless coefficient. We will see that none of the *lines* reveals a common factor, except in the last or final equation, which will display the factor (ac').

Then, this factor indicates nothing else but the presence of a solution of the form (279) and (287). It does not indicate that the final equation can be obtained with fewer coefficients, but rather something else worthy of attention: The chosen polynomial multipliers are worthless for elimination purposes. Here is why:

Assume we use, for example, the three multipliers

$$Dx + F, \ D'x + F', \ A''x^2 + B''xy + D''x + E''y + F''$$

and we find $e'' = 0$; that is, the third equation is simply $d''x + f'' = 0$; then the final equation would be $0 = 0$ with these polynomials, which brings no new information.

The reason is that these three polynomials, whose general form is

$$Dx + Ey + F, \ D'x + E'y + F', \ A''x^2 + B''xy + C''y^2 + D''x + E''y + F''$$

have been reduced to their simplest form only by implicitly assuming that the terms Ey and $E'y$ could be cancelled in the first two polynomials using the third equation. However, this assumption is justified as long as e'' is not zero, but it collapses when $e'' = 0$; in that case, there are no terms left in the equation $d''x + f'' = 0$, and it can be used only to cancel terms in x. The first two polynomial multipliers must then be $Ey + F$ and $E'y + F'$, instead of $Dx + F$ and $D'x + F'$. The third polynomial is then $B''xy + C''y^2 + D''x + E''y + F''$.

By the way, this does not prevent the other factor from being the true final equation, once the final equation is computed with the original polynomials, and the factor e'' is then removed. The true final equation escapes this polynomial multiplier form only if, before proceeding to computations, we have accounted for $e'' = 0$ in the equation $d''x + e''y + f'' = 0$, that is, when this equation is used as $d''x + f'' = 0$.

A similar process applies to the case $(ac') = 0$.

Thus, we see the meaning of the factor when it occurs in the last line only: It shows that the form used for the polynomial multipliers is inappropriate when this factor is zero, and another form must be used; this is always easy to do.

(373.) We now look for the equation resulting from the elimination of x and y in the following three equations:

$$
\begin{aligned}
axy + bx + cy + d &= 0, \\
a'xy + b'x + c'y + d' &= 0, \\
a''xy + b''x + c''y + d'' &= 0.
\end{aligned}
$$

From (359), I consider polynomial multipliers of the form $(x^2, y^2)^4$.

Applying the same methods as those of (359) to these polynomials, we can suppress the dimensions 4, 3 and 2, because each of the corresponding coefficients is zero. Thus the simplest polynomial multiplier is of the form $(x^1, \ y^1)^1$ for each equation.

At this point, the number of useless coefficients is 1, because one term can always be cancelled in the first polynomial using the last two equations, without introducing new coefficients.[11]

---

[11] If we had used the form $(x^1, y^1)^2$ for each polynomial multiplier, we would have found (359) that this form could reduce to $(x^1, y^1)^1$ with one arbitrary equation in the sum equation. This is in agreement with our current discussion.

Multiplying each equation by polynomials of the form $Ax + By + C$, and adding the three products together, the sum equation has the form

$$Aa \quad x^2y \quad + \quad Ba \quad xy^2 = 0.$$

$$+ \quad Ab \quad x^2 \quad + \quad Ac \quad xy \quad + \quad Bc \quad y^2$$
$$+ \quad Bb$$
$$+ \quad Ca$$

$$+ \quad Ad \quad x \quad + \quad Bd \quad y$$
$$+ \quad Cb \quad\quad\quad + \quad Cc$$

$$+ \quad Cd$$

Since there is one useless coefficient, I form the arbitrary equation $Ac + A'C' + A''c'' = 0$, or $Bb + B'b' + B''b'' = 0$, or $Ca + C'a' + C''a'' = 0$, or, etc. Using the first of these equations, together with the two equations provided by the terms in x^2y and x^2 (the unknowns are A, A', A'' only), I conclude $A = 0$, $A' = 0$, $A'' = 0$. Thus I should really compute $BB'B''CC'C''$ only.

Using the terms xy^2, xy, y^2, x and y and the term without x nor y successively, I get

First line $aB'B''CC'C''$
Second line $(ab')B''CC'C'' + aB'B''aC'C''$
Third line $(ab'c'')CC'C'' + (ac')B''aC'C''$
Fourth line $(ab'c'')bC'C'' - (ac')B''(ab')C''$
Fifth line $(ab'c'').(bc')C'' - (ac'd'').(ab')C''$,

where the fifth line is obtained by rejecting the term where B'' remains, because it cannot influence the final equation.

Sixth line $(ab'c'').(bc'd'') - (ac'd'').(ab'd'') = 0$,

which is the final equation.

(374.) Assume a, b, c and d have dimensions 0, 1, 1 and 2 in z, respectively, and that the same holds for a', b', c', d', and a'', b'', c'', d''; the degree of the final equation in z is $0 + 1 + 1 + 1 + 1 + 2 = 6$. But the three proposed equations are of the form $(x^1, y^1, z^2)^2 = 0$, which must indeed (62) lead to a sixth-degree final equation.

(375.) Consider now three equations of the form

$$ax^2 \quad + \quad bxy = 0.$$
$$+ \quad cx \quad + \quad dy$$
$$+ \quad e$$

We can pick each polynomial multiplier of the form $(x^4, y^2)^4$. But reasoning as have done earlier in (349 ff), we can use the simpler form $(x^4, y)^4$, then the simpler form $(x^2, y)^2$, then finally $(x^2)^2$.

The number of useless coefficients is zero, since no terms can be excluded from this polynomial multiplier without introducing new ones.

Let us therefore multiply each equation by a polynomial of the form $Ax^2 + Bx + C$. The sum equation has the form

$$Aa \quad x^4 \quad + \quad Ab \quad x^3 y = 0.$$

$$\begin{aligned} &+ \quad Ac \quad x^3 \quad + \quad Ad \quad x^2 y \\ &+ \quad Ba \qquad\quad + \quad Bb \end{aligned}$$

$$\begin{aligned} &+ \quad Ae \quad x^2 \quad + \quad Bd \quad xy \\ &+ \quad Bc \qquad\quad + \quad Cb \\ &+ \quad Ca \end{aligned}$$

$$\begin{aligned} &+ \quad Be \quad x \quad + \quad Cd \quad y \\ &+ \quad Cc \end{aligned}$$

$$+ \quad Ce$$

Thus we get

First line $\quad aA'A''$

Second line $\quad (ab')A''BB'B''$

Third line $\quad (ab'c'')BB'B'' - (ab')A''aB'B''$

Fourth line $\quad [(ab'c'')bB'B'' - (ab'd'')aB'B'' + (ab')A''(ab')B'']CC'C''$

Fifth line $\quad [(ab'c'').(bc')B'' - (ab'd'').(ac')B'' + (ab'e').(ab')B'']CC'C''$
$\qquad\qquad +[(ab'c'')bB'B'' - (ab'd'')aB'B'']aC'C'',$

where the terms in A'' have been rejected since they do not appear in the subsequent lines and thus cannot influence the final equation.

Sixth line $\quad [(ab'c'').(bc'd'') - (ab'd'').(ac'd'') + (ab'e').(ab'd'')]CC'C''$
$\qquad\qquad +[(ab'c'').(bd')B'' - (ab'd'').(ad')B'']aC'C''$
$\qquad\qquad -[(ab'c'').(bc')B'' - (ab'd'').(ac')B'' + (ab'e').(ab')B'']bC'C'',$

where the terms in $B'B''$ have been rejected since they cannot influence the final equation.

Seventh line $\quad [(ab'c'').(bc'd'') - (ab'd'').(ac'd'') + (ab'e').(ab'd'')]cC'C''$
$\qquad\qquad\quad +[(ab'c'').(bd'e'') - (ab'd'').(ad'e'')]aC'C''$
$\qquad\qquad -[(ab'c'').(bc'e'') - (ab'd'').(ac'e'') + (ab'e').(ab'e'')]bC'C'',$

where the terms in B'' have been rejected.

Eighth line
$[(ab'c'').(bc'd'') - (ab'd'').(ac'd'') + (ab'e').(ab'd'')](cd')C''$
$+[(ab'c'').(bd'e'') - (ab'd'').(ad'e'')](ad')C''$
$-[(ab'c'').(bc'e'') - (ab'd'').(ac'e'') + (ab'e').(ab'e'')](bd')C''$

Ninth line or
Final equation
$$\left.\begin{aligned} [(ab'c'').(bc'd'') - (ab'd'').(ac'd'') + (ab'e').(ab'd'')](cd'e'') \\ +[(ab'c'').(bd'e'') - (ab'd'').(ad'e'')](ad'e'') \\ -[(ab'c'').(bc'e'') - (ab'd'').(ac'e'') + (ab'e').(ab'e'')](bd'e'') \end{aligned}\right\} = 0.$$

The final equation is devoid of any superfluous terms.

(376.) Assume the three equations have the form $(x^2, y^1, z^2)^2 = 0$, when they are expanded from two to three unknowns. From (62), we know that the degree of the final equation must be $8 - 1 = 7$. This is the same as the degree of the final equation we have just computed; in this case indeed, the dimensions of a, b, c, d, e are 0, 0, 1, 1, 2, respectively. The same holds for a', b', c', d', e' and for a'', b'', c'', d'', e'', from which it is easy to conclude that each term in the final equation of the form

$$(ab'c'').(bc'd'').(cd'e'')$$

must have dimension $0 + 0 + 1 + 0 + 1 + 1 + 1 + 1 + 2 = 7$.

If the three equations in x, y and z have the form $[x, (y, z)^1]^2$, then, from (131), the final equation must have degree 4. This is also what is given by the final equation we have just found, because then a, b, c, d, e have dimensions 0, 0, 1, 0, 1, respectively, and the same holds for a', b', c', d', e' and a'', b'', c'', d'', e''; thus each term in the above final equation has dimension $0 + 0 + 1 + 0 + 1 + 0 + 1 + 0 + 1 = 4$.

(377.) In (320) and (321), we have given the final equation in x, resulting from three equations of the form $[x, (y, z)^1]^2 = 0$; we have also said that it was harder to get the equation in y and z, because it is more complex, and that we would delay giving this expression. The final equation above provides the simplest possible such expression.

Assume in general that the three proposed equations have the form $[x, (y, z)^1]^t = 0$. Putting them in the form of equations in two unknowns, we only have to solve for

$$\begin{aligned} ay + bz + c &= 0, \\ a'y + b'z + c' &= 0, \\ a''y + b''z + c'' &= 0, \end{aligned}$$

which is $(ab'c'') = 0$. Thus the final equation for the third unknown is very easy to find as long as two of the unknowns do not exceed degree 1 when taken separately or together.

The final equation relative to one or the other unknowns can be obtained by writing the equations as

$$(x^t, y^1)^t = 0.$$

Following the same argument as in the previous example, we find that the simplest polynomial multipliers leading to the final equation have the form

$$\begin{aligned} (x)^{t'+t''-2} &\quad \text{for the first equation,} \\ (x)^{t+t''-2} &\quad \text{for the second equation} \\ (x)^{t+t'-2} &\quad \text{for the third equation;} \end{aligned}$$

that is, these polynomials only contain one unknown.

(378.) Let us now assume that the three proposed equations can be recast as equations in two unknowns like $(x, y)^2 = 0$.

The polynomial multiplier for each equation reduces (350) to the form $(x, y)^2$ and even (351) to the form $(x^2, y^1)^2$, with one arbitrary equation in whichever dimension is most convenient.

Assume that the three proposed equations have the form

$$\begin{aligned} ax^2 &\;+\; bxy \;+\; cy^2 = 0 \\ +\; dx &\;+\; ey \\ +\; f \end{aligned}$$

and that we multiply each equation by a polynomial of the form

$$\begin{aligned} Ax^2 &\;+\; Bxy \\ +\; Dx &\;+\; Ey \\ +\; F. \end{aligned}$$

The sum equation is of the form

$$\begin{array}{llllllll}
Aa & x^4 &+& Ab & x^3y &+& Ae & x^2y^2 &+& Bc & xy^3 = 0, \\
&&+& Ba &&+& Bb
\end{array}$$

$$\begin{array}{llllllll}
+ & Ad & x^3 &+& Ae & x^2y &+& Be & xy^2 &+& Ec & y^3 \\
+ & Da &&+& Bd \\
&&&+& Db &&+& Dc \\
&&&+& Ea &&+& Eb
\end{array}$$

$$\begin{array}{llllll}
+ & Af & x^2 &+& Bf & xy &+& Ee & y^2 \\
+ & Dd &&+& De &&+& Fc \\
+ & Fa &&+& Ed \\
&&&+& Fb
\end{array}$$

$$\begin{array}{llll}
+ & Df & x &+& Ef & y \\
+ & Fd &&+& Fe
\end{array}$$

$$+ \; Ff$$

The number of useless coefficients is one. Since this coefficient can be taken in the first (highest) dimension, I create the arbitrary equation $Ba + B'a' + B''a'' = 0$; I could make many other assumptions, but this one considerably simplifies the computations.

The question reduces to computing the value of

$$AA'A''BB'B''DD'D''EE'E''FF'F''.$$

We have already given many examples about how to do so and will not go into details for this case; rather we will pursue our computations until we reach the line where the factor of the final equation appears and will only then give the final result.

We choose to successively go through the equations provided by the terms x^4, x^3y, the equation $Ba + B'a' + B''a'' = 0$, and then x^2y^2 and xy^3. We get

First line	...	$aA'A''$
Second line	...	$(ab')A''BB'B''$
Third line	...	$-(ab')A''aB'B''$
Fourth line	...	$-(ab'c'')aB'B'' + (ab')A''(ab')B''$
Fifth line	...	$[-(ab'c'').(ac')B'' - (ab')A''(ab'c'')]DD'D''$, etc.

We therefore see that all the following lines have the common factor $(ab'c'')$, and it will therefore be a factor in the final equation. Taking this factor away for simplicity, we need to compute

$$-[(ac')B'' + (ab')A'']DD'D''EE'E''FF'F''$$

using the factors x^3, x^2y, etc. We get the final equation (A):

$$
\left.
\begin{aligned}
&[(ad'e'').[(ab'e'') + (ac'd'')] - (ab'd'').[(bd'e'') - (ac'f'')] - (ab'f'').(ab'e'')] \\
&\times[(bc'e'').(de'f'') + (bc'f'').(cd'f'')] \\
&+[(ab'c'').[(ab'f'') - (ad'e'')] + (ac'd'')^2 - (ab'd'').(bc'd'')] \\
&\times[(bd'f'').(ce'f'') - (cd'f'')^2 - (cd'e'').(de'f'')] \\
&-[(ac'e'').[(ad'e'') - (ab'f'')] + (ac'd'').(ac'f'') - (ab'd'').(cd'e'')] \\
&\times[(ac'e'').(de'f'') + (ac'f'').(cd'f'')] \\
&+[(ad'f'').[(ab'e'') + (ac'd'')] - (ab'd'').(bd'f'') - (ab'f'')^2] \\
&\times[(bc'd'').(ce'f'') - (bc'e'').(be'f'') + (bc'f'')^2] \\
&-[(ac'e'').(ad'f'') - (ab'f'').(ac'f'') - (ab'd'').(cd'f'')] \\
&\times[(ac'd'').(ce'f'') + (ac'f'').(bc'f'') - (ac'e'').(be'f'')] \\
&+[(ae'f'').[(ab'e'') + (ac'd'')] - (ac'f'').(ab'f'') - (ab'd'').(be'f'')] \\
&\times[(ae'f'').(bc'e'') - (ac'f'').(bc'f'')] \\
&+[(ac'e'').(ae'f'') - (ac'f'')^2 - (ab'd'').(ce'f'')].[(ac'f'')^2 - (ac'e'').(ae'f'')] \\
&+[(ab'c'').(ae'f'') + (ac'd'').(ac'f'') - (ab'd'').(bc'f'')] \\
&\times[(ab'f'').(ce'f'') + (cd'e'').(ae'f'') - (ac'f'').(cd'f'')] \\
&+(ab'c'').(ce'f'').[(ab'f'').(ae'f'') - (ab'd'').(de'f'') - (ac'f'').(ad'f'')] \\
&-(ad'f'').(ce'f'').[(ab'c'').(ac'f'') + (ac'd'').(ac'e'') - (ab'd'').(bc'e'')].
\end{aligned}
\right\} = 0.
$$

This is the final equation resulting from three equations in three unknowns, irrespective of the degree of these equations, as long as the degree of two of the unknowns does not exceed 2.

(379.) When computing the eighth *line*, we find the following terms:

$$(ab'e'').(ac')D'' - (ac'e'').(ab')D''.$$

These have been replaced by $(ab'c'').(ae')D''$, since from (221), we have

$$(ac'e'').(ab') - (ab'e'').(ac') + (ab'c'').(ae') = 0.$$

This substitution leads to the term $(ab'c'').(ae'e'')$ in the eleventh line, which is nothing but

$$(ab'c'')[(ae - a'e)e'' - (ae'' - a''e)e' + (a'e'' - a''e')e]$$

and thus it is zero.

(380.) Assume $c = c' = c'' = 0$; then each quantity such as $(ab'c'')$, $(ac'd'')$, $(cd'e'')$, etc. where c, c' or c'' is present, is zero.

Assume we set to zero all terms where any one of the quantities c, c' or c'' reaches more than one dimension in equation (A). Then the equation reduces to

$$[(ad'e'').(ab'e'') - (ab'd'').(bd'e'') - (ab'f'').(ab'e'')].(bc'e'').(de'f'')$$
$$- \quad [(ad'f'').(ab'e'') - (ab'd'').(bd'f'') - (ab'f'')^2](bc'e'').(be'f'')$$
$$+ \quad [(ae'f'').(ab'e'') - (ab'd'').(be'f'')].(ae'f'').(bc'e'') = 0.$$

It is now clear that the first element of this equation is zero since $c = c' = c'' = 0$. But since the entire equation admits the factor $(bc'e'')$, we clearly also have (B)

$$[(ad'e'').(ab'e'') - (ab'd'').(bd'e'') - (ab'f'').(ab'e'')].(de'f'')$$
$$- \quad [(ad'f'').(ab'e'') - (ab'd'').(bd'f'') - (ab'f'')^2].(be'f'')$$
$$+ \quad [(ae'f'').(ab'e'') - (ab'd'')(be'f'')](ae'f'') = 0.$$

When changing d into c, e into d and f into e, this equation is entirely equivalent to the one given in (375), and it is easy to show this must be true.

(381.) Assume $a = a' = a'' = 0$ in equation (B). We first cancel the terms where a, a', a'' must reach more than one dimension, to get

$$-(ab'd'').(bd'e'').(de'f'') + (ab'd'').(be'f'').(bd'f'') = 0,$$

or, if we cancel the factor $(ab'd'')$, we get (C):

$$(be'f'').(bd'f'') - (bd'e'').(de'f'') = 0.$$

This equation is the same as the one found in (373), if we replace b by a, d by b, e by c, and f by d. And this must be true indeed.

(382.) Assume $b = b' = b'' = 0$ in equation (C), and assume we first cancel the terms where b, b', b'' exceed the first dimension. We get $-(bd'e'').(de'f'') = 0$, or, suppressing the factor $-(bd'e'')$, we get $(de'f'') = 0$; This is indeed the condition equation resulting from the three equations

$$\begin{aligned} dx + ey + f &= 0 \\ d'x + e'y + f' &= 0, \\ d''x + e''y + f'' &= 0. \end{aligned}$$

(383.) Let us now examine the factor $(ab'c'')$ found during the computation of equation (A).

This factor only indicates a pathological solution of the form given in (279) and (287), as we have already said.

Indeed, assume that, using the last two among the three proposed equations, we determine the values of y^2 and xy, and we substitute them in the third equation to conclude the value of x^2; we get

$$(ab'c'')x^2 + (bc'd'')x + (bc'e'')y + (bc'f'') = 0.$$

Assume now that this value of x^2 is substituted in any one of the three proposed equations. I claim it will satisfy all three when $(ab'c'') = 0$.

Indeed, we then have

$$(bc'd'')x + (bc'e'')y + (bc'f'') = 0,$$

Thus $x^2 = \frac{0}{0}$; this value satisfies each of the three proposed equations. This is a solution of the kind mentioned in (279) and (287).

(384.) However, by (370), $(ab'c'') = 0$ also indicates that we may reach the final equation with fewer coefficients, since this factor occurs before the final equation.

Indeed, assume that we form an arbitrary equation, following these thoughts. For example, we form the equation $Bb + B'b' + B''b'' = 0$ in addition to the previous arbitrary equation $Ba + B'a' + B''a''$, which was formed when writing the general solution. This equation leads to $B = 0$, $B' = 0$, $B'' = 0$ when combined with the equation given by the term xy^3 in the sum equation. When we compute $AA'A''DD'D''EE'E''FF'F''$, we see that, although there is one fewer coefficient than there are equations, we can still complete the elimination process. Indeed, one of the three equations

$$Aa + A'a' + A''a'' = 0, \quad Ab + A'b' + A''b'' = 0, \quad Ac + A'c' + A''c'' = 0$$

provided by the terms x^4, x^3y, x^2y^2 is always true when we assume $(ab'c'') = 0$. Equivalently, the condition equation for these three equations is precisely $(ab'c'') = 0$.

Thus we can use three polynomial multipliers of the form $Ax^2 + Dx + Ey + F$ to reach the proper final equation in this case. The computation of the lines requires omitting one of the three equations

$$Aa + A'a' + A''a'' = 0, \quad Ab + A'b' + A''b'' = 0, \quad Ac + A'c' + A''c'' = 0.$$

By the way, we will examine this factor further later.

(385.) Concluding about three equations in three unknowns, we must warn the reader about a simpler but illusory solution.

Assume we form three other equations from the three proposed equations, such that each one of them contains only one of the three quantities x^2, xy and y^2. We get the following three equations:

$$
\begin{aligned}
(ab'c'')x^2 + (bc'd'')x + (bc'e'')y + (bc'f'') &= 0,\\
(ab'c'')xy - (ac'd'')x - (ac'e'')y - (ae'f'') &= 0,\\
(ab'c'')y^2 + (ab'd'')x + (ab'e'')y + (ab'f'') &= 0.
\end{aligned}
$$

When $(ab'c'') = 0$, these equations become

$$
\begin{aligned}
(bc'd'')x + (bc'e'')y + (bc'f'') &= 0,\\
(ac'd'')x + (ac'e'')y + (ac'f'') &= 0,\\
(ab'd'')x + (ab'e'')y + (ab'f'') &= 0.
\end{aligned}
$$

Thus it appears we could reach the final equation much faster than described above, since we only need to substitute the values of x and y found in two of these equations into the third one.

This solution would be illusory, however, and would lead to a trivial equation.

Indeed, the first two equations lead to

$$[(bc'd'').(ac'e'') - (ac'd'').(bc'e'')]x + (bc'f'').(ac'e'') - (bc'e'').(ac'f'') = 0.$$

We easily see from the theorems given in (221) that

$$(bc'd'').(ac'e'') - (ac'd'').(bc'e'') = 0$$
$$\text{and} \quad (bc'f'').(ac'e'') - (bc'e'').(ac'f'') = 0.$$

The same would be true of the equation leading to the value of y. The same would also hold by combining the first of these equations with the third, or the second with the third. Thus, these three equations are redundant and express no more information than two of them do.

Considerations about the factor in the final equation obtained by using the second method

(386.) The first method we have given to reach the final equation cannot yield any factor that may alter the degree of the final equation: This factor can only be a function of the coefficients of the proposed equations. These factors are important, since they indicate cases when the degree of the final equation may be lowered.

The second method, which uses transformed equations such that there is one fewer unknown than there are equations, is rarely devoid of any factor. Since the coefficients of the various unknowns to be eliminated are functions of the unknown that appears in the final equation, the apparent degree of this final equation can very often differ from the true degree.

However, the computations involved in this second method are considerably shorter than those involved in the first method. This advantage justifies the drawback of generating superflous factors. But we must find the means to identify these factors in the final equation if, as we believe, they cannot be avoided in general.

We have already stated in (339), and will prove in the sequel, that these factors never occur in equations in two unknowns when they are recast as equations in a single unknown. The same is not true when the number of unknowns exceeds two. In general, the factor becomes all the more complicated, as far as its length and the number of letters are concerned, when the number of unknowns is large.

(387.) At first, it appears that, since the methods given in the first part of this work allow one to determine the degree of the final equation, we only have to look for the commensurate dividers in the final equation, that the superfluous factor should be among these dividers, and that its degree is determined by the difference between the true degree and the apparent degree given by the second elimination method.

This is true; however, using such a method to look for the superfluous factor would result in computations whose nature would be infinitely more painstaking than the elimination process proposed in the first method: The advantages promised by the second method would completely vanish. We also must add that the method aimed at finding the commensurate dividers is a trial-and-error method, which could not be useful when dealing with such complex quantities as those of interest here. Our goal is to reach the final equation without any superfluous factor, not by a trial-and-error method such as the commensurate dividers method, but by a more rigorous process. Here is one that may be used in general.

(388.) The process we have presented involves forming a number of arbitrary equations, beyond those resulting from the cancellation of terms in the sum equation. These arbirtrary equations can always be chosen according to different choices, and both the superfluous factor and the apparent final equation will depend upon these choices. Thus the final equation can always be considered to be the product of two factors. One is the true final equation and does not change with the arbitrary equations. The other is the superfluous factor and changes with these arbitrary equations.

Thus we can get two apparent final equations by changing one or more arbitrary equations and computing the corresponding final equation. These two equations will contain the true final equation as a common factor. Thus the remaining work will be to compute the greatest common divider among these two apparent final equations.

(389.) The computation of the final equation is rather arduous in and of itself and we should avoid repeating this task if possible. This can always be done by observing the following.

When computing the *lines* to reach the apparent final equation, we use one fewer arbitrary equation than required; we then pursue the computations excluding the last line, as if this arbitrary equation did not exist.

When computing the last line, that is, when computing the apparent final equation, we then use the last arbitrary equation; but we use this equation differently: If we use two different equations in the last line, we reach two apparent final equations. The final equation must be a common factor of these two.

(390.) Consider, for example, the computation of the final equation resulting from the three equations (378) of the form $(x, y)^2 = 0$. We have used the arbitrary equation $Ba + B'a' + B''a'' = 0$, but we have used it as early as the third *line*.

If the factor, which we know is $(ab'c'')$, had not appeared in the computations as easily as we have seen, I would have proceeded with the computation of the lines by using the terms x^4, x^3y, x^2y^2, xy^3, x^3, x^2y, etc. up until the last line excluded, and without considering the arbitrary equation.

When I reach this line, I use the arbitrary equation $Ba + B'a' + B''a'' = 0$ to get a first apparent final equation; then, I use another arbitrary equation along with the same second-to-last line, to get a second apparent final equation. It then becomes obvious that only computations relative to the last *line* need changing, rather than all computations leading to the apparent final equation.

(391.) It is not always enough to create a different arbitrary equation to avoid encountering the same factor in the second apparent equation and in the first. For example, if I used the arbitrary equation $Bb + B'b' + B''b'' = 0$ as the second arbitrary equation, the second apparent final equation would have the same factor as the first. Consequently, this method would not be appropriate to identify the final equation with no factor.

This drawback can always be easily prevented by picking this arbitrary equation in the lower dimensions of the sum equation. Considering our example, I could use the equation $Be + B'e' + B''e'' = 0$ as the arbitrary equation leading to the second apparent final equation.

About the means to recognize which coefficients in the proposed equations can appear in the factor of the apparent final equation

(392.) The method we have just presented can be used very successfully to obtain the factor of the final equation, or, rather, to obtain the final equation without this factor. However, it would be quite advantageous to be able to determine this factor independently from the computations required by this method. We will now present observations that, we believe, can clarify this issue; we believe these observations can also be useful for other research efforts in analysis.

(393.) A part of our presentation will rely on finding the number of terms in the polynomial $[u, (x \ldots n-1)^B \ldots n]^T$. We could simply give this expression here and send the reader back to the methods presented in the first part of this book to obtain the expression of the number of terms in an arbitrary polynomial. But we believe there is some value in applying the methods given in the first book to this polynomial. We will first give the way to find this expression, and we will then compute the expression itself.

(394.) In (75), we have found the expression of the number of terms in the polynomial $[(u^A, x^{\overset{A}{\prime}})^B, y \ldots n]^T$. If we set $A = \underset{\prime}{A} = B$, we get

$$N[(u,x)^B, y \ldots n]^T = N(u \ldots n)^T - N(u \ldots n)^{T-B-1} - N(u \ldots n)^{T-B-1}$$
$$+ N(u \ldots n)^{T-B-2} - N(u)^{B-1} \times N(u \ldots n-1)^{T-B-1}.$$

But since

$$N(u \ldots n)^{T-B-1} - N(u \ldots n)^{T-B-2} = N(u \ldots n-1)^{T-B-1},$$

a shorter expression is

$$N[(u,x)^B, y \ldots n]^T, \text{ or } N[u, (x \ldots 2)^B \ldots n]^T$$
$$= N(u \ldots n)^T - N(u \ldots n)^{T-B-1} - N(u)^B \times N(u \ldots n-1)^{T-B-1}.$$

To compute the expression of $N[u, (x \ldots 3)^B \ldots n]^T$, I assume this polynomial is ordered with respect to one of the three letters entering in the expression $(x \ldots 3)^B$; we choose z, for example. Denoting by s the exponent of z in an arbitrary term, each term is of the form $z^s[u \ldots (x \ldots 2)^{B-s} \ldots n-1]^{T-s}$. We therefore have to compute the sum of $N[u \ldots (x \ldots 2)^{B-s} \ldots n-1]^{T-s}$, from $s = 0$ to $s = B$.

According to what we have just presented,

$$N[u \ldots (x \ldots 2)^{B-s} \ldots n-1]^{T-s} = N(u \ldots n)^{T-s} - N(u \ldots n)^{T-B-1}$$
$$- N(u)^{B-s} \times N(u \ldots n-1)^{T-B-1},$$

whose sum (70) from $s = 0$ to $s = B$ is

$$N(u \ldots n)^T - N(u \ldots n)^{T-B-1} - N(u)^B \times N(u \ldots n-1)^{T-B-1}$$
$$- N(u \ldots 2)^B \times N(u \ldots n-2)^{T-B-1}.$$

(395.) We now use this expression to get $N[u \ldots (x \ldots 4)^B \ldots n]^T$; as before, we order this polynomial with respect to one of the four variables whose degree must not exceed the dimension B; choose z, for example. Assuming the polynomial is ordered with respect to z, any term in this polynomial can be represented as $z^s[u \ldots (x \ldots 3)^{B-s} \ldots n-1]^{T-s}$. Thus we need to compute the sum of $N[u \ldots (x \ldots 3)^{B-s} \ldots n-1]^{T-s}$ from $s = 0$ to $s = B$.

According to what was just found, we have

$$N(u \ldots (x \ldots 3)^{B-s} \ldots n-1)^{T-s} = N(u \ldots n-1)^{T-s} - N(u \ldots n-1)^{T-B-1}$$
$$- N(u)^{B-s} \times N(u \ldots n-2)^{T-B-1} - N(u \ldots 2)^{B-s} \times N(u \ldots n-3)^{T-B-1},$$

whose sum from $s = 0$ to $s = B$ is

$$N(u \ldots n)^T - N(u \ldots n)^{T-B-1} - N(u)^B \times N(u \ldots n-1)^{T-B-1}$$
$$- N(u \ldots 2)^B \times N(u \ldots n-2)^{T-B-1} - N(u \ldots 3)^B \times N(u \ldots n-3)^{T-B-1}.$$

Thus, in general,

$$N[u \ldots (x \ldots n-1)^B \ldots n]^T = N(u \ldots n)^T - N(u \ldots n)^{T-B-1}$$
$$- N(u \ldots 1)^B \times N(u \ldots n-1)^{T-B-1}$$
$$- N(u \ldots 2)^B \times N(u \ldots n-2)^{T-B-1} - N(u \ldots 3)^B \times N(u \ldots n-3)^{T-B-1}$$
$$- N(u \ldots 4)^B \times N(u \ldots n-4)^{T-B-1} \ldots - N(u \ldots n-2)^B \times N(u \ldots 2)^{T-B-1}.$$

(396.) Thus, and following all what was said in the first book, the degree of the final equation resulting from the elimination of $n-1$ of the unknowns in n equations of the form $[u \ldots (x \ldots n-1)^b]^t = 0$ is obtained by differentiating the quantity

$$N(u \ldots n)^T - N(u \ldots n)^{T-B-1} - N(u)^B \times N(u \ldots n-1)^{T-B-1}$$
$$- N(u \ldots 2)^B \times N(u \ldots n-2)^{T-B-1} - N(u \ldots 3)^B \times N(u \ldots n-3)^{T-B-1} \ldots$$
$$\ldots \ldots - N(u \ldots n-2)^B \times N(u \ldots 2)^{T-B-1}$$

n times and varying T by the successive increments t, t', t'', t''', etc., and varying B by the successive increments b, b', b'', b''', etc.

(397.) Thus, for example, in the case of two equations we obtain

$$D = tt' - (t-b).(t'-b').$$

Considering three equations, we obtain

$$D = tt't'' - (t-b).(t'-b').(t''-b'') - b(t'-b').(t''-b'')$$
$$- b'(t-b).(t''-b'') - b''(t-b).(t'-b').$$

Considering four equations, we get

$$D = tt't''t''' - (t-b).(t'-b').(t''-b'').(t'''-b''') - b(t'-b').(t''-b'').(t'''-b''')$$
$$- b'(t-b).(t''-b'').(t'''-b''') - b''(t-b).(t'-b').(t'''-b''')$$
$$- b'''(t-b).(t'-b').(t''-b'') - bb'(t''-b'').(t'''-b''') - bb''(t'-b').(t'''-b''')$$
$$- bb'''(t'-b').(t''-b'') - b'b''(t-b).(t'''-b''') - b'b'''(t-b).(t''-b'')$$
$$- b''b'''(t-b).(t'-b'),$$

and so on.

(398.) We will be able to estimate the difference between the degree of the final equation using this method and the degree of the true final equation by comparing these formulas. This difference will indicate the dimension of the factor in the final equation and which coefficients can enter this factor.

(399.) Assume the n equations $[u \ldots (x \ldots n-1)^b \ldots n]^t = 0$ are complete, when they are written as equations in $n-1$ unknowns. If u is the unknown to appear in the final equation, the coefficients of the unknowns x, y, z, etc. (to be eliminated) are functions of u and known quantities. Assume these functions of u have dimensions denoted by p, p', p'', etc., respectively, in the terms of highest dimension. Then these functions of u will have dimensions $p+q-1$, $p'+q-1$, $p''+q-1$, etc. in the qth dimension, in descending order from the highest dimension.

For example, consider three equations of the form

$$
\begin{array}{llll}
ax^3 & +\ bx^2y & +\ cxy^2 & +\ dy^3 = 0, \\
+\ ex^2 & +\ fxy & +\ gy^2 & \\
+\ hx & +\ ky & & \\
+\ l & & &
\end{array}
$$

$$
\begin{array}{lll}
e'x^2 & +\ f'xy & +\ g'y^2 = 0, \\
+\ h'x & +\ k'y & \\
+\ l' & &
\end{array}
$$

$$
\begin{array}{ll}
h''x & +\ k''y = 0. \\
+\ l'' &
\end{array}
$$

If a, b, c, d have dimension p, e', f', g' have dimension p', and h'', k'' have dimension p'', then e, f and g will have dimension $p+1$, h', k' will have dimension $p'+1$, h and k will have dimension $p+2$, l' will have dimension $p'+2$, and finally l will have dimension $p+3$.

Likewise, assume the dimensions of the undetermined coefficients in the highest dimension of the polynomial multipliers are P, P', P'', etc., respectively. Then those in the Qth dimension (beginning from the highest dimension) will be $P+Q-1$, $P'+Q-1$, $P''+Q-1$, etc., respectively.

(400.) Considering an arbitrary dimension K in the sum equation, we now want to know the dimension of the known coefficient in that dimension, which also stands next to the unknown of dimension Q, Q' or Q'' in one of the polynomial multipliers. Denote this dimension r. We have $r+P+Q-1 = P+p+K-1$, and consequently $r = p+K-Q$; thus it follows that if K is smaller than Q, the undetermined coefficient will not be present in the dimension K of the sum equation. But if, for other reasons, we can assume it is indeed present, it will be supposed to have a known coefficient with dimension $p+K-Q$. We formulate similar statements relative to Q', Q'', etc.: $r = p'+K-Q'$, $r = p''+K-Q''$, etc.

(401) This being said, we must remember we must compute the product of all undetermined coefficients remaining in the polynomial multipliers to reach the final equation. Then, going through all equations provided both

by the cancellation of terms in the sum equation and by the arbitrary equations whose usefulness and usage we have taught, we successively swap each undetermined coefficient against the corresponding known coefficient in the current equation.

Thus any term in the final equation must be the product of as many known coefficients as there remains undetermined coefficients in all polynomial multipliers.

(402.) By the way, the sum of the dimension of the known coefficient with the degree of the unknowns standing next to it is constant; the same occurs for each undertermined coefficient of each polynomial multiplier. Thus we can easily conclude that the total dimension of the product of known terms that contribute to each term of the final equation is the same for each term. That is, each term in the final equation is such that: (i) It is the product of the same number of known coefficients and (ii) the sum of the dimensions of all known coefficients is the same for each term in the final equation.

(403.) We first consider equations in two unknowns, which we write as equations in one unknown.

Let A, B, C, D, etc., A', B', C', D', etc. be the undetermined coefficients of the two polynomial multipliers, which, from (346), must be of the form

$$(x \ldots 1)^{t'-1}, \quad (x \ldots 1)^{t-1}.$$

Assume we have written the product $AA'BB'CC'DD'$, etc. and that we successively use the equations

$$Aa + A'a' = 0, \quad Ab + A'b' + Ba + B'a' = 0, \quad \text{etc.}$$

to get the final equation. Since our concern is to obtain one term of the final equation only, we can swap only one undetermined coefficient with the corresponding determined coefficient, when using each one of these equations.

Without loss of generality, we may assume $t' < t$.

Assume we replace A, B, C, D, etc. by their corresponding determined coefficient, using the equations provided by the dimensions $t+t'-1$, $t+t'-2$, $t + t' - 3$ in the sum equation, successively. From what was said in (400), the dimension of the known coefficients used for substitution purposes is always p. Since there are t' coefficients A, B, C, D, etc., the total resulting dimension is pt'.

Assume we replace $A'B'C'D'$, etc. by the corresponding determined coefficients in the sum equation, starting from the dimension $t-1$. We see from (400) that the dimensions of the determined coefficients that we use for the substitution are $p' + t'$. And since there are t of them, the total resulting dimension is $p't + tt'$.

Thus the total dimension of each term in the final equation is $tt' + p't + pt'$.

From what was said in (397), and remembering (i) that the variable t we then used should be replaced here by $t + p$, (ii) that the variable we then called t' should be replaced by $t' + p'$, (iii) that the variables we then called b and b' should be replaced by t and t', the degree of the final equation is

$$D = (t + p).(t' + p') - (t + p - t).(t' + p' - t) = tt' + p't + pt',$$

which is the same result as that obtained with the elimination method described in that section.

(404.) Thus the current elimination method does not change the degree of the final equation for equations in two unknowns; therefore it introduces no new factor.

(405.) Let us now turn our attention to equations in three unknowns, which we first rewrite as equations in two unknowns.

If t, t', t'' are the degrees of these equations, the polynomial multiplier of the first equation (350) is of the form $(x\ldots 2)^{t'+t''-2}$; that of the second equation is of the form $(x\ldots 2)^{t+t''-2}$; that of the third equation is of the form $(x\ldots 2)^{t+t'-2}$.

This form gives

$$N(x\ldots 2)^{t-2} + N(x\ldots 2)^{t'-2} + N(x\ldots 2)^{t''-2}$$

arbitrary equations to construct in the sum equation: Starting from the first, that is, the highest dimension and operating in descending order, the number of arbitrary equations is

$$N(x\ldots 1)^{t-2} + N(x\ldots 1)^{t'-2} + N(x\ldots 1)^{t''-2}.$$

In the second dimension, the number of such arbitrary equations is

$$N(x\ldots 1)^{t-3} + N(x\ldots 1)^{t'-3} + N(x\ldots 1)^{t''-3},$$

and so on.

The form of these polynomial multipliers can still be reduced relative to the specific exponents of x and y, as we have seen in (351 ff). These can indeed be less than $t'+t''-2$, etc. However, this matter would lead us to too many details and we leave this form at that level of generality. The result is that, when using a simpler multiplier form, the resulting factor will have lower dimension than the one we are about to determine. However, we will see that if we know the dimension of the factor in the general form, we will always know the form taken by this factor when the polynomial multipliers are simpler.

(406.) We determine the undetermined coefficients of terms that may be cancelled in the polynomial multipliers by assigning other arbitrary equations rather than setting these coefficients to zero; we create as many of these arbitrary equations as possible in each dimension of the sum equation, avoiding assigning equations to lower dimensions when they can be assigned to higher dimensions.

At the same time, we create these arbitrary equations using the same approach followed until now; that is, all similar coefficients are simultaneously present. For example, if the total coefficient of an arbitrary term in the sum equation is

$$Ac + A'c' + A''c'' + Bb + B'b' + B''b'' + Ca + C'a' + C''a''$$

and we must create an arbitrary equation from one part of this term, we will write any of the arbitrary equations

$$Ac + A'c' + A''c'' = 0, \text{ or } Bb + B'b' + B''b'' = 0, \text{ or } Ca + C'a' + C''a'' = 0,$$
$$\text{or } Bb + B'b' + B''b'' + Ca + C'a' + C''a'' = 0,$$

or, etc. We will not write $Bb + B'b' = 0$, not because it is forbidden, but because maintaining symmetry has been shown to facilitate computations; we will make the same implicit assumption to obtain the characteristics of the factor.

(407.) This being said, we can assume with no loss of generality that $t > t' > t''$. Consider all equations provided by (i) the cancellation of terms and (ii) the arbitrary equations in the sum equation; assume that we use a number of equations equal to the number of terms in each dimension of the first polynomial multiplier and that we replace each undetermined coefficient in this polynomial by its determined coefficient in the currently used equation, successively.

From (400), it is easy to see that, denoting by q the order of the dimension of the sum equation to which this equation belongs, the undetermined coefficient of an arbitrary term bearing the same order in the first polynomial multiplier will have a known coefficient whose dimension is p. Thus, once all $N(x \dots 2)^{t'+t''-2}$ undetermined coefficients in the first multiplier have been replaced, the product of the determined coefficients that replace these undetermined coefficients will have dimension $pN(x \dots 2)^{t'+t''-2}$.

(408.) Using the same approach, the product of the $N(x \dots 2)^{t+t''-2}$ known coefficients replacing the undetermined coefficients in the second polynomial multiplier will have dimension $p'N(x \dots 2)^{t+t''-2}$.

(409.) However, the highest dimension in the sum equation and many of the following dimensions do not provide enough equations to determine the coefficients of the terms in the dimensions of same order in the third polynomial multiplier. We must therefore examine how many equations remain to be used in each dimension of the sum equation.

(410.) The investigation we are about to begin considers two general cases: $t' + t'' - t > 0$ and $t' + t'' - t < 0$. Consider the first case.

Starting from the highest dimension in the sum equation, and in a number of successive dimensions in descending order, the dimension of the order q in the sum equation provides $(x \dots 1)^{t+t'+t''-1-q}$ equations.

The number of arbitrary equations in that same dimension is

$$N(x \dots 1)^{t-1-q} + N(x \dots 1)^{t'-1-q} + N(x \dots 1)^{t''-1-q},$$

such that the total number of equations in the qth dimension of the sum equation is

$$N(x \dots 1)^{t+t'+t''-1-q} + N(x \dots 1)^{t-1-q} + N(x \dots 1)^{t'-1-q}$$
$$+ N(x \dots 1)^{t''-1-q}.$$

However, a number

$$N(x \dots 1)^{t'+t''-1-q} + N(x \dots 1)^{t'+t''-1-q}$$

of these equations is used by the coefficients, whose number is the same in the first two polynomial multipliers. Thus only

$$N(x \dots 1)^{t+t'+t''-1-q} + N(x \dots 1)^{t-1-q} + N(x \dots 1)^{t'-1-q} + N(x \dots 1)^{t''-1-q}$$
$$- N(x \dots 1)^{t'+t''-1-q} - N(x \dots 1)^{t+t''-1-q} = t + t' - 2q$$

equations remain to be used in the qth dimension in the sum equation, and the number of coefficients in the dimension with the same number in the third polynomial multiplier is $N(x\ldots1)^{t+t'-1-q} = t + t' - q$.

This argument holds from $q = 1$ to $q = t''$. Assume now $q = t'' + q'$; then the number of remaining equations is given by

$$N(x\ldots1)^{t+t'-1-q'} + N(x\ldots1)^{t-t''-1-q'} + N(x\ldots1)^{t'-t''-1-q'}$$
$$-N(x\ldots1)^{t'-1-q'} - N(x\ldots1)^{t-1-q'} = t + t' - 2t'' - q'.$$

The corresponding number of coefficients in the third polynomial multiplier is $N(x\ldots1)^{t+t'-t''-1-q'} = t + t' - t'' - q'$.

These expressions are valid from $q' = 1$ to $q' = t' - t''$.

Writing $q' = t' - t'' + q''$, the number of remaining equations is

$$N(x\ldots1)^{t+t''-1-q''} + N(x\ldots1)^{t-t'-1-q''} - N(x\ldots1)^{t''-1-q''}$$
$$-N(x\ldots1)^{t+t''-t'-1-q''} = t - t'';$$

the corresponding number of coefficients in the third polynomial multiplier is $N(x\ldots1)^{t-1-q''} = t - q''$.

These expressions hold from $q'' = 1$ to $q'' = t - t'$. Now write $q'' = t - t' + q'''$. The number of remaining equations is

$$N(x\ldots1)^{t'+t''-1-q'''} - N(x\ldots1)^{t'+t''-t-1-q'''} - N(x\ldots1)^{t''-1-q'''} = t - t'' + q'''.$$

The corresponding number of coefficients in the third polynomial multiplier is $N(x\ldots1)^{t'-1-q'''} = t' - q'''$.

These expressions hold from $q''' = 1$ to $q''' = t' + t'' - t$.

Let us thus write $q''' = t' + t'' - t + q^{IV}$. The number of remaining equations becomes

$$N(x\ldots1)^{t-1-q^{IV}} - N(x\ldots1)^{t-t'-1-q^{IV}} = t';$$

The corresponding number of coefficients in the third polynomial multiplier is $N(x\ldots1)^{t-t''-1-q^{IV}} = t - t'' - q^{IV}$.

These expressions hold from $q^{IV} = 1$ to $q^{IV} = t - t'$.

Now introduce $q^{IV} = t - t' + q^{V}$; the number of remaining equations is $N(x\ldots1)^{t'-1-q^{V}} = t' - q^{V}$; the corresponding number of coefficients in the third polynomial multiplier is $N(x\ldots1)^{t'-t''-1-q^{V}} = t' - t'' - q^{V}$.

These expressions are true from $q^{V} = 1$ to $q^{V} = t' - t''$.

Then introduce $q^{V} = t' - t'' + q^{VI} = t'' - q^{VI}$. The corresponding number of coefficients in the third polynomial multiplier is zero, and this holds from $q^{VI} = 1$ to $q^{VI} = t''$, where the sum equation is finally exhausted.

Since we will need to compare many of these expressions, we bring them together in the following table for convenience.

Ranges	Number of equations	Number of coefficients
From $q = 1$ to $q = t''$.	$t + t' - 2q$	$t + t' - q$
From $q' = 1$ to $q' = t' - t''$	$t + t' - 2t'' - q'$	$t + t' - t'' - q'$
From $q'' = 1$ to $q'' = t - t'$	$t - t''$	$t - q''$
From $q''' = 1$ to $q''' = t' + t'' - t$	$t - t'' + q'''$	$t' - q'''$
From $q^{IV} = 1$ to $q^{IV} = t - t'$	t'	$t - t'' - q^{IV}$
From $q^V = 1$ to $q^V = t' - t''$	$t' - q^V$	$t' - t'' - q^V$
From $q^{VI} = 1$ to $q^{VI} = t''$	$t'' - q^{VI}$	0

(411.) We now consider the case $t' + t'' - t < 0$.

As before, the number of remaining equations in the qth dimension of the sum equation is $t + t' - 2q$, and the corresponding number of coefficients in the third polynomial multiplier is $t + t' - q$; these figures hold from $q = 1$ to $q = t''$.

Introducing q' such that $q = t'' + q'$, the number of remaining equations is

$$N(x \dots 1)^{t+t'-1-q'} + N(x \dots 1)^{t-t''-1-q'} + N(x \dots 1)^{t'-t''-1-q'}$$
$$-N(x \dots 1)^{t'-1-q'} - N(x \dots 1)^{t-1-q'} = t + t' - 2t'' - q'.$$

The corresponding number of coefficients in the third polynomial multiplier is $N(x \dots 1)^{t+t'-t''-1-q'} = t + t' - t'' - q'$. These expressions hold for $q' = 1$ to $q' = t' - t''$.

Introducing q'' such that $q' = t' - t'' + q''$, the number of remaining equations is

$$N(x \dots 1)^{t+t''-1-q''} + N(x \dots 1)^{t-t'-1-q''} - N(x \dots 1)^{t''-1-q''}$$
$$-N(x \dots 1)^{t+t''-t'-1-q''} = t - t''.$$

The corresponding number of coefficients in the third polynomial multiplier is $N(x \dots 1)^{t-1-q''} = t - q''$; these expressions hold from $q'' = 1$ to $q'' = t''$.

Then introducing q''' such that $q'' = t'' + q'''$, the number of remaining equations is

$$N(x \dots 1)^{t-1-q'''} + N(x \dots 1)^{t-t'-t''-1-q'''} - N(x \dots 1)^{t-t'-1-q'''}$$
$$= t - t'' - q'''.$$

The corresponding number of coefficients, in the third polynomial multiplier, is $N(x \dots 1)^{t-t''-1-q'''} = t - t'' - q'''$, and these expressions hold from $q''' = 1$ to $q''' = t - t' - t''$.

Introducing q^{IV} such that $q''' = t - t' - t'' + q^{IV}$, the number of remaining equations is

$$N(x \dots 1)^{t'+t''-1-q^{IV}} - N(x \dots 1)^{t''-1-q^{IV}} = t'.$$

The corresponding number of coefficients in the third polynomial multiplier is $N(x \dots 1)^{t'-1-q^{IV}} = t' - q^{IV}$. These expressions hold from $q^{IV} = 1$ to $q^{IV} = t''$.

Introducing q^V such that $q^{IV} = t'' + q^V$, the number of remaining equations is $N(x\ldots1)^{t'-1-q^V} = t' - q^V$; the corresponding number of coefficients in the third polynomial multiplier is $N(x\ldots1)^{t'-t''-1-q^V} = t' - t'' - q^V$. These expressions hold from $q^V = 1$ to $q^V = t' - t''$.

Finally, introducing q^{VI} such that $q^V = t' - t'' + q^{VI}$, the number of remaining equations is $N(x\ldots1)^{t''-1-q^{VI}} = t'' - q^{VI}$. The corresponding number of coefficients in the third polynomial multiplier is zero. These expressions hold from $q^{VI} = 1$ to $q^{VI} = t''$, where the sum equation is finally exhausted. We now summarize all these results for the case $t' + t'' - t < 0$ in the following table.

Ranges	Number of equations	Number of coefficients
From $q = 1$ to $q = t''$	$t + t' - 2q$	$t + t' - q$
From $q' = 1$ to $q' = t' - t''$	$t + t' - 2t'' - q'$	$t + t' - t'' - q'$
From $q'' = 1$ to $q'' = t''$	$t - t''$	$t - q''$
From $q''' = 1$ to $q''' = t - t' - t''$	$t - t'' - q'''$	$t - t'' - q'''$
From $q^{IV} = 1$ to $q^{IV} = t''$	t'	$t' - q^{IV}$
From $q^V = 1$ to $q^V = t' - t''$	$t' - q^V$	$t' - t'' - q^V$
From $q^{VI} = 1$ to $q^{VI} = t''$	$t'' - q^{VI}$	0

(412.) We now look at how this enumeration can help us evaluate the total dimension of the product of the known coefficients used to substitute the undetermined coefficients in the third polynomial multiplier. Consider first the case $t' + t'' - t > 0$.

From (410), the number of equations used to substitute the same number of coefficients in the dimension with the same order in the third polynomial multiplier is $t + t' - 2q$, where q ranges from $q = 1$ to $q = t''$.

However, the dimension of each undetermined coefficient in this dimension is $P'' + q - 1$, and the dimension of each coefficient in the corresponding dimension in the sum equation is $P'' + p'' + q - 1$. Thus each known coefficient which, in that dimension of the sum equation, influences an undetermined coefficient whose dimension bears the same order, must have dimension p''. Exchanging undetermined coefficients against the $t + t' - 2q$ known coefficients thus results in creating a dimension $p''(t + t' - 2q)$. Consequently, from $q = 1$ to $q = t''$, all these exchanges will result in a dimension equal to $\int p''(t + t' - 2q) = p''t''(t + t' - t'' - 1)$.

In each dimension from $q = 1$ to $q = t''$, q undetermined coefficients remain to be exchanged, since only $t + t' - 2q$ coefficients out of $t + t' - q$ coefficients have already been replaced.

From $q' = 1$ to $q' = t' - t''$, there are $t + t' - 2t'' - q'$ equations and $t + t' - t'' - q'$ coefficients.

However, the dimension of each of these coefficients is $P'' + t'' + q' - 1$, and the dimension of the coefficients of the terms in the sum equation in the corresponding dimension is $P'' + p'' + t'' + q' - 1$; thus the known coefficient used for substitution against each undetermined coefficient contributes the dimension p''; since there are $t + t' - 2t'' - q'$ such exchanges, the resulting dimension is $p''(t + t' - 2t'' - q')$ and, summing from $q' = 1$ to $q = t' - t''$, results in $\int p''(t + t' - 2t'' - q') = p''[(t' - t'').(t + t' - 2t'') - \frac{(t' - t'').(t' - t'' + 1)}{2}]$. In each dimension from $q' = 1$ to $q' = t' - t''$, t'' known coefficients remain to be exchanged.

From $q'' = 1$ to $q'' = t - t'$, the number of equations in each dimension is $t - t''$, and the corresponding number of undetermined coefficients is $t - q''$. However, each of these coefficients has dimension $P' + t' + q'' - 1$; thus each exchange of undetermined coefficient contributes a dimension p''. Thus $t - t''$ such exchanges will contribute a dimension $p''(t - t'')$. Since q'' ranges from $q'' = 1$ to $q'' = t - t'$, the total contributed dimension is $p''(t - t'')(t - t')$. In each dimension from $q'' = 1$ to $q'' = t - t'$, there will remain $t'' - q''$ coefficients to be exchanged.

From $q''' = 1$ to $q''' = t' + t'' - t$, there are $t - t'' + q'''$ equations and $t' - q'''$ coefficients in each dimension.

Among these $t - t'' + q'''$ equations, we first use q''' of them to replace the q undetermined coefficients remaining in each dimension from $q = 1$ to $q = t''$.

The dimension of each of these coefficients is $P'' + q - 1$, and the dimension of the coefficient for each term in the sum equation leading to the equation used to substitute the variable is $P'' + p'' + t - 1 + q'''$. Thus exchanging each coefficient will result in the added dimension $p'' + t + q''' - q = p'' + t$; Indeed, when picking the equations and coefficients at an equal distance from $q''' = 1$ and $q = 1$ we get $q''' = q$.

Thus exchanging q coefficients leads to the dimension $(p'' + t)q$. Adding these from $q = 1$ to $q = q''' = t' + t'' - t$ leads to the dimension

$$\int (p'' + t)q = \frac{(p'' + t).(t' + t'' - t).(t' + t'' - t + 1)}{2}.$$

Since among all coefficients from $q = 1$ to $q = t' + t'' - t$, we have only exchanged those remaining coefficients from $q = 1$ to $q = t''$, we still need to exchange the undetermined coefficients of each dimension from $q = t' + t'' - t + 1$ to $q = t''$. Introducing $\underset{\prime}{q}$ such that $q = t' + t'' - t + \underset{\prime}{q}$, this is equivalent to exchanging the undetermined coefficients from $\underset{\prime}{q} = 1$ to $\underset{\prime}{q} = t - t'$.

In order to fulfill this objective, I use the same number of equations occuring from $q^{IV} = 1$ to $q^{IV} = t - t'$. Since each undetermined coefficient in the qth dimension has dimension $P'' + q - 1 = P'' + t' + t'' - t + \underset{\prime}{q} - 1$,

and since the dimension of the coefficient in each term of the sum equation used for the exchange is $P'' + p'' + t' + t'' + q^{IV} - 1$, each exchange will contribute the dimension $p'' + t + q^{IV} - \underset{\prime}{q} = p'' + t$; thus $t' + t'' - t - \underset{\prime}{q}$ such

coefficients provide the dimension $(p'' + t).(t' + t'' - t + \underset{\prime}{q})$; adding these from

$\underset{\prime}{q} = q^{IV} = t - t'$ to $\underset{\prime}{q}$ or q^{IV} results in the dimension

$$\int_{\prime} (p'' + t).(t' + t'' - t + q)$$

$$= (p'' + t).(t' + t'' - t).(t - t') + \frac{(p''+t).(t-t').(t-t'+1)}{2}.$$

This being done: (1) There remains no coefficient to be exchanged from $q = 1$ to $q = t''$. (2) There remains no equations from $q' = 1$ to $q' = t' - t''$; but there remains t'' undetermined coefficients to exchange. (3) There remains no equations from $q'' = 1$ to $q'' = t - t'$; however, $t'' - q''$ coefficients remain to be exchanged. (4) From $q''' = 1$ to $q''' = t' + t'' - t$, $t - t''$ equations remain and $t' - q'''$ coefficients remain to be exchanged. (5) From $q^{IV=1}$ to $q^{IV} = t - t'$, the number of remaining equations is

$$t' - (t' + t'' - t - \underset{\prime}{q}) \text{ or } t - t'' - \underset{\prime}{q} \text{ or } t - t'' - q^{IV}$$

and so is the number of coefficients to exchange. Beyond this, the same number of equations and coefficients remains as in (410).

Let us now use the $t - t''$ remaining equations from $q''' = 1$ to $q''' = t' + t'' - t$, to substitute the same number of undetermined coefficients in the corresponding dimensions.

The dimension of each of these coefficients is $P'' + t - 1 + q'''$. The dimension of the coefficient of each term in the sum equation whose dimension bears the same order is $P'' + p'' + t - 1 + q'''$. Thus, the substitution will contribute the dimension p''. For the $t - t''$ coefficients, this contribution is $p''(t - t'')$, and from $q''' = 1$ to $q''' = t' + t'' - t$, the contribution will be the dimension $p''(t - t'').(t' + t'' - t)$.

Thus there remains no equations from $q''' = 1$ to $q''' = t' + t'' - t$, and there remains only $t' + t'' - t - q'''$ coefficients.

We just saw that there remains $t - t'' - q^{IV}$ equations and the same number of coefficients from $q^{IV} = 1$ to $q^{IV} = t - t'$.

However, the dimension of each of these coefficients is $P'' + t' + t'' - 1 + q^{IV}$; the dimension of the corresponding coefficients in the terms of the sum equation is $P'' + p'' + t' + t'' - 1 + q^{IV}$. Thus each substitution contributes the dimension p''. For $t - t'' - q^{IV}$ coefficients, the contributed dimension is $p''(t - t'' - q^{IV})$; From $q^{IV} = 1$ to $q^{IV} = t - t'$, this dimension is $p''(t - t'').(t - t') - \frac{p''(t-t').(t-t'+1)}{2}$.

From $q^V = 1$ to $q^V = t' - t''$, we have $t' - q'$ equations and $t' - t'' - q^V$ undetermined coefficients. Let us use these equations to substitute these coefficients.

The dimension of each coefficient is $P'' + t + t'' - 1 + q^V$, and the dimension of each of the corresponding coefficients of the terms in the sum equation is $P'' + p'' + t + t'' - 1 - q^V$ for each coefficient. Thus the substitution of each coefficient contributes the dimension p''. A number $t' - t'' - q^V$ of these coefficients contributes the dimension $p''(t' - t'' - q^V)$; thus from $q^V = 1$ to $q^V = t' - t''$, the contributed dimension will be

$$p''(t' - t'').(t' - t'') - \frac{p''(t' - t'').(t' - t'' + 1)}{2} = \frac{p''(t' - t'').(t' - t'' - 1)}{2}.$$

Thus there still remain t'' equations from $q^{\mathrm{V}} = 1$ to $q^{\mathrm{V}} = t'-t''$. However, we have seen that the number of remaining coefficients from $q' = 1$ to $q' = t'-t''$ is t''. We can therefore use these equations to substitute these coefficients.

However, each of these coefficients has dimension $P''+t''+q'-1$, and the coefficient of each term of the sum equation, which provides the equation used for substitution, has dimension $P'' + p'' + t + t'' + q^{\mathrm{V}} - 1$; thus each substitution contributes the dimension $p'' + t + q^{\mathrm{V}} - q' = p'' + t$. Thus, t'' coefficients will contribute the dimension $(p''+t)t''$. From $q' = 0$ (or $q^{\mathrm{V}} = 1$) to q' or $q^{\mathrm{V}} = t' - t''$, the contribution will be $(p'' + t)t''(t' - t'')$.

There remain $t'' - q''$ coefficients from $q'' = 1$ to $q'' = t - t'$, and there remain $t'' - q^{\mathrm{IV}}$ equations from $q^{\mathrm{VI}} = 1$ to $q^{\mathrm{VI}} = t''$; thus there remains the same number of equations for each dimension.

However, the dimension of each of these coefficients is $P''+t'+q''-1$, and the dimension of the coefficients of the term in the sum equation that will contribute the equation used for the substitution is $P''+p''+t+t'+q^{\mathrm{VI}}-1$; thus substituting each undetermined coefficient contributes the dimension $p'' + t$; A number $t'' - q''$ of these substitutions contributes the dimension $(p''+t).(t''-q'')$. Thus from $q'' = 1$ to $q'' = t-t'$, the contributed dimension is $(p'' + t)t''(t - t') - \frac{(p''+t).(t-t').(t-t'+1)}{2}$.

We have now used all equations from $q^{\mathrm{VI}} = 1$ to $q^{\mathrm{VI}} = t - t'$ to perform these substitutions. Thus the only remaining equations occur from $q^{\mathrm{VI}} = t - t' + 1$ to $q^{\mathrm{VI}} = t''$. Writing $q^{\mathrm{VI}} = t - t' + \underset{\prime}{q}$, the number of remaining equations from $\underset{\prime}{q} = 1$ to $\underset{\prime}{q} = t' + t'' - t$ is $t' + t'' - t - q'''$.

However, we have previously seen that $t' + t'' - t - q'''$ coefficients remain from $q''' = 1$ to $q''' = t' + t'' - t$. We now substitute them.

The dimension of each coefficient is $P'' + t + q''' - 1$; the coefficient of the term in the sum equation that contributes the equation used for the substitution has dimension $P'' + p'' + 2t + \underset{\prime}{q} - 1$; thus each substitution contributes the dimension $p'' + t + \underset{\prime}{q} - q''' = p'' + t$. Thus, substituting $t + t'' - t - q'''$ such coefficients contributes the dimension $(p'' + t).(t' + t'' - t - q''')$. The total contribution from $q''' = 1$ to $q''' = t' + t'' - t$ is

$$(p'' + t).(t' + t'' - t).(t' + t'' - t) - \frac{(p'' + t).(t' + t'' - t).(t' + t'' - t + 1)}{2}$$
$$= \frac{(p'' + t).(t' + t'' - t).(t' + t'' - t - 1)}{2}.$$

Bringing together all the results we have found, the dimension contributed by substituting each undetermined coefficient of the third polynomial multiplier with the corresponding known coefficient in the equation provided by the sum equation reduces to

$$\frac{p''(t + t').(t + t' - 1)}{2} + tt't'' = p''N(x\ldots 2)^{t+t'-2} + tt't''.$$

(413.) Thus, from (407) and (408), the total dimension, or the degree of the final equation resulting from the second elimination method, is

$$pN(x\dots2)^{t'+t''-2} + p'N(x\dots2)^{t+t''-2} + p''N(x\dots2)^{t+t'-2} + tt't''$$

when $t' + t'' - t > 0$.

Before we examine the significance of this result, let us examine the case $t' + t'' - t < 0$.

(414.) As in the previous case, we first have $t + t' - 2q$ equations and $t + t' - q$ coefficients in each dimension from $q = 1$ to $q = t''$. By reasoning as in (412), the substitution of $t+t'-2q$ coefficients contributes the dimension $p''t''(t+t'-t''-1)$. The number of remaining coefficients in each dimension from $q = 1$ to $q = t''$ is q.

Likewise, we have $t+t'-2t''-q'$ equations and $t+t'-t''-q'$ coefficients in each dimension from $q' = 1$ to $q' = t'-t''$. The substitution of $t+t'-2t''-q'$ coefficients contributes the dimension

$$p'' \left[(t' - t'').(t+t'-2t'') - \frac{(t'-t'').(t'-t''+1)}{2} \right] ;$$

there remains t'' non substituted coefficients in each dimension from $q' = 1$ to $q' = t' - t''$.

There are $t - t''$ equations and $t - q''$ coefficients from $q'' = 1$ to $q'' = t''$. The dimension of each coefficient is $P'' + t' + q'' - 1$, and the dimension of the coefficient of the term in the sum equation leading to the equation used in the substitution is $P'' + p'' + t' + q'' - 1$. Thus each substitution contributes the dimension p''. Substituting $(t - t'')$ coefficients contributes the dimension $p''(t - t'')$. Thus, adding these contributions from $q'' = 1$ to $q'' = t''$ leads to the dimension $\int p''(t - t'') = p''t''(t - t'')$.

Thus $t'' - q''$ non substituted coefficients remain in each dimension from $q'' = 1$ to $q'' = t''$. But there is exactly the same number of equations in each dimension from $q^{VI} = 1$ to $q^{VI} = t''$; we can use these equations to perform the substitutions.

The dimension of each coefficient is $P'' + t' + q'' - 1$; the dimension of the coefficient of each term in the sum equation, which provides the equation used for substitution, is $P'' + p'' + t + t' + q^{VI} - 1$; each substitution thus contributes the dimension $(p'' + t)$. Substituting $t'' - q''$ coefficients thus contributes the dimension $(p'' + t).(t'' - q'')$. Summing from $q'' = 1$ to $q'' = t''$ results in the dimension

$$\int (p'' + t).(t'' - q'') = (p'' + t)t''t'' - \frac{(p'' + t)t''(t'' + 1)}{2} = \frac{(p'' + t)t''(t'' - 1)}{2}.$$

There are $t - t'' - q''$ equations and the same number of coefficients from $q''' = 1$ to $q''' = t - t' - t''$. The dimension of each coefficient is $P'' + t' + t'' + q''' - 1$ and the dimension of the coefficient in each term of the sum equation providing the equation used for substitution is $P'' + p'' + t' + t'' + q''' - 1$; thus each equation contributes the dimension p''. Since there are $t - t'' - q'''$ coefficients, their combined contributed dimension is $p''(t - t'' - q''')$. From

$q''' = 1$ to $q''' = t - t' - t''$, the total dimension is

$$p''(t - t'').(t - t' - t'') - \frac{p''(t - t' - t'').(t - t' - t'' + 1)}{2}.$$

There are t' equations and $t' - q^{\text{IV}}$ coefficients from $q^{\text{IV}} = 1$ to $q^{\text{IV}} = t''$; each coefficient has dimension $P'' + t + q^{\text{IV}} - 1$, and the coefficient of each term in the corresponding dimension of the sum equation is $P'' + p'' + t + q^{\text{IV}} - 1$; thus each substitution contributes the dimension p''; $t' - q^{\text{IV}}$ coefficients thus provide the dimension $p''(t' - q^{\text{IV}})$; summing from $q^{\text{IV}} = 1$ to $q^{\text{IV}} = t''$ leads to the total contributed dimension: $\int p''(t' - q^{\text{IV}}) = p''t't'' - \frac{p''t''(t'' + 1)}{2}$.

At this point q^{IV} equations will therefore remain in each dimension from $q^{\text{IV}} = 1$ to $q^{\text{IV}} = t''$. But we have also seen earlier that q coefficients remain in each dimension from $q = 1$ to $q = t''$. Thus we can use these equations to substitute these coefficients. Each coefficient has dimension $P'' + q - 1$; the coefficient of the term in the sum equation that leads to the equation for the substitution has dimension $P'' + p'' + t + q^{\text{IV}} - 1$. Thus each substitution contributes the dimension $p'' + t$. With q such substitutions in each dimension, we get a contributed dimension equal to $(p'' + t)q$. Summing from $q = 1$ to $q = t''$ leads to the total dimension $\frac{(p'' + t)t''(t'' + 1)}{2}$.

From $q^{\text{V}} = 1$ to $q^{\text{V}} = t' - t''$, there are $t' - q^{\text{V}}$ equations and $t' - t'' - q^{\text{V}}$ coefficients; each coefficient has dimension $P'' + t + t'' + q^{\text{V}} - 1$, and the coefficient of each term with corresponding dimension in the sum equation is $P'' + p'' + t + t'' + q^{\text{V}} - 1$. Thus each substitution contributes the dimension p''. Since there are $t' - t'' - q^{\text{V}}$ coefficients, the resulting contributed dimension is $p''(t' - t'' - q^{\text{V}})$. Summing from $q^{\text{V}} = 1$ to $q^{\text{V}} = t' - t''$ yields the total dimension

$$p''(t' - t'').(t' - t'') - \frac{p''(t' - t'').(t' - t'' + 1)}{2} = \frac{p''(t' - t'').(t' - t'' - 1)}{2}.$$

From $q^{\text{V}} = 1$ to $q^{\text{V}} = t' - t''$, t'' equations remain in each dimension. However, we have seen above that there are t'' coefficients in each dimension from $q' = 1$ to $q' = t' - t''$. We can therefore use these equations to substitute these coefficients. Each coefficient has dimension $P'' + t'' + q' - 1$; the coefficient of the term of the sum equation leading to the equation used for the substitution has dimension $P'' + p'' + t + t'' + q^{\text{V}} - 1$. Thus each substitution contributes the dimension $p'' + t$; hence the contribution from t'' coefficients is $(p'' + t)t''$; adding the contributions from $q' = 1$ to $q' = t' - t''$ leads to the total dimension $(p'' + t)t''(t' - t'')$.

Bringing all these results together, we get, for the case $t' + t'' - t < 0$, the same result as for the previous case. That is, the degree of the final equation is given again by

$$pN(x \dots 2)^{t' + t'' - 2} + p'N(x \dots 2)^{t + t'' - 2} + p''N(x \dots 2)^{t + t' - 2} + tt't''.$$

(415.) Thus, in general, the degree of the final equation reached via our second method is always

$$pN(x \dots 2)^{t' + t'' - 2} + p'N(x \dots 2)^{t + t'' - 2} + p''N(x \dots 2)^{t + t' - 2} + tt't''.$$

(416.) From what was said in (397), and remarking that what we then called t is now $t + p$ and what we then called b is now t, the true degree of the final equation is

$$(t + p).(t' + p').(t'' + p'') - pp'p'' - tp'p''$$
$$-t'pp'' - t''pp'$$
$$= tt't'' + pt't'' + p't t'' + p''tt'.$$

Thus the factor introduced in the final equation by the second method has degree

$$p[N(x \ldots 2)^{t'+t''-2} - t't''] + p'[N(x \ldots 2)^{t+t''-2} - tt'']$$
$$+p''[N(x \ldots 2)^{t+t'-2} - tt']$$
$$= p\left(\frac{t'^2+t''^2-t'-t''}{2}\right) + p'\left(\frac{t^2+t''^2-t-t''}{2}\right) + p''\left(\frac{t^2+t'^2-t-t'}{2}\right).$$

(417.) Thus:

1. If the proposed equations are complete, the second elimination method does not alter the degree of the final equation since then $p = p' = p'' = 0$.

2. The same holds true, for the same reason, if the equations are incomplete and the unknowns to be eliminated, when combined pairwise, do not exceed a total dimension smaller than that of the equation.

(418.) In all other cases, the factor will contain the unknown relative to which the final expression is written. Consequently it would hide the true degree of the final equation; however, we have the means to determine the degree of this factor.

(419.) Moreover, we will always be able to determine the coefficients of the proposed equations that, alone, can enter this factor. This will considerably simplify the necessary work to find it. Before showing how to determine which coefficients can, alone, enter the factor, we need to say a word about the general expression of the degree of the final equation obtained with the second method.

(420.) Looking back at what we said in (403) about equations in two unknowns rewritten as equations in one unknown, we see that the expression for the degree of the final equation found using the second method is

$$pN(x \ldots 1)^{t'-1} + p'N(x \ldots 1)^{t-1} + tt'.$$

We just saw that, when considering equations in three unknowns recast as equations in two unknowns, the degree of the final equation is

$$pN(x \ldots 2)^{t'+t''-2} + p'N(x \ldots 2)^{t+t''-2} + p''N(x \ldots 2)^{t+t'-2} + tt't''.$$

We must therefore conclude that the degree of the final equation for equations in four unknowns recast as equations in three unknowns must be

$$pN(x \ldots 3)^{t'+t''+t'''-3} + p'N(x \ldots 3)^{t+t''+t'''-3}$$
$$+p''N(x \ldots 3)^{t+t'+t'''-3} + p'''N(x \ldots 3)^{t+t'+t''-3} + tt't''t''',$$

and this is indeed what happens by reasoning about these equations as we did for the preceding cases.

Comparing with the actual degree of the final equation found in (397), we can always determine the dimension of the factor; considering the cases mentioned in (417), the factor does not affect the degree of the final equation.

(421.) We now easily see what the degree of the final equation is, using the second method, for an arbitrary number of unknowns.

(422.) The degree of the factor of the final equation is in general expressed as a function of t, t', t'', etc., whose elements are multiplied by p, p' p'', and so on, and this expression becomes zero when $p = p' = p'' =$ etc. $= 0$; it is easy to conclude from this that the factor can be composed of nothing else but coefficients of the terms in the highest dimension.

Indeed, only those can, when taken in arbitrary combinations, yield zero dimension when $p = 0$, $p' = 0$, etc. The coefficients of the terms in the lower dimensions all have their dimension greater than zero; thus it is impossible for the dimension of the factor to become zero if this factor included any one of these coefficients.

For example, considering the three equations

$$ax^2 + bxy + cy^2 = 0,$$
$$+dx + ey$$
$$+f$$

$$a'x^2 + b'xy + c'y^2 = 0,$$
$$+d'x + e'y$$
$$+f'$$

$$a''x^2 + b''xy + c''y^2 = 0,$$
$$+d''x + e''y$$
$$+f''$$

the factor can only contain the letters a, b, c; a', b', c'; a'', b'', c''.

(423.) This observation leads us to exclude many letters right away when computing the factor and that makes it easier to find. It may often be found much more easily than using the method of the common divider discussed in (388). Consider indeed the example given by the three equations above. We immediately see that the factor can be nothing but $(ab'c'')^2$. Indeed, using the formula

$$p\left(\frac{t'^2 + t''^2 - t' - t''}{2}\right) + p'\left(\frac{t^2 + t''^2 - t - t''}{2}\right) + p''\left(\frac{t^2 + t'^2 - t - t'}{2}\right)$$

and assuming $p = p' = p'' = 1$, $t = t' = t'' = 2$, the dimension of the factor is 6. This is the dimension of $(ab'c'')^2$ when we assume a, b, c, a', b', c', a'', b'', c'' all have dimension 1.

(424.) Such is the dimension of the factor when the arbitrary equations are formed so as not to cancel any coefficients of the polynomial multipliers. If we had followed the simpler and more natural approach to forming the arbitrary equations so as to minimize the number of coefficients in the polynomial multipliers, the dimension of the factor would have gone down as the number of cancelled coefficients grew larger. It is always possible, using our general formula that gives this dimension and the number of cancelled coefficients, to determine the dimension of this factor.

For example, consider the three equations above. We know from (349 ff) that we can cancel one term in each polynomial multiplier; then the dimension of the factor is only 3; that is, the factor is only $(ab'c'')$, as we have already seen in (278).

(425.) The general expression we have found for the dimension of the factor in the final equation required the assumption that all possible arbitrary equations provided by each dimension of the sum equation have been formed. As seen in (236), it is possible that fewer of these equations be formed in the higher dimensions, where the leftover is formed in the lower dimensions instead. By doing this, the number of letters entering the factor does not increase, but its degree relative to the unknown in the final equation would increase. In other terms, the degree of the final equation would be affected, even in the case of complete equations.

(426.) Thus the expression we have given for the general dimension of the factor is the simplest possible, among all expressions where the arbitrary equations are not used to cancel any term in the polynomial multipliers. It also leads to the lowest possible dimension, when the arbitrary equations are used to cancel all possible terms that may be cancelled in the polynomial multipliers.

Determining the factor of the final equation:
How to interpret its meaning

(427.) We have said in (339) that the factor in the final equation resulting from our second method indicates solutions of the type described in (279) and (287). It has an even deeper significance in the General Theory of Equations. This significance and the factor itself will occur simultaneously.

(428.) We have just seen that the factor can only be composed of coefficients of the terms in the highest dimension of each equation. Let us therefore assume that the coefficient of each term in the lower dimensions is zero; the final equation must contain the solution for all arbitrary values of the coefficients of the proposed equations, including the solution to this specific case. However, the final equation is composed of two factors: One is the factor itself, denoted by F; the other, denoted by E, is the true final equation. Among these two factors, the factor E becomes zero if all coefficients in the lower dimensions are zero. Indeed, if any term in E remained nonzero, this term would only be composed of the coefficients of the terms in the highest dimension in the proposed equations: Thus all terms of E would not be homogenous functions, or with the same dimension as the coefficients in the proposed equations, which is not possible.

Since all terms of E become zero when assuming that the coefficients in the lower dimensions are zero, the solution to this specific case can only be borne by the factor F; that is, the equation $F = 0$ is the sought result.

(429.) At this stage, let us reformulate our problem. Our problem is to determine the condition(s) such that the new equations obtained by setting the highest dimension to zero are all simultaneously satisfied. That is, the factor F is the condition equation, or one of the condition equations, or one of their products, so that the new equations can hold simultaneously.

(430.) There is no doubt this factor can be divided by one or more condition equations mentioned above. The number of these equations can always be reduced to two; however, a much larger number of varieties can occur. The factor, however, may itself contain other factors than those condition equations: Indeed, those arbitrary equations not used to cancel any terms in the polynomial multipliers necessarily lead to increasing the total dimension of the factor with no particular relevance to those condition equations.

(431.) The preceding dicussion leads us to see that it is almost impossible to determine this factor directly, although we have determined that it can only contain the coefficients in the highest dimensions of the proposed equations. However, all of what we have just said offers a general and simple method to find this factor in each case. Here it is:

Since the factor only contains coefficients in the highest dimension of each proposed equation, it follows that this factor is not sensitive to whether the other coefficients are zero or not. However, simply setting these coefficients to zero would annihilate the final equation. This problem can be avoided as follows: Pick the coefficient in the lowest dimension of the equation and assume it is infinitesimally small instead of zero. We then keep only the lowest-order terms in the final equation; assuming that $n-1$ coefficients are zero, the equation can be divided by the nth coefficient.

We can perform the same operation for the coefficient of each term in the second dimension (dimension 1), the third dimension (dimension 2), etc., successively, up to and excluding the highest dimension. This process leads us to the factor, without looking for a complicated divider. We will not give examples of this method, since such examples are already provided in (381), (382) and (383).

(432.) Thus, although our elimination method is in general unable to avoid appending a factor to the final equation, we see that (1) this factor is not unrelated to the original problem, and (2) it can always be found and therefore extracted from the final equation. This step is absolutely necessary: Any equation where a factor is left is completely useless.

About the factor that arises when going from the general final equation to final equations of lower degrees

(433.) So far we have given the fastest method to construct the general formulas resulting from an arbitrary number of equations whose number of unknowns is the number of equations minus one.

We have also given means to obtain the factor that alienates this final equation and, consequently, to find the most reduced final equation.

In order to compute the final equations resulting from equations with lower degrees from the final equation, we only need to set all coefficients in the highest dimensions of some of the proposed equations to zero.

For example, we have found in (378) the simplest final equation resulting from the three equations

$$\begin{aligned} ax^2 &+ bxy + cy^2 = 0, \\ &+ dx + ey \\ &+ f \end{aligned}$$

$$\begin{aligned} a'x^2 &+ b'xy + c'y^2 = 0, \\ &+ d'x + e'y \\ &+ f' \end{aligned}$$

$$\begin{aligned} a''x^2 &+ b''xy + c''y^2 = 0. \\ &+ d''x + e''y \\ &+ f'' \end{aligned}$$

We can get the final equation resulting from the three equations

$$\begin{aligned} ax^2 &+ bxy + cy^2 = 0, \\ &+ dx + ey \\ &+ f \end{aligned}$$

$$\begin{aligned} a'x^2 &+ b'xy + c'y^2 = 0, \\ &+ d'x + e'y \\ &+ f' \end{aligned}$$

$$\begin{aligned} &\quad\; d''x + e''y = 0, \\ &+ f'' \end{aligned}$$

by setting $a'' = 0$, $b'' = 0$, $c'' = 0$ in the general final equation.

This process cancels a large number of terms and yields a much simpler equation. However, it is not, by far, the simplest such equation. This equation is affected by a factor, and this factor is all the more complex if there are more equations and their degrees are higher; and it cannot be identified by inspection.

It is nevertheless important to rid the final equation from this factor, which no method can prevent and which is an intrinsic issue tied to the elimination process.

In general, the factor affecting the true final equation must be extracted from all computed final equations: This is not only because this factor adds much useless complexity, but also because, in some cases, it completely invalidates the final equation.

Indeed, every time the coefficients of the proposed equation satisfy the necessary conditions, such that the condition equation resulting from setting this factor to zero can occur, the final equation will reduce to $0 = 0$, or, in other words, it will mean nothing.

Looking for this factor is thus absolutely necessary: Failing to do so would invalidate many of the benefits of the general elimination formulas, including the ability to find formulas for equations of lower degrees.

Looking for the factors is no problem when only two equations are involved: The factor is a monomial and very easy to detect.

Consider, for example, the final equation found in (348) for two equations of the form

$$ax^3 + bx^2 + cx + d = 0.$$

Assume we set $a' = 0$ in the final equation to get the final equation corresponding to the two equations

$$ax^3 + bx^2 + cx + d = 0,$$
$$b'x^3 + c'x + d' = 0.$$

We obtain

$$[ab'(bc') - a^2c'^2 + a^2b'd'].(cd') - [ab'(bd') - a^2c'd'].(bd')$$
$$+[ab'(cd') - a^2d'^2]ad' = 0,$$

which can obviously be divided by a: the resulting quotient is the final equation that could be obtained directly, following the method given in (346).

When there is more than one unknown, the factor is not a monomial anymore, and many superfluous efforts would be required to find it, unless it could be determined a priori. Let us now show how this can be done.

(434.) Let us assume that the proposed equations are originally of the form given in (396), and assume that, when they are recast as equations in one less unknown than there are equations, these equations are complete, with degrees t, t', t'', etc., respectively. Under such assumptions, the terms p, p', p'', etc. (which are the dimension of the coefficients in the terms with highest dimension of each equation) must be such that $t+p$, $t'+p'$, $t''+p''$, etc. replace what we have called, in (396), t, t', t''; and t, t', t'' replace what we then called b, b', b'', etc.

Then, for two equations, the degree of the final equation is expressed as $tt'+pt'+p't$; for three equations, $tt't''+pt't''+p'tt''+p''tt'$; for four equations, $tt't''t''' + pt't''t''' + p'tt''t''' + p''tt't''' + p'''tt't''$, and so on.

Assume now that each coefficient in the highest dimension in one of the equations, say with degree t, is zero. Then t becomes $t-1$, and p becomes $p+1$.

The expression for the degree of the final equation then becomes

For two equations	...	$tt' - t' + pt' + t' + p't - p'$,
or	...	$tt' + pt' + p't - p'$;
For three equations	...	$tt't'' + pt't'' + p'(tt'' - t'') + p''(tt' - t')$,
For four equations	...	$tt't''t''' + pt't''t''' + p'(tt''t''' - t''t''')$
		$+p''(tt't''' - t't''') + p'''(tt't'' - t't'')$,

and so on. The degree of the final equation will therefore be reduced as follows:

For two equations	...	p',
for three equations	...	$p't'' + p''t'$,
for four equations	...	$p't''t''' + p''t't''' + p'''t't''$.

However, setting each coefficient to zero in the highest dimension in the equation of degree t only destroys some terms in the final equation, but does not change its dimension. Thus it can be divided by a factor whose dimension is

For two equations	...	p',
For three equations	...	$p't'' + p''t'$,
For four equations	...	$p't''t''' + p''t't''' + p'''t't''$,

and so on.

However, we can see that (i) t and p do not enter these expressions; thus the factor must be completely independent of the equation of degree t; that is, it contains none of the coefficients of this equation; (ii) the dimension of this factor becomes zero when p', p'' and p''' are zero; thus, it only contains the factors in the highest dimensions in the equations with degree t', t'', t''', etc.; (iii) this factor is therefore the same as if the coefficients in the lower dimensions of the equations were zero; (iv) thus it necessarily is the condition equation that must be satisfied so that the equations arising from the highest dimension in each equation t', t'', t'' are simultaneously satisfied.

What we have said about the equation with degree t is applicable to all others; thus we conclude as follows:

Assume we have found the most reduced general final equation resulting from an arbitrary number n of equations in $n - 1$ unknowns and degrees t, t', t'', t''', etc., and we want to compute the most reduced final equation when the degree of one of the equations drops by one unit. We must first compute the value of this final equation when each coefficient, in the highest dimension of the equation whose degree is dropped, is set to zero. This equation must then be divided by the factor determined by computing the necessary condition equation that must be satisfied so that all equations formed from the highest dimension of each proposed equation, except the equation of interest, simultaneously hold.

(435.) Thus, we now see how, given the most reduced final equation for a given set of equations, we can extract the most reduced final equation for all sets of equations with lower degrees.

For example, by (433), given the two equations

$$ax^3 + bx^2 + cx + d = 0,$$
$$a'x^3 + b'x^2 + c'x + d' = 0,$$

and their final equation, we can obtain the final equation for the two equations

$$ax^3 + bx^2 + cx + d = 0,$$
$$b'x^2 + c'x + d' = 0$$

as follows. First set $a' = 0$; then divide the resulting equation by a. However, $a = 0$ is the necessary condition equation for the equation $ax^3 = 0$ to hold. This equation is indeed formed from the highest dimension of the other equation.

Similarly, consider the three equations

$$ax^2 \quad + \quad bxy \quad + \quad cy^2 = 0,$$
$$+ \quad dx \quad + \quad ey$$
$$+ \quad f$$

$$a'x^2 \quad + \quad b'xy \quad + \quad c'y^2 = 0,$$
$$+ \quad d'x \quad + \quad e'y$$
$$+ \quad f'$$

$$a''x^2 \quad + \quad b''xy \quad + \quad c''y^2 = 0,$$
$$+ \quad d''x \quad + \quad e''y$$
$$+ \quad f''$$

and the resulting final equation (378). Assume $a'' = 0$, $b'' = 0$, $c'' = 0$; the resulting reduced final equation can be divided by $(ab').(bc') - (ac')^2$, which is precisely the necessary condition equation so that the two equations

$$ax^2 + bxy + cy^2 = 0,$$
$$a'x^2 + b'xy + c'y^2 = 0$$

can hold simultaneously.

Determination of the factor mentioned above

(436.) The factor we are looking for is therefore the condition equation that must hold so that the equations formed from the highest dimension of the $n - 1$ among the n proposed equations hold.

However, will the factor be this equation itself, as we have said, or is it only included as a factor in this equation, as we have pointed out may happen for the factor of the general final equation?

It turns out that the factor is the equation itself, as we have suggested.

Indeed, assume it is only a factor of this condition equation. Then, its dimension would be smaller than the dimension of this equation. However, it must be precisely the same: From (434), the factor's dimension is

For two equations   ... p',
For three equations   ... $p't'' + p''t'$,
For four equations   ... $p't''t''' + p''t't''' + p'''t't''$.

I claim that the dimension of the condition equation is precisely that dimension in all cases.

Indeed, consider the case of n equations: The condition equation results from $n-1$ equations resulting from each one of the highest dimensions of the $n-1$ proposed equations. Although these equations contain $n-1$ unknowns, they should really be seen as equations in $n - 2$ unknowns since they only consist of the highest dimensions. Thus, given $n - 1$ equations in $n - 2$ unknowns, we can get the factor by following the method we have taught to compute the final equation. Let us compute the dimension of this final equation for 1, 2, 3, etc. equations, corresponding to 2, 3, etc. proposed equations.

Let p', p'', p''', p^{IV}, etc. be the dimensions of the unknown present in the coefficients of the proposed equations, in their highest dimension. For one equation, where there is no unknown left, the dimension is p'.

For two equations, where one unknown remains, our method gives $p't'' + p''t'$ as the dimension of the final equation. Indeed, all known coefficients in each equation share the same dimension in general. Since the final equation results from substituting the unknown coefficients with known ones in their product, its dimension is

$$p'N(x \ldots 1)^{t''-1} + p''N(x \ldots 1)^{t'-1} = p't'' + p''t'.$$

We have seen, from (404), that the final equation resulting from two equations does not have any factor; thus, the actual dimension of the final equation is indeed $p't'' + p''t'$.

Now considering three equations (in two apparent unknowns), our method finds the degree of the final equation to be

$$p'N(x \ldots 2)^{t''+t'''-2} + p''N(x \ldots 2)^{t'+t'''-2} + p'''N(x \ldots 2)^{t'+t''-2}.$$

From (415), however, we have seen that this final equation must contain a factor whose dimension is

$$p'N(x \ldots 2)^{t''+t'''-2} + p''N(x \ldots 2)^{t'+t'''-2} + p'''N(x \ldots 2)^{t'+t''-2}$$
$$-p't''t''' - p''t'''t' - p'''t't'';$$

thus, the dimension of the true final equation is $p't''t''' + p''t't''' + p'''t't''$ and so on.

Hence, the factor of interest is exactly the condition equation resulting from the $n-1$ equations formed with each highest-dimensional term of $n-1$ out of the n proposed equations.

(437.) This condition equation or factor can be obtained by multiplying all equations by the highest dimension of the appropriate polynomial multipliers, which can be determined from (340 ff).

Assume, for example, that we have three proposed equations. Considering the reduced final equation resulting from the three equations

$$(x, y)^t = 0, \ (x, y)^{t'} = 0, \ (x, y)^{t''} = 0,$$

we want to compute the final equation for the three equations

$$(x, y)^{t-1} = 0, \ (x, y)^{t'} = 0, \ (x, y)^{t''} = 0.$$

To compute the factor of this final equation after setting to zero all coefficients in the highest dimension of the first equation, we compute the condition equation resulting from the two equations formed with the highest dimension of the equation $(x, y)^{t'} = 0$ and $(x, y)^{t''} = 0$. We get this equation by multiplying the first equation with the highest dimension of the polynomial multiplier $(x, y)^{t''-1}$ and the second equation with the highest dimension of the polynomial multiplier $(x, y)^{t'-1}$: We then proceed with the computation of the *lines*, following what was done up to now.

(438.) However, from all that we have said so far, we see that if the total number of proposed equations exceeds three, our method will add a factor to this condition equation. Since the coefficients of the terms leading to this condition equation all have the same dimension, it seems this new factor would be hard to find. Here is how we can resolve this apparent difficulty.

(439.) Consider, for example, four equations, all of which have degree 3. We must find the condition equation corresponding to the three equations

$$\left.\begin{array}{l} ax^3 + bx^2y + cx^2z + dxy^2 + exyz \\ \quad + fxz^2 + gy^3 + hy^2z + kyz^2 + lz^3 \end{array}\right\} = 0,$$

$$\left.\begin{array}{l} a'x^3 + b'x^2y + c'x^2z + d'xy^2 + e'xyz \\ \quad + f'xz^2 + g'y^3 + h'y^2z + k'yz^2 + l'z^3 \end{array}\right\} = 0,$$

$$\left.\begin{array}{l} a''x^3 + b''x^2y + c''x^2z + d''xy^2 + e''xyz \\ \quad + f''xz^2 + g''y^3 + h''y^2z + k''yz^2 + l''z^3 \end{array}\right\} = 0.$$

This condition equation is no different from the one resulting from the three other equations

$$
\begin{array}{rcccccccc}
& ax^3 & + & bx^2y & + & dxy^2 & + & gy^3 = 0, \\
+ & cx^2 & + & exy & + & hy^2 & & \\
+ & fx & + & ky & & & & \\
+ & l & & & & & & \\
\end{array}
$$

$$
\begin{array}{rcccccccc}
& a'x^3 & + & b'x^2y & + & d'xy^2 & + & g'y^3 = 0, \\
+ & c'x^2 & + & e'xy & + & h'y^2 & & \\
+ & f'x & + & k'y & & & & \\
+ & l' & & & & & & \\
\end{array}
$$

$$
\begin{array}{rcccccccc}
& a''x^3 & + & b''x^2y & + & d''xy^2 & + & g''y^3 = 0. \\
+ & c''x^2 & + & e''xy & + & h''y^2 & & \\
+ & f''x & + & k''y & & & & \\
+ & l'' & & & & & & \\
\end{array}
$$

These equations are nothing but the preceding equations with $z = 1$.

In general, we can assume that one of the unknowns is equal to 1 in each of the equations created from the highest dimensions: These equations can then be treated exactly like we did so far, to get both the final equation and its factor.

About equations where the number of unknowns is less than the number of equations by two units

(440.) When the number of equations exceeds the number of unknowns by two units, we obtain two equations among the known coefficients of their terms: These equations may be more or less convoluted depending upon the method used to obtain them.

(441.) Not only can these equations be more or less convoluted, but also things do not happen as in the case when the number of equations exceeds the number of unknowns by one unit. In the latter case, we are sure the equation has a factor when this equation is more complicated than expected, and we have methods to isolate this factor.

In the current case, the two condition equations may appear as more complex than they really are, and trying to identify a factor in each condition equation would be a vain task. There would be none, or, if there were one, eliminating it would not necessarily simplify the equations as much as possible.

(442.) For clarity of exposition, assume the two reduced condition equations are

$$\begin{aligned} E &= 0, \\ E' &= 0. \end{aligned}$$

Multiplying the first by a and the second by a', I form the equation

$$aE + a'E' = 0.$$

Multiplying the first by b and the second by b', I form the equation

$$bE + b'E' = 0.$$

It is easy to see that these two equations can be reduced, since they can be changed in two other equations with a factor each. But obviously none has any factor, and it would be vain to try and simplify the equations by looking for the factor that makes them complicated.

(443.) Consider again the two equations

$$\begin{aligned} aE + a'E' &= 0, \\ bE + b'E' &= 0. \end{aligned}$$

I can simplify them by multiplying the first equation by m. Adding the product to the second equation, I get $(ma + b)E + (ma' + b')E' = 0$. By forcing $ma' + b' = 0$ or $m = -\frac{b'}{a'}$, I get the equation $(ma + b)E = 0$, which is the same as $-\frac{(ab')}{a'}E = 0$ or $(ab')E = 0$. Dividing by (ab') yields $E = 0$. A similar process leads to $E' = 0$. However, we cannot always use such simple means.

(444.) Nonetheless, we propose to give means to reach the two final equations or two condition equations which have been reduced as much as possible. However, we cannot reach this goal immediately, and we believe we may not be able to reach that goal in general.

Indeed, finding the two simplest final equations resulting from an arbitrary number of equations whose number of unknowns is less than the number of equations by two units is a specific case of the following, much more general question. Find the means to satisfy a given number of equations containing the same number of unknowns minus two. This more general question must present several cases which are not relevant to the first question. This is how we have seen that the condition equation resulting from n equations in $n-1$ unknowns contains a factor which contains the solution to the question when all lower dimensions are missing from the proposed equations.

(445.) Our object is to give a method to answer simply and completely the following question: *What are the condition equations that contain all possible solutions to a given number of equations containing two fewer unknowns than there are equations?* We will then show how to bring these equations to the lowest possible dimension, that is, how to rid them from the pathological solutions they may contain.

(446.) We begin by searching the simplest form of the polynomial multipliers that must be used to reach these two final equations.

Form of the simplest polynomial multipliers used to reach the two condition equations resulting from n equations in $n-2$ unknowns

(447.) Assume there is only one unknown and, consequently, three equations whose degrees are t, t', t'' for the first, second and third equation, respectively. Assume also that $t > t' > t''$.

In general we can use the following polynomial multipliers (227):

First equation	...	$(x\ldots 1)^{T+t'+t''}$,
Second equation	...	$(x\ldots 1)^{T+t+t''}$,
Third equation	...	$(x\ldots 1)^{T+t+t'}$.

In order to know the simplest form of these polynomial multipliers, we notice that with such multipliers, the sum equation is of the form $(x\ldots 1)^{T+t+t'+t''}$, and the difference between the number of terms in the highest dimension in this equation and the number of useful coefficients in the highest dimension of each polynomial multiplier is

$$d^3[N(x\ldots 0)^{T+t+t'+t''}]\ldots \begin{pmatrix} T+t+t'+t'' \\ t,t',t'' \end{pmatrix}.$$

This quantity is zero as long as $T+t+t'+t'' > t+t'+t''$. Thus, we can set all coefficients that would raise the sum equation beyond $t+t'+t''-1$ to zero. In other words, the three polynomial multipliers must have the following forms:

First equation	...	$(x\ldots 1)^{t'+t''-1}$,
Second equation	...	$(x\ldots 1)^{t+t''-1}$,
Third equation	...	$(x\ldots 1)^{t+t'-1}$

if they are not to contain any superfluous terms.

(448.) This form, however, can be further reduced. To do so, I assume it can be reduced to

$$(x\ldots 1)^{t'+t''-q},$$
$$(x\ldots 1)^{t+t''-q},$$
$$(x\ldots 1)^{t+t'-q}.$$

Then the difference between the number of terms in the highest dimension in the sum equation and the number of useful coefficients in the highest dimension of each polynomial multiplier is not

$$d^3 N(x\ldots 0)^{t+t'+t''-q}\ldots \begin{pmatrix} t+t'+t''-q \\ t,t',t'' \end{pmatrix}$$

anymore. To determine what it is, I rewrite

$$d^3 N(x\ldots 0)^{t+t'+t''-q} \ldots \begin{pmatrix} t+t'+t''-q \\ t,t',t'' \end{pmatrix}$$

as the equivalent quantity

$$d^3 N(x\ldots 0)^{t+t'+t''-q} \ldots \begin{pmatrix} t+t'+t''-q \\ t,t',t'' \end{pmatrix}$$
$$= ddN(x\ldots 0)^{t+t'+t''-q} \ldots \begin{pmatrix} t+t'+t''-q \\ t',t'' \end{pmatrix}$$
$$-dN(x\ldots 0)^{t'+t''-q} \ldots \begin{pmatrix} t'+t''-q \\ t'' \end{pmatrix} + N(x\ldots 0)^{t''-q} - N(x\ldots 0)^{-q}.$$

I notice that (1) the expression $N(x\ldots 0)^{-q}$ does not make sense because of its negative exponent $-q$. Thus, the true expression for the number of terms sought is

$$ddN(x\ldots 0)^{t+t'+t''-q} \ldots \begin{pmatrix} t+t'+t''-q \\ t',t'' \end{pmatrix}$$
$$-dN(x\ldots 0)^{t'+t''-q} \ldots \begin{pmatrix} t'+t''-q \\ t'' \end{pmatrix} + N(x\ldots 0)^{t''-q},$$

at least as long as q is not greater than t''.

From $q = 1$ to $q = t''$, the first two terms in this expression are zero each, and the last term $N(x\ldots 0)^{t''-q}$ is always 1.

Hence, in each dimension of the sum equation from $t+t'+t''-1$ to $t+t'$, the number of terms in each dimension exceeds the number of corresponding useful coefficients in the polynomial multipliers by 1. Thus, from (325), the form of the polynomial multipliers can be lowered by the quantity t''. The form of the polynomial multipliers is therefore allowed to be

First equation	...	$(x\ldots 1)^{t'-1}$,
Second equation	...	$(x\ldots 1)^{t-1}$,
Third equation	...	$(x\ldots 1)^{t+t'-t''-1}$,

with t'' arbitrary coefficients or equations available.

(449.) I now want to determine whether this form can still be reduced. I write

First equation	...	$(x\ldots 1)^{t'-q'}$,
Second equation	...	$(x\ldots 1)^{t-q'}$,
Third equation	...	$(x\ldots 1)^{t+t'-t''-q'}$;

that is, I introduce q' such that $q = t'' + q'$ in the above expressions.

Then the expression for the difference between the number of terms in the highest dimension of the sum equation and the number of useful coefficients in the highest dimension of the polynomial multipliers reduces to

$$ddN(x\ldots 0)^{t+t'-q'} \ldots \begin{pmatrix} t+t'-q' \\ t',t'' \end{pmatrix} - dN(x\ldots 0)^{t'-q'} \ldots \begin{pmatrix} t'-q' \\ t'' \end{pmatrix},$$

whose two terms are zero as long as $t' - q'$ is strictly greater than t''. Thus the number of terms in the highest dimension of the sum equation and the number of useful coefficients provided by the highest dimension of each polynomial multiplier is the same. We can therefore assume each of these coefficients, from $q' = 0$ to $q' = t' - t''$, to be zero.

Thus, the polynomial multipliers can be reduced to the following form

First equation	...	$(x \ldots 1)^{t''-1}$,
Second equation	...	$(x \ldots 1)^{t-t'+t''-1}$,
Third equation	...	$(x \ldots 1)^{t-1}$.

(450.) In order to determine whether this form can be further reduced, I assume the form

First equation	...	$(x \ldots 1)^{t''-q''}$,
Second equation	...	$(x \ldots 1)^{t-t'+t''-q''}$,
Third equation	...	$(x \ldots 1)^{t-q''}$;

that is, I introduce q'' such that $q' = t' - t'' + q''$ in the previous form.

Then the difference between the number of terms in the highest dimension of the sum equation and the number of useful coefficients in the highest dimension of each polynomial multiplier becomes

$$ddN(x \ldots 0)^{t+t''-q''} \ldots \left(\begin{array}{c} t+t''-q'' \\ t',t'' \end{array} \right) - dN(x \ldots 0)^{t''-q''} \ldots \left(\begin{array}{c} t''-q'' \\ t'' \end{array} \right),$$

that is,

$$ddN(x \ldots 0)^{t+t''-q''} \ldots \left(\begin{array}{c} t+t''-q'' \\ t',t'' \end{array} \right) - N(x \ldots 0)^{t''-q''} + N(x \ldots 0)^{-q''}.$$

Because of its negative exponent $-q''$, the expression $N(x \ldots 0)^{-q''}$ cannot occur and the expression becomes

$$ddN(x \ldots 0)^{t+t''-q''} \ldots \left(\begin{array}{c} t+t''-q'' \\ t',t'' \end{array} \right) - N(x \ldots 0)^{t''-q''}.$$

(451.) Two cases may arise: Either $t' + t'' < t$ or $t' + t'' > t$. Let us examine the first case. The first term in the expression

$$ddN(x \ldots 0)^{t+t''-q''} \ldots \left(\begin{array}{c} t+t''-q'' \\ t',t'' \end{array} \right) - N(x \ldots 0)^{t''-q''}$$

is zero as long as $t + t'' - q''$ is greater than $t' + t''$, that is, until $q'' = t - t'$. Thus, if $t' + t'' < t$ or $t - t' > t''$, the expression

$$ddN(x \ldots 0)^{t+t''-q''} \ldots \left(\begin{array}{c} t+t''-q'' \\ t',t'' \end{array} \right) - N(x \ldots 0)^{t''-q''}$$

will be negative and equal to -1, from $q'' = 1$ to $q'' = t''$: If no arbitrary equations were available, the form of the polynomial multipliers could not be lowered since the number of terms in the highest dimension of the sum equation is smaller than the number of useful coefficients provided by the highest dimension of each polynomial multiplier.

But we have seen above that we have one arbitrary equation or coefficient available from $q = 1$ to $q = t''$. If we use them from $q'' = 1$ to $q'' = t''$, each coefficient in the corresponding dimensions of the polynomial multipliers can be assumed to be zero. Thus, when $t' + t'' < t$, the form of the polynomial multipliers can be reduced to

First equation   ... $(x \ldots 1)^{-1}$,
Second equation   ... $(x \ldots 1)^{t-t'-1}$,
Third equation   ... $(x \ldots 1)^{t-t''-1}$.

Since the form $(x \ldots 1)^{-1}$ is that of a ficitious polynomial multiplier, the simplest final equation will result only from a combination of the second and third equations, without the first.

(452.) We now conclude about the simplest polynomial multipliers for the second and third equations.

Assume the polynomial multipliers are of the form

Second equation   ... $(x \ldots 1)^{t-t'-q'''}$,
Third equation   ... $(x \ldots 1)^{t-t''-q'''}$;

that is, let us introduce q''' such that $q'' = t'' + q'''$.

Then the difference between the number of terms in the highest dimension of the sum equation and the number of useful coefficients in the highest dimension of each polynomial multiplier is $ddN(x \ldots 0)^{t-q'''} \ldots \left(\begin{array}{c} t - q''' \\ t', t'' \end{array} \right)$, which is zero as long as $t - q'''$ is greater than $t' + t''$, in other words, from $q''' = 1$ to $q''' = t - t' - t''$. Hence, we can still set to zero each of the coefficients in the highest dimensions of the polynomial multipliers from $q''' = 1$ to $q''' = t - t' - t''$: Thus, the form of the polynomial multipliers can be reduced to

Second equation   ... $(x \ldots 1)^{t''-1}$,
Third equation   ... $(x \ldots 1)^{t'-1}$.

This is the simplest expression. Indeed, if we introduce q^{IV} such that $q''' = t - t' - t'' + q^{IV}$, the quantity $ddN(x \ldots 0)^{t-q'''} \ldots \left(\begin{array}{c} t - q''' \\ t', t'' \end{array} \right)$ becomes

$$ddN(x \ldots 0)^{t-q'''} \ldots \left(\begin{array}{c} t - q''' \\ t', t'' \end{array} \right)$$
$$= dN(x \ldots 0)^{t'+t''-q^{IV}} \ldots \left(\begin{array}{c} t' + t'' - q^{IV} \\ t'' \end{array} \right) - N(x \ldots 0)^{t''-q^{IV}} + N(x \ldots 0)^{-q^{IV}}.$$

Because of the negative exponent $-q^{IV}$, this expression reduces to

$$dN(x \ldots 0)^{t'+t''-q^{IV}} \ldots \left(\begin{array}{c} t' + t'' - q^{IV} \\ t'' \end{array} \right) - N(x \ldots 0)^{t''-q^{IV}}.$$

Since $dN(x\ldots0)^{t'+t''-q^{\mathrm{IV}}}\left(\dfrac{t'+t''-q^{\mathrm{IV}}}{t''}\right)=0$, the expression reduces to $-N(x\ldots0)^{t''-q^{\mathrm{IV}}}=-1$. Thus, the number of terms in the highest dimension of the sum equation is smaller than the number of useful coefficients, and the coefficients in the higher dimensions of the polynomial multipliers cannot be assumed to be zero, unless there remain some arbitrary equations. But none remain.

Let us note that this last form is in perfect agreement with that was said in (346).

Thus, when $t'+t''<t$, combining the three proposed equations gives the simplest final equation or condition equation in the form of a combination of the second and third equations only.

(453.) However, two condition equations must exist. Thus, still considering the case $t>t'+t''$, we must determine the form of the polynomial multipliers that will provide this second equation.

Let us start our examination again, from the form

$$(x\ldots1)^{t''-q''},$$
$$(x\ldots1)^{t-t'+t''-q''},$$
$$(x\ldots1)^{t-q''}.$$

Instead of using all available arbitrary equations, assume we set one of them aside. Then we can still assume the coefficients of the polynomial multipliers to be zero, from $q''=1$ to $q''=t''-1$. The form of the polynomial multipliers then reduces to

First equation	...	$(x\ldots1)^0,$
Second equation	...	$(x\ldots1)^{t-t'},$
Third equation	...	$(x\ldots1)^{t-t''}.$

There remain $t-t'-t''+1$ arbitrary equations beyond the arbitrary equation we have set aside. Since, by hypothesis, we do not use this equation in the highest dimension of the sum equation, we will have more useful coefficients than terms to cancel in this dimension; thus, the dimension cannot be lowered any further.

(454.) Let us now examine whether the resulting final equation is simpler than that obtained by combining the first and third equations.

The three polynomial multipliers provide $1+t-t'+1+t-t''+1$ coefficients.

Among these, $t-t'-t''+2$ are arbitrary. Assuming they are each set to zero, the elimination will therefore be performed using $t+1$ coefficients. Thus, the number of known coefficients entering each term of the final equation is $t+1$.

If we were to combine the first and third equations, the appropriate polynomial multipliers (346) would be $(x\ldots1)^{t''-1}$ and $(x\ldots1)^{t-1}$, such that the number of known coefficients entering each term of the final equation would be $t + t''$. Therefore, when $t > t' + t''$, the following multipliers:

$$(x \ldots 1)^0,$$
$$(x \ldots 1)^{t-t'},$$
$$(x \ldots 1)^{t-t''}$$

are the ones leading to the simplest final equation after the final equation resulting from the combination of the second and third equations.

(455.) Consider now the case $t < t' + t''$.

We begin with the form from (450):

$$(x \ldots 1)^{t''-q''},$$
$$(x \ldots 1)^{t-t'+t''-q''},$$
$$(x \ldots 1)^{t-q''}.$$

Recall the difference between the number of terms in the highest dimension of the sum equation and the number of useful coefficients in the highest dimension of each polynomial multiplier is

$$ddN(x \ldots 0)^{t+t''-q''} \ldots \begin{pmatrix} t+t''-q'' \\ t', t'' \end{pmatrix} - N(x \ldots 0)^{t''-q''}.$$

However, $ddN(x \ldots 0)^{t+t''-q''} \ldots \begin{pmatrix} t+t''-q'' \\ t', t'' \end{pmatrix}$ can no longer be zero from $q'' = 1$ to $q'' = t''$, when $t < t' + t''$. It can be zero only from $q'' = 1$ to $q'' = t - t'$. Throughout this interval, we also have $-N(x \ldots 0)^{t''-q''} = -1$.

Thus, assuming we use one of the t'' available arbitrary equations for each dimension of the sum equation from $q'' = 1$ to $q'' = t - t'$, the difference between the number of terms in the highest dimension of the sum equation and the number of useful coefficients in the highest dimensions of each polynomial multiplier is zero from $q'' = 1$ to $q'' = t - t'$. Thus we can assume each coefficient of the polynomial multipliers to be zero from $q'' = 1$ to $q'' = t - t'$.

The form of the polynomial multipliers is then

First equation	...	$(x \ldots 1)^{t'+t''-t-1},$
Second equation	...	$(x \ldots 1)^{t''-1},$
Third equation	...	$(x \ldots 1)^{t'-1}.$

(456.) To determine whether this form can be further reduced I write it as

First equation	...	$(x \ldots 1)^{t'+t''-t-q'''},$
Second equation	...	$(x \ldots 1)^{t''-q'''},$
Third equation	...	$(x \ldots 1)^{t'-q'''}.$

The expression of the difference between the number of terms in the highest dimension of the sum equation and the number of useful coefficients in the highest dimension of the polynomial multipliers becomes

$$ddN(x \ldots 0)^{t'+t''-q'''} \ldots \begin{pmatrix} t'+t''-q''' \\ t', t'' \end{pmatrix} - N(x \ldots 0)^{t'+t''-t-q'''},$$

that is,

$$dN(x\ldots0)^{t'+t''-q'''}\ldots\left(\begin{array}{c}t+t''-q'''\\t''\end{array}\right)-N(x\ldots0)^{t''-q'''}$$
$$+N(x\ldots0)^{-q'''}-N(x\ldots0)^{t'+t''-t-q'''}.$$

Since $N(x\ldots0)^{-q'''}$ cannot occur, this expression reduces to

$$dN(x\ldots0)^{t'+t''-q'''}\ldots\left(\begin{array}{c}t+t''-q'''\\t''\end{array}\right)-N(x\ldots0)^{t''-q'''}-N(x\ldots0)^{t'+t''-t-q'''},$$

that is, $0-1-1=-2$.

Thus, the number of useful coefficients in the q'''th dimension of the polynomial multipliers exceeds the number of terms to be cancelled in the q'''th dimension of the sum equation. It would, therefore, not have been possible to lower the form of the polynomial multipliers if we had not had a number of arbitrary equations available.

Among the t'' available arbitrary equations, we have already used $t-t'$ of them. Thus $t'+t''-t$ of them remain. We must use two of these for each dimension, and we will therefore be able to lower the form of the polynomial multipliers from $q''=1$ to $q'''=\frac{t'+t''-t-\alpha}{2}$, where α is 0 or 1 depending upon whether $t'+t''-t$ is even or odd. When this quantity is odd, one arbitrary equation remains available.

Hence, when $t<t'+t''$, the simplest form for the polynomial multipliers is as follows

First equation   ... $(x\ldots1)^{\frac{t'+t''-t+\alpha}{2}-1}$,

Second equation   ... $(x\ldots1)^{\frac{t+t''-t'+\alpha}{2}-1}$,

Third equation   ... $(x\ldots1)^{\frac{t+t'-t''+\alpha}{2}-1}$,

where α is 0 or 1, depending upon whether $t'+t''-t$ is even or odd. In the latter case one arbitrary equation remains.

(457.) The final equation found by using these polynomial multipliers is always the simplest. It is simpler than that given by the combination of any two of the three proposed equations.

Indeed, if we were to combine the two lowest equations, the resulting literal degree of the final equation would be $t'+t''$. Using the polynomial multipliers we have just identified, the literal degree of the final equation is $\frac{t+t'+t''+\alpha}{2}<t'+t''$, since $t<t'+t''$ and α cannot exceeed 1.

(458.) We now turn our attention to the second condition equation. We consider the following form for the three polynomial multipliers:

First equation   ... $(x\ldots1)^{\frac{t'+t''-t+\alpha}{2}}$,

Second equation   ... $(x\ldots1)^{\frac{t+t''-t'+\alpha}{2}}$,

Third equation   ... $(x\ldots1)^{\frac{t+t'-t''+\alpha}{2}}$.

Again, α is zero or 1 depending upon whether $t'+t''-t$ is even or odd. Three arbitrary equations will remain in the second case, and two will remain in the first. Indeed, we can use the $t' + t'' - t$ available arbitrary equations (456) as desired in the sum equation. Thus, we can use two of these per dimension from $q''' = 1$ to $q''' = \frac{t'+t''-t-\alpha}{2} - 1$.

(459.) The new form of the polynomial multipliers always gives a simpler final equation than that obtained from the combination of any two of the three proposed equations, except when t is greater than $t' + t'' - \alpha - 6$. In that case, the second condition equation can be chosen from the combination of the second and third equations.

(460.) Obtaining the simplest final equation after the one resulting from the first form does not always require us to use the second form we just gave for the three polynomial multipliers. When $t' + t'' - t$ is odd, the first form provides the two condition equations. Indeed, since one arbitrary equation must be formed, two final equations can be obtained from two different forms of that arbitrary equation. We now apply this to specific examples.

(461.) Consider the following three equations:

$$\begin{aligned} ax + b &= 0, \\ a'x + b' &= 0, \\ a''x + b'' &= 0. \end{aligned}$$

In this case, $t' + t'' - t$ is odd, so $\alpha = 1$ and the form of the polynomial multipliers is $(x \ldots 1)^0$. Hence, each equation must be multiplied by a single undetermined coefficient. Since in (456) there is one arbitrary coefficient or equation, we set this coefficient to zero to get the simplest possible equation. Then the combination of any two equations leads to the simplest equation.

Denoting by A, A', A'' the respective multipliers of these equations, we obtain the final equation

$$(ab') = 0$$

by setting $A'' = 0$. We get the final equation

$$(ab'') = 0$$

by setting $A' = 0$, and we get the final equation

$$(a'b'') = 0$$

by setting $A = 0$. When two of these equations hold, the third necessarily holds too.

(462.) Assume the three proposed equations are

$$\begin{aligned} ax^2 + bx + c &= 0, \\ a'x^2 + b'x + c' &= 0, \\ a''x^2 + b''x + c'' &= 0. \end{aligned}$$

This time $t' + t'' - t$ is even and we have $\alpha = 0$: the three simplest polynomial multipliers have the form $(x \ldots 1)^0$; that is, they are A, A', A'' with no arbitrary coefficient or equation.

The sum equation is therefore of the form

$$Aax^2 + Abx + Ac = 0.$$

We can therefore compute $AA'A''$ as follows:

> First line   ... $aA'A''$,
> Second line   ... $ab'A''$,
> Third line   ... $(ab'c'')$.

The final equation is therefore $(ab'c'') = 0$ or

$$(ab' - a'b)c'' - (ab'' - a''b)c' + (a'b'' - a''b')c = 0.$$

This is the simplest equation that may be formed.

To get the second condition equation, we can refer to (458) and use the polynomial multiplier form $(x \ldots 1)^1$, to obtain two arbitrary coefficients, to be set to zero. Assuming we take both coefficients in the same polynomial multiplier, we only have to combine two equations. Indeed, we have seen in (346) that two equations with that degree must have polynomial multipliers of the form $(x \ldots 1)^1$, with no arbitrary coefficient. Many other choices can be made for the arbitrary coefficients, but they will not lead to anything simpler.

Thus, the two final equations are $(ab'c'') = 0$ combined with any one of the following three equations:

$$
\begin{aligned}
(ab').(bc') - (ac')^2 &= 0, \\
(ab'').(bc'') - (ac'')^2 &= 0, \\
(a'b'').(b'c'') - (a'c'')^2 &= 0.
\end{aligned}
$$

If the first equation $(ab'c'') = 0$ is satisfied along with any one of the three remaining equations, then the other two are automatically satisfied.

(463.) For our third example, we assume the following three equations:

$$
\begin{aligned}
ax^3 + bx^2 + cx + d &= 0, \\
a'x^3 + b'x^2 + c'x + d' &= 0, \\
a''x^3 + b''x^2 + c''x + d'' &= 0.
\end{aligned}
$$

In this case $t + t'' - t$ is odd. Consequently $\alpha = 1$ and the form of the three polynomial multipliers (456) is $(x \ldots 1)^1$ with one arbitrary coefficient or equation.

Let $Ax + B$, $A'x + B'$, $A''x + B''$ be the three polynomial multipliers; then the sum equation has the following form:

$$
\begin{array}{cccccccc}
Aa & x^4 & + & Ab & x^3 & + & Ac & x^2 & + & Ad & x & +Bd = 0, \\
 & & + & Ba & & + & Bb & & + & Bc & &
\end{array}
$$

I first compute $AA'A''BB'B''$ without accounting for the arbitrary equation and I get as follows:

First line   ... $aA'A''$,

Second line ... $(ab')A''BB'B'' + aA'A''aB'B''$,

Third line   ... $(ab'c'')BB'B'' - (ab')A''bB'B''$
                          $+(ac')A''aB'B'' + aA'A''(ab')B''$,

Fourth line ... $(ab'c'')cB'B'' - (ab'd'')bB'B''$
                          $+(ab')A''(bc')B'' + (ac'd'')aB'B''$
                          $-(ac')A''(ac')B'' + (ad')A''(ab')B'' + aA'A''(ab'c'')$,

Fifth line   ... $(ab'c'').(cd')B'' - (ab'd'').(bd')B''$
                         $-(bc'd'').(ab')A'' + (ac'd'').(ad')B''$
                         $+(ac'd'').(ac')A'' - (ab'd'').(ad')A''$.

At this point, we can use the arbitrary equation to write

$Ab + A'b' + A''b'' = 0$, or $Ac + A'c' + A''c'' = 0$, or $Ad + A'd' + A''d'' = 0$, or $Ba + B'a' + B''a'' = 0$, or $Bb + B'b' + B''b'' = 0$, or $Bc + B'c' + B''c'' = 0$.

We can then compute a sixth line according to any one of these arbitrary equations, leading us to the first final equation. Computing another sixth line using another one of these arbitrary equations, I get a second final equation.

For example, using the arbitrary equations $Ab + A'b' + A''b'' = 0$ and $Ac + A'c' + A''c'' = 0$ successively, I obtain the following two final equations:

$$-(ac'd'').(ab'c'') + (ab'd'')^2 = 0,$$
$$-(bc'd'').(ab'c'') + (ab'd'').(ac'd'') = 0.$$

These two equations, as simple as they may seem, are not the simplest possible ones. Indeed, the arbitrary coefficient could have been assumed to be zero or determined by any other arbitrary equation. If the arbitrary coefficient is zero, then the literal dimension of the final equation must be smaller by one unit. Thus the literal dimension of the two equations is too high by one unit, although none of them accepts any divider.

Thus, instead of proceeding with the computation of the sixth line using any one of the arbitrary equations given above, I use one of the following arbitrary equations:

$$A'' = 0, \ A' = 0, \ A = 0, \ B'' = 0, \ B' = 0, \ B = 0.$$

I first remark that, for example, $(ac')A''$ is shorthand for

$$(ac')A'' - (ac'')A' + (a'c'')A.$$

Together with the equation $A'' = 0$ or $A + 0A' + 1A'' = 0$, it yields $(ac')A'' = (ac')$.

We can use this observation to combine the fifth line found above with $A'' = 0$ and $A' = 0$, successively. We then get the following two final equations:

$$-(bc'd'').(ab') + (ac'd'').(ac') - (ab'd'').(ad') = 0$$
$$\text{and} \ \ +(bc'd'').(ab'') - (ac'd'').(ac'') + (ab'd'').(ad'') = 0.$$

Many other equations can be formed, but they will not be simpler. Given any two of these equations, the others will always be redundant.

(464.) To determine the relation of the two equations we have found to the two preceding ones, we compute the equation leading to $A = 0$, which is

$$-(bc'd'').(a'b'') + (ac'd'').(a'c'') - (ab'd'').(a'd'') = 0.$$

Consider the last three equations; multiply the first by b'', the second by b' and the third by b. Subtract the second product from the sum of the two others, to get

$$-(ac'd'').(ab'c'') + (ab'd'')^2 = 0.$$

Likewise, multiply the first equation by c'', the second by c' and the third by c, and subtract the second product from the sum of the other two to get

$$-(bc'd'').(ab'c'') + (ab'd'').(ac'd'') = 0.$$

Thus the two equations

$$-(ac'd'').(ab'c'') + (ab'd'')^2 = 0$$
$$-(bc'd'').(ab'c'') + (ab'd'').(ac'd'') = 0$$

do not express anything more than the three equations

$$-(bc'd'').(ab') + (ac'd'').(ac') - (ab'd'').(ad') = 0,$$
$$+(bc'd'').(ab'') - (ac'd'').(ac'') + (ab'd'').(ad'') = 0,$$
$$-(bc'd'').(a'b'') + (ac'd'').(a'c'') - (ab'd'').(a'd'') = 0.$$

(465.) Assume we now have two unknowns, and therefore four equations whose degrees are t, t', t'', t''' for the first, second, third and fourth equations, respectively. With no loss of generality, we assume $t > t' > t'' > t'''$.

Any one of the following five general cases can arise:

$$
\begin{array}{llll}
t' > t'' + t''', & t > t' + t''', & t > t' + t'', & t > t' + t'' + t''', \\
t' > t'' + t''', & t > t' + t''', & t > t' + t'', & t < t' + t'' + t''', \\
t' > t'' + t''', & t > t' + t''', & t < t' + t'', & t < t' + t'' + t''', \\
t' > t'' + t''', & t < t' + t''', & t < t' + t'', & t < t' + t'' + t''', \\
t' < t'' + t''', & t < t' + t''', & t < t' + t'', & t < t' + t'' + t'''.
\end{array}
$$

The last four cases subdivide into many others, whose details would require too much effort. Thus we will only examine the first case in detail and consider only one of the subcases for the fifth case.

Consider the first case first.

(466.) Assume that we choose polynomial multipliers of the form

$$(x\ldots 2)^{T+t'+t''+t'''},\ (x\ldots 2)^{T+t+t''+t'''},\ (x\ldots 2)^{T+t+t'+t'''},\ (x\ldots 2)^{T+t+t'+t''}$$

according to what was said in (224). The difference between the number of terms in the highest dimension of the sum equation and the number of useful coefficients in the highest dimension of the polynomial multipliers is

$$d^4 N(x\ldots 1)^{T+t+t'+t''+t'''} \ldots \binom{T+t+t'+t''+t'''}{t,t',t'',t'''}.$$

However, this expression is zero as long as the exponent of each polynomial expression it contains is no less than -1. Since the smallest of these polynomials is $(x \ldots 1)^T$ with $T+1$ terms, we easily see that all coefficients of the highest dimension of the polynomial multipliers can be set to zero from any arbitrary $T > 0$ until $T = -1$. Then

$$d^4 N (x \ldots 1)^{T+t+t'+t''+t'''} \ldots \binom{T+t+t'+t''+t'''}{t,t',t'',t'''} = 0.$$

The polynomial multipliers therefore reduce to

First equation	...	$(x \ldots 1)^{t'+t''+t'''-1}$,
Second equation	...	$(x \ldots 1)^{t+t''+t'''-1}$,
Third equation	...	$(x \ldots 1)^{t+t'+t''-1}$,
Fourth equation	...	$(x \ldots 1)^{t+t'+t''-1}$.

I now determine whether this form can be further reduced by writing it as

First equation	...	$(x \ldots 1)^{t'+t''+t'''-1-q}$,
Second equation	...	$(x \ldots 1)^{t+t''+t'''-1-q}$,
Third equation	...	$(x \ldots 1)^{t+t'+t''-1-q}$,
Fourth equation	...	$(x \ldots 1)^{t+t'+t''-1-q}$.

Then the difference between the number of terms in the highest dimension of the sum equation and the number of useful coefficients in the highest dimension of the polynomial multipliers becomes

$$d^4 N (x \ldots 1)^{t+t'+t''+t'''-1-q} \ldots \binom{t+t'+t''+t'''-1-q}{t,t',t'',t'''}.$$

Since polynomials with negative exponent must be rejected from this expression, I rewrite it as

$$d^3 N (x \ldots 1)^{t+t'+t''+t'''-1-q} \ldots \binom{t+t'+t''+t'''-1-q}{t',t'',t'''}$$
$$-dd N (x \ldots 1)^{t'+t''+t'''-1-q} \ldots \binom{t'+t''+t'''-1-q}{t'',t'''}$$
$$+d N (x \ldots 1)^{t''+t'''-1-q} \ldots \binom{t''+t'''-1-q}{t'''} - N (x \ldots 1)^{t'''-1-q}$$

after rejecting the term $N(x \ldots 1)^{-1-q}$.

The first two terms are obviously zero as long as $q < t'$. The third term reduces to t''' as long as q remains smaller than t''. The fourth term reduces to $t''' - q$ as long as q is smaller than t'''. So the total expression reduces to q.

Since the number of useful coefficients is smaller than the number of terms to be cancelled (see (325)), we can set each coefficient, from $q = 1$ to $q = t'''$, to zero. For each value of q in this interval, we will have q arbitrary equations available.

(467.) Let us introduce q' such that $q = t''' + q'$. The difference between the number of terms to be cancelled and the number of useful coefficients becomes

$$d^3 N(x \ldots 1)^{t+t'+t''-1-q'} \ldots \begin{pmatrix} t+t'+t''-1-q' \\ t', t'', t''' \end{pmatrix}$$

$$-ddN(x \ldots 1)^{t'+t''-1-q'} \ldots \begin{pmatrix} t'+t''-1-q' \\ t'', t''' \end{pmatrix}$$

$$+dN(x \ldots 1)^{t''-1-q'} \ldots \begin{pmatrix} t''-1-q' \\ t''' \end{pmatrix}.$$

The first two terms are zero as long as $q' < t' - t'''$; the third term is positive and equals t''' as long as $q' < t'' - t'''$. Thus, since the number of useful coefficients is smaller than the number of terms to be cancelled from $q' = 1$ to $q' = t'' - t'''$, we can set each coefficient to zero from $q' = 1$ to $q' = t'' - t'''$. Each value of q' in this inteval also generates t''' available arbitrary equations.

(468.) Let us introduce q'' such that $q' = t'' - t''' + q''$. The expression for the difference between the number of terms to be cancelled and the number of useful coefficients is

$$d^3 N(x \ldots 1)^{t+t'+t'''-1-q''} \ldots \begin{pmatrix} t+t'+t'''-1-q'' \\ t', t'', t''' \end{pmatrix}$$

$$-ddN(x \ldots 1)^{t'+t'''-1-q''} \ldots \begin{pmatrix} t'+t'''-1-q'' \\ t'', t''' \end{pmatrix}$$

$$+N(x \ldots 1)^{t'''-1-q''}.$$

The first two terms are each zero as long as $q'' < t' - t''$. Since $t' > t'' + t'''$, this expression reduces to $N(x \ldots 1)^{t'''-1-q''}$ or $t''' - q''$ from $q'' = 1$ to $q'' = t'''$. We can assume again that all coefficients are zero from $q'' = 1$ to $q'' = t'''$. For each value of q'' in this interval, $t''' - q''$ arbitrary equations become available.

(469.) Let us now introduce q''' such that $q'' = t''' + q'''$. The expression of the difference between the number of terms to cancel and the number of useful coefficients is

$$d^3 N(x \ldots 1)^{t+t'-1-q'''} \ldots \begin{pmatrix} t+t'-1-q''' \\ t', t'', t''' \end{pmatrix}$$

$$-ddN(x \ldots 1)^{t'-1-q'''} \ldots \begin{pmatrix} t'-1-q''' \\ t'', t''' \end{pmatrix}.$$

This quantity is zero as long as $t' - q''' > t'' + t'''$ or $q''' < t' - t'' - t'''$; thus, we can assume all coefficients to be zero from $q''' = 1$ to $q''' = t' - t'' - t'''$.

(470.) Let us introduce q^{IV} such that $q''' = t' - t'' - t''' + q^{\mathrm{IV}}$. The expression of the difference between the number of terms to be cancelled and the number of useful coefficients is

$$d^3 N(x \ldots 1)^{t+t''+t'''-1-q^{\mathrm{IV}}} \ldots \begin{pmatrix} t+t''+t'''-1-q^{\mathrm{IV}} \\ t', t'', t''' \end{pmatrix}$$

$$-ddN(x \ldots 1)^{t''+t'''-1-q^{\mathrm{IV}}} \ldots \begin{pmatrix} t''+t'''-1-q^{\mathrm{IV}} \\ t'', t''' \end{pmatrix},$$

that is,

$$d^3 N(x \ldots 1)^{t+t''+t'''-1-q^{IV}} \ldots \begin{pmatrix} t+t''+t'''-1-q^{IV} \\ t',t'',t''' \end{pmatrix}$$

$$-dN(x \ldots 1)^{t''+t'''-1-q^{IV}} \ldots \begin{pmatrix} t''+t'''-1-q^{IV} \\ t''' \end{pmatrix}$$

$$+N(x \ldots 1)^{t'''-1-q^{IV}}$$

once the term $N(x \ldots 1)^{-1-q^{IV}}$ has been eliminated.

Since we assume $t > t'+t''$, the first term is zero; the second term is $-t'''$; the third term, $t''' - q^{IV}$, can exist only until $q^{IV} = t'''$. Thus, this difference reduces to $-q^{IV}$.

If no arbitrary equations were available, the number of useful coefficients would exceed the number of terms to be cancelled and these coefficients could not be set to zero. However, from $q = 1$ to $q = t'''$, we have q arbitrary equations available. Since q and q^{IV} take the same values, we can use these arbitrary equations from $q^{IV} = 1$ to $q^{IV} = t'''$ and assume that the coefficients are zero within this interval.

(471.) Let us now introduce q^V such that $q^{IV} = t''' + q^V$. The expression of the difference between the number of terms to cancel and the number of useful coefficients is

$$d^3 N(x \ldots 1)^{t+t''-1-q^V} \ldots \begin{pmatrix} t+t''-1-q^V \\ t',t'',t''' \end{pmatrix}$$

$$-dN(x \ldots 1)^{t''-1-q^V} \ldots \begin{pmatrix} t''-1-q^V \\ t''' \end{pmatrix}.$$

Since we assume $t > t'+t''$, the first term is zero as long as $q^V < t''-t'''$; the second term is t'''. Thus, with no available arbitrary equations, we could not assume any of the coefficients were zero. However, since we have t''' available arbitrary equations in each dimension from $q' = 1$ to $q' = t''-t'''$, and since q' and q^V take the same values within the same interval, we can set all coefficients to zero, from $q^V = 1$ to $q^V = t''-t'''$.

(472.) Let us now introduce q^{VI} such that $q^V = t''-t'''+q^{VI}$. The difference between the number of terms to cancel and the number of useful coefficients is

$$d^3 N(x \ldots 1)^{t+t'''-1-q^{VI}} \ldots \begin{pmatrix} t+t'''-1-q^{VI} \\ t',t'',t''' \end{pmatrix}$$

$$-dN(x \ldots 1)^{t'''-1-q^{VI}} \ldots \begin{pmatrix} t'''-1-q^{VI} \\ t''' \end{pmatrix};$$

that is,

$$d^3 N(x \ldots 1)^{t+t'''-1-q^{VI}} \ldots \begin{pmatrix} t+t'''-1-q^{VI} \\ t',t'',t''' \end{pmatrix} - N(x \ldots 1)^{t'''-1-q^{VI}}.$$

Since $t > t'+t''+t'''$, the first term is zero as long as q^V is no greater than t'''; the second term is $-t'''+q^{VI}$; that is, it is negative. It would not be possible to set any coefficient to zero if there were no arbitrary equations left. However, there are $t''' - q''$ available arbitrary equations from $q'' = 1$ to $q'' = t'''$, and q'' and q^{VI} have the same values over the same intervals. Hence, we can use these arbitrary equations from $q^{VI} = 1$ to $q^{VI} = t'''$. Consequently, all coefficients in this interval are zero.

(473.) Let us now introduce q^{VII} such that $q^{VI} = t''' + q^{VII}$. The difference between the number of terms to cancel and the number of useful coefficients is

$$d^3 N(x\ldots 1)^{t-1-q^{VII}} \cdots \left(\begin{array}{c} t - 1 - q^{VII} \\ t', t'', t''' \end{array} \right).$$

This quantity is zero as long as $t - q^{VII} > t' + t'' + t'''$ or $q^{VII} < t - t' - t'' - t'''$; Thus we can still assume that all the coefficients in the polynomial multipliers from $q^{VII} = 1$ to $q^{VII} = t - t' - t'' - t'''$ are zero.

This is the end of the reduction process when $t' > t'' + t'''$, $t > t' + t'''$, $t > t' + t''$, and $t > t' + t'' + t'''$. Indeed, the expression

$$d^3 N(x\ldots 1)^{t-1-q^{VII}} \cdots \left(\begin{array}{c} t - 1 - q^{VII} \\ t', t'', t''' \end{array} \right)$$

can only take negative values (after eliminating the term

$$N(x\ldots 1)^{t-t'-t''-t'''-1-q^{VII}}$$

when $q^{VII} > t - t' - t'' - t'''$). Thus, we have more useful coefficients than terms to be cancelled in the highest dimension. Since no arbitrary equations remain available, we cannot assume any of the coefficients are zero.

(474.) Let us now examine the final form of the polynomial multipliers.

Since our statements hold until $q^{VII} = t - t' - t'' - t'''$, it follows that setting $q^{VII} = t - t' - t'' - t''' + 1$ leads to the last reducible form, and this form is the simplest possible.

However, the form of the sum equation before the last reduction is $(x\ldots 2)^{t-1-q^{VII}}$. Thus, it becomes $(x\ldots 2)^{t'+t''+t'''-2}$ after this last reduction. The form of the polynomial multipliers is therefore

First equation	...	$(x\ldots 2)^{t'+t''+t'''-t-2}$,
Second equation	...	$(x\ldots 2)^{t''+t'''-2}$,
Third equation	...	$(x\ldots 2)^{t'+t'''-2}$,
Fourth equation	...	$(x\ldots 2)^{t'+t''-2}$.

But the polynomial $(x\ldots 2)^{t'+t''+t'''-t-2}$ has zero terms. Hence, for the case when $t' > t'' + t'''$, $t > t' + t'''$, $t > t' + t''$, $t > t' + t'' + t'''$, the simplest condition equation is obtained by combining the last three equations without involving the first one.

The second simplest condition equation can be obtained by combining the last two equations with the first.

(475.) We now examine the case $t' < t'' + t'''$, $t < t' + t'''$, $t < t' + t''$, $t < t' + t'' + t'''$.

What we have said about the first case still holds until $q'' = t' - t''$. From $q'' = 1$ to $q'' = t' - t''$, $t''' - q''$ arbitrary equations are made available. Thus, it is possible to cancel all dimensions of the polynomial multipliers within this interval.

(476.) Let us introduce q''' such that $q'' = t' - t'' + q'''$. The expression of the difference between the number of terms to cancel and the number of useful coefficients is

$$d^3 N(x \ldots 1)^{t+t''+t'''-1-q'''} \ldots \binom{t+t''+t'''-1-q'''}{t', t'', t'''}$$
$$-dd N(x \ldots 1)^{t''+t'''-1-q'''} \ldots \binom{t''+t'''-1-q'''}{t'', t'''}$$
$$+N(x \ldots 1)^{t''+t'''-t'-1-q'''},$$

which reduces to

$$d^3 N(x \ldots 1)^{t+t''+t'''-1-q'''} \ldots \binom{t+t''+t'''-1-q'''}{t', t'', t'''}$$
$$-d N(x \ldots 1)^{t''+t'''-1-q'''} \ldots \binom{t''+t'''-1-q'''}{t'''}$$
$$+N(x \ldots 1)^{t'''-1-q'''} + N(x \ldots 1)^{t''+t'''-t'-1-q'''}.$$

Since $t < t' + t''$, this expression holds only until $q''' = t - t'$, and its value is $-t''' + t''' - q''' + t'' + t''' - t' - q'''$ or $t'' + t''' - t' - 2q'''$.

Two cases arise at this point: Either $t'+t''+t'''-2t > 0$ or $t'+t''+t'''-2t < 0$. We will only look at the first case. We can cancel all the dimensions of the polynomial multipliers within the interval from $q''' = 1$ to $q''' = t - t'$.

(477.) Introduce q^{IV} such that $q''' = t - t' + q^{IV}$. The difference between the number of terms to be cancelled and the number of useful coefficients becomes

$$d^3 N(x \ldots 1)^{t'+t''+t'''-1-q^{IV}} \ldots \binom{t'+t''+t'''-1-q^{IV}}{t', t'', t'''}$$
$$-d N(x \ldots 1)^{t'+t''+t'''-t-1-q^{IV}} \ldots \binom{t'+t''+t'''-t-1-q^{IV}}{t'''}$$
$$+N(x \ldots 1)^{t'+t'''-t-1-q^{IV}} + N(x \ldots 1)^{t''+t'''-t-1-q^{IV}}.$$

This expression reduces to

$$dd N(x \ldots 1)^{t'+t''+t'''-1-q^{IV}} \ldots \binom{t'+t''+t'''-1-q^{IV}}{t'', t'''}$$
$$-d N(x \ldots 1)^{t''+t'''-1-q^{IV}} \ldots \binom{t''+t'''-1-q^{IV}}{t'''}$$
$$-d N(x \ldots 1)^{t'+t''+t'''-t-1-q^{IV}} \ldots \binom{t'+t''+t'''-t-1-q^{IV}}{t'''}$$
$$+N(x \ldots 1)^{t'''-1-q^{IV}} + N(x \ldots 1)^{t'+t'''-t-1-q^{IV}}$$
$$+N(x \ldots 1)^{t''+t'''-t-1-q^{IV}}.$$

This expression holds from $q^{IV} = 1$ to $q^{IV} = t'' + t''' - t$ and takes the value $-2t''' + t''' - q^{IV} + t' + t''' - t - q^{IV} + t'' + t''' - t - q^{IV}$ or $t' + t'' + t''' - 2t - 3q^{IV}$.

This quantity is positive until some value of q^{IV} and becomes negative before $q^{IV} = t'' + t''' - t$.

Let us now add q^{IV} of the arbitrary equations that we have available from $q = 1$ to $q = t'''$ to this expression for each dimension from $q^{IV} = 1$ to a given rank. Then the expression becomes $t' + t'' + t''' - 2t - 2q^{IV}$, whose sum is $(t' + t'' + t''' - 2t)q^{IV} - q^{IV}(q^{IV} + 1)$. This sum becomes zero when $q^{IV} + 1 = t' + t'' + t''' - 2t$ or when $q^{IV} = t' + t'' + t''' - 2t - 1$. Thus, we can use this part of the arbitrary equations to suppress all the dimensions of the polynomial multipliers from $q^{IV} = 1$ to $q^{IV} = t' + t'' + t''' - 2t - 1$.

(478.) Let us introduce $\underset{\prime}{q}$ such that $q^{IV} = t' + t'' + t''' - 2t - 1 + q$. Then the expression $t' + t'' + t''' - 2t - 3q^{IV}$ becomes $-2(t' + t'' + t''' - 2t) + 3 - 3\underset{\prime}{q}$. This expression holds from $\underset{\prime}{q} = 1$ to $\underset{\prime}{q} = t'' + t''' - t - (t' + t'' + t''' - 2t - 1)$, that is, until $\underset{\prime}{q} = t - t' + 1$.

Introduce

1. $\underset{\prime\prime}{q}$ such that $q = t' + t'' + t''' - 2t - 1 + \underset{\prime\prime}{q}$, and let us use, for each dimension from $q = 1$, the q remaining arbitrary equations, from $q = t' + t'' + t''' - 2t$ or $\underset{\prime\prime}{q} = 1$ to $q = t'''$.

2. $\underset{\prime\prime\prime}{q}$ such that $q''' = t - t' + 1 - \underset{\prime\prime\prime}{q}$, and let us use, for each dimension from $\underset{\prime}{q} = 1$ to $t'' + t''' - t' - 2q'''$, the $t' + t'' + t''' - 2t - 2 + 2\underset{\prime\prime\prime}{q}$ arbitrary equations remaining from $q''' = 1$ to $q''' = t - t'$.

Then, for each dimension from $\underset{\prime}{q} = 1$ to $\underset{\prime}{q} = t - t'$, we have an excess of $2(t' + t'' + t''' - 2t) - 3 + 3\underset{\prime}{q}$ useful coefficients. There will also be $2(t' + t'' + t''' - 2t) - 3 - \underset{\prime\prime}{q} + 2\underset{\prime\prime\prime}{q} = 2(t' + t'' + t''' - 2t) - 3 + 3\underset{\prime}{q}$ available arbitrary equations.

Hence, we can cancel all dimensions of the polynomial multipliers from $\underset{\prime}{q} = 1$ to $\underset{\prime}{q} = t - t'$.

We have therefore exhausted the expression $t' + t'' + t''' - 2t - 3q^{IV}$ from $q^{IV} = 1$ to $q^{IV} = t'' + t''' - t - 1$. In the dimension $q^{IV} = t'' + t''' - t$, there remains an excess of $t' - 2t'' - 2t''' + t$ useful coefficients. This excess can be absorbed by the remaining available arbitrary equations.

(479.) At this point we are left with (1) q available arbitrary equations in each dimension from $q = t' + t'' + t''' - 2t - 1 + t - t' + 1$ to $q = t'''$, that is, from $q = t'' + t''' - t$ to $q = t'''$; (2) t''' arbitrary equations from $q' = 1$ to $q' = t'' - t'''$; (3) $t''' - q''$ arbitrary equations in each dimension from $q'' = 1$ to $q'' = t' - t''$.

(480.) To determine whether the degree of the polynomial multipliers can be lowered further, we introduce q^V such that $q^{IV} = t'' + t''' - t - q^V$. The difference between the number of terms to be cancelled and the number of useful coefficients can be expressed as

$$ddN(x\ldots 1)^{t+t'-1-q^{\mathrm{V}}}\ldots\binom{t+t'-1-q^{\mathrm{V}}}{t'',t'''}$$

$$-dN(x\ldots 1)^{t-1-q^{\mathrm{V}}}\ldots\binom{t-1-q^{\mathrm{V}}}{t'''}$$

$$-dN(x\ldots 1)^{t'-1-q^{\mathrm{V}}}\ldots\binom{t'-1-q^{\mathrm{V}}}{t''}$$

$$+N(x\ldots 1)^{t-t''-1-q^{\mathrm{V}}}+N(x\ldots 1)^{t'-t''-1-q^{\mathrm{V}}}.$$

This expression holds from $q^{\mathrm{V}}=1$ to $q^{\mathrm{V}}=t'-t''$, and it takes the value $-2t'''+t-t''-q^{\mathrm{IV}}+t'-t''-q^{\mathrm{IV}}$ or $-2t'''-2t''+t+t'-2q^{\mathrm{IV}}$.

We have seen there remains an excess of $-2t'''-2t''+t+t'$ useful coefficients from the previous reductions. This quantity also corresponds to the value of $-2t'''-2t''+t+t'-2q^{\mathrm{IV}}$ when $q^{\mathrm{IV}}=0$. Thus, for each dimension from $q^{\mathrm{V}}=0$ to $q^{\mathrm{V}}=t'-t''$, there is an excess of $-2t'''-2t''+t+t'-2q^{\mathrm{V}}$ useful coefficients.

This excess can be absorbed by noting that (1) there are q available arbitrary equations in each dimension from $q=t''+t'''-t$ to $q=t'''$. Introducing $\underset{\mathrm{IV}}{q}$ such that $q=t''+t'''-t+\underset{\mathrm{IV}}{q}$, we have $t''+t'''-t+\underset{\mathrm{IV}}{q}$ available arbitrary equations in every dimension from $\underset{\mathrm{IV}}{q}=0$ to $\underset{\mathrm{IV}}{q}=t-t'$, and consequently the same is true until $\underset{\mathrm{IV}}{q}=t'-t''$; (2) we have $t'''-q''$ available arbitrary equations in every dimension from $q''=1$ to $q''=t'-t''$. Introducing $\underset{\mathrm{V}}{q}$, such that $q''=t'-t''-\underset{\mathrm{V}}{q}$, the number of available arbitrary equations is $t'''+t''-t'+\underset{\mathrm{V}}{q}$. Summing the number of these arbitrary equations yields $2t'''+2t''-t'-t+\underset{\mathrm{IV}}{q}+\underset{\mathrm{V}}{q}=2t'''+2t''-t-t'+2q^{\mathrm{V}}$ arbitrary equations in each dimension from $q^{\mathrm{V}}=0$ to $q^{\mathrm{V}}=t'-t''$. This number is the same as the number of useful coefficients.

(481.) Thus, we can cancel all dimensions of the polynomial multipliers corresponding to $q^{\mathrm{V}}=1$ to $q^{\mathrm{V}}=t-t''$.

Introduce q^{VI} such that $q^{\mathrm{V}}=t'-t''+q^{\mathrm{VI}}$. The difference between the number of terms to be cancelled and the number of useful coefficients is

$$ddN(x\ldots 1)^{t+t''-1-q^{\mathrm{VI}}}\ldots\binom{t+t''-1-q^{\mathrm{VI}}}{t'',t'''}$$

$$-dN(x\ldots 1)^{t+t''-t'-1-q^{\mathrm{VI}}}\ldots\binom{t+t''-t'-1-q^{\mathrm{VI}}}{t'''}$$

$$-dN(x\ldots 1)^{t''-1-q^{\mathrm{VI}}}\ldots\binom{t''-1-q^{\mathrm{VI}}}{t'''}+N(x\ldots 1)^{t-t'-1-q^{\mathrm{VI}}}.$$

This expression holds from $q^{\mathrm{VI}}=t-t'$ and is equal to $-2t'''+t-t'-q^{\mathrm{VI}}$ or $-(2t'''+t'-t+q^{\mathrm{VI}})$.

However,

1. There remain q available arbitrary equations from $q = t' + t''' - t$ to $q = t'''$. Introducing $\underset{\text{VI}}{q}$ such that $q = t' + t''' - t - 1 + \underset{\text{VI}}{q}$, it also means there remain $t' + t''' - t - 1 + \underset{\text{VI}}{q}$ available arbitrary equations for each dimension from $\underset{\text{VI}}{q} = 1$ to $\underset{\text{VI}}{q} = t - t' + 1$.

2. There remain t''' arbitrary equations from $q' = 1$ to $q' = t'' - t'''$. Let us assume that $t'' - t''' > t - t'$.

Bringing together these two sets of arbitrary equations yields a total of $2t''' + t' - t - 1 + \underset{\text{VI}}{q} = 2t''' + t' - t - 1 + q^{\text{VI}}$ available arbitrary equations from $q^{\text{VI}} = 1$ to $q^{\text{VI}} = t - t'$. In other words, there is one less available arbitrary equation than there are useful coefficients. However, there also remain t''' available arbitrary equations coming from the set of equations available from $q = 1$, which we used from $q = 1$ to $q = t'''$.

Using $t - t'$ of these last equations, we then have as many arbitrary equations as there are useful coefficients, from $q^{\text{VI}} = 1$ to $q^{\text{VI}} = t - t'$. Thus we can set the coefficients of the polynomial multipliers from $q^{\text{VI}} = 1$ to $q^{\text{VI}} = t - t'$ to zero. At this point, we are left with (1) $t''' + t' - t$ arbitrary equations, (2) t''' arbitrary equations from $q' = t - t' + 1$ to $q' = t'' - t'''$.

(482.) Let us introduce q^{VII} such that $q^{\text{VI}} = t - t' + q^{\text{VII}}$. The difference between the number of terms to cancel and the number of useful coefficients becomes

$$ddN(x \ldots 1)^{t' + t'' - t - 1 - q^{\text{VII}}} \ldots \left(\begin{array}{c} t' + t'' - q^{\text{VII}} \\ t'', t''' \end{array} \right)$$

$$-dN(x \ldots 1)^{t'' - 1} \ldots \left(\begin{array}{c} t'' - 1 \\ t''' \end{array} \right)$$

$$-dN(x \ldots 1)^{t' + t'' - t - 1 - q^{\text{VII}}} \ldots \left(\begin{array}{c} t' + t'' - t - 1 - q^{\text{VII}} \\ t''' \end{array} \right)$$

This expression holds from $q^{\text{VII}} = 1$ to $q^{\text{VII}} = t' + t'' - t''' - t$, where it reduces to $-2t'''$.

However, we have just seen that t''' arbitrary equations remain from $q' = t - t' + 1$ to $q' = t'' - t'''$, that is, throughout $t'' + t' - t''' - t$ dimensions. Assuming we use $2t'''$ such equations from $q^{\text{VII}} = 1$, we can set to zero all coefficients of the polynomial multipliers from $q^{\text{VII}} = 1$ to $q^{\text{VII}} = \frac{t' + t'' - t''' - t}{2}$, or more exactly until $q^{\text{VII}} = \frac{t' + t'' - t''' - t - \alpha}{2}$, where α is zero or 1, depending upon whether $t' + t'' - t''' - t$ is even or odd. In the latter case, t''' arbitrary equations remain, beyond the $t''' + t' - t$ others that remain for both cases.

Since $t''' + t' - t$ and $2t''' + t' - t$ are both smaller than $2t'''$, it is not possible to lower the form of the polynomial multipliers beyond $q^{\text{VII}} = \frac{t' + t'' - t''' - t - \alpha}{2}$; thus, $q^{\text{VII}} = \frac{t' + t'' - t''' - t - \alpha}{2} + 1$ determines the simplest form of the multipliers.

The sum equation will therefore be of the form

$$(x \ldots 1)^{\frac{t' + t'' + t''' + t + \alpha}{2} - 1}.$$

Consequently, the form of the polynomial multipliers is as follows:

First equation $\qquad \ldots \quad (x \ldots 1)^{\frac{t'+t''+t'''-t+\alpha}{2}-1}$,

Second equation $\quad \ldots \quad (x \ldots 1)^{\frac{t''+t'''+t-t'+\alpha}{2}-1}$,

Third equation $\qquad \ldots \quad (x \ldots 1)^{\frac{t'+t'''+t-t''+\alpha}{2}-1}$,

Fourth equation $\quad \ldots \quad (x \ldots 1)^{\frac{t'+t''+t-t'''+\alpha}{2}-1}$.

In these expressions, α is zero or 1 depending upon whether $t'+t''-t'''-t$ is even or odd. $t'''+t'-t$ arbitrary equations remain to be formed in the first case and $2t'''+t'-t$ in the second case.

(483.) Such is the form of the polynomial multipliers when $t' < t'' + t'''$, $t < t'' + t'''$, $t < t' + t'''$, $t < t' + t''$, $t < t' + t'' + t'''$, $2t < t' + t'' + t'''$, and $t'' - t''' > t - t'$.

We will not give much attention to the other cases. What we have said illustrates what must be done to cover these cases well enough. A lot of care must be exercised to use the available arbitrary equations, but they can always be distributed such that, combined with the quantities q, q', q'', etc., they lead to the lowest possible dimension for the polynomial multipliers.

(484.) We now turn our attention to applications. First consider four equations of the form

$$ax + by + c = 0.$$

Therefore, we have $t = t' = t'' = t''' = 1$ and $t'' - t''' = t - t'$. These equations satisfy all other conditions of the case we just investigated. Moreover, $t' + t'' - t''' - t = 0$. Consequently, $\alpha = 0$.

The common form for the four polynomial multipliers is therefore $(x \ldots 2)^0$, with only one arbitrary equation.

We may put this equation to best use by setting one coefficient to zero. Doing this for two different coefficients can lead to two of the simplest condition equations by combining any three of the four proposed equations in two different ways.

Combining the first three equations to create the sum equation leads to the condition equation

$$(ab'c'') = 0.$$

Likewise, we can combine the first two equations with the fourth one, leading to the condition equation

$$(ab'c''') = 0.$$

Combining the first with the last two would lead to

$$(ab''c''') = 0.$$

Finally, the combination of the last three equations leads to

$$(ab'c'') = 0.$$

However, if the first two equations hold, then the last two are automatically satisfied.

(485.) Indeed, the first and second equations are shorthand for

$$(ab')c'' - (ab'')c' + (a'b'')c = 0,$$
$$(ab')c''' - (ab''')c' + (a'b''')c = 0.$$

Multiplying the first equation by $(a'b''')$ and the second by $(a'b'')$ and subtracting the second product from the first yields the following equation:

$$(ab').(a'b''')c'' - (ab').(a'b'')c''' - [(ab'').(a'b''') - (ab''').(a'b'')]c' = 0 \ldots \text{(A)}.$$

Likewise, multiplying the first equation by (ab''') and the second by (ab''), and subtracting the second from the first yields the following equation:

$$(ab').(ab''')c'' - (ab').(ab'')c''' + [(a'b'').(ab''') - (a'b''').(ab'')]c = 0 \ldots \text{(B)}.$$

From (220), we have

$$(ab'').(a'b''') - (ab''').(a'b'') - (ab').(a''b''') = 0.$$

The two equations (A) and (B) thus become

$$(ab').(a'b''')c'' - (ab').(a'b'')c''' - (ab').(a''b''')c' = 0,$$
$$(ab').(ab''')c'' - (ab').(ab'')c''' - (ab').(a''b''')c = 0.$$

These two equations can be divided by (ab') and become the following two equations

$$\begin{cases} -(a'b'')c''' + (a'b''')c'' - (a''b''')c' = 0, \\ -(ab'')c''' + (ab''')c'' - (a''b''')c = 0, \end{cases}$$

or

$$\begin{cases} -(a'b''c''') = 0 \\ -(ab''c''') = 0. \end{cases}$$

The second and first equations are precisely the third and fourth equations among the four introduced above.

Thus, if two of these equations hold, the other two hold as well.

Note that these kinds of verifications require the theorems pointed out in (220). It would be quite difficult to find one's own way in the required computations without these results, even for low-order equations.

(486.) Assume now that the four equations are of the form

$$ax^2 + bxy + cy^2$$
$$+dx + ey$$
$$+f.$$

We have $t = t' = t'' = t'''$, $t'' - t''' = t - t'$, and all the other conditions elicited in (475 ff) hold. Moreover, $t' + t'' - t''' - t = 0$; consequently $\alpha = 0$.

The common form of the polynomial multipliers is therefore $(x \ldots 2)^1$ and there are two arbitrary equations in the sum equation.

To get the simplest pair of condition equations, we will assume that two of the twelve resulting undetermined coefficients are equal to zero. In order not to lose any of the advantages brought forward by our method, we will introduce these equations only after computing the tenth line.

We assume we have multiplied each one of the proposed equations by a polynomial of the form $Ax + By + C$ to get a sum equation of the form

$$
\begin{aligned}
Aa \quad & x^3 & + \quad Ab \quad & x^2 y & + \quad Ac \quad & xy^2 & + \quad Bc \quad & y^3 = 0. \\
+ \quad & Ba & + \quad Bb & & & & & \\[6pt]
& & + \quad Ad \quad & x^2 & + \quad Ae \quad & xy & + \quad Be \quad & y^2 \\
& & + \quad Ca & & + \quad Bd & & + \quad Cc & \\
& & & & + \quad Cb & & & \\[6pt]
& & + \quad Af \quad & x & + \quad Bf \quad & y & & \\
& & + \quad Cd & & + \quad Ce & & & \\[6pt]
& & + \quad Cf & & & & &
\end{aligned}
$$

Two arbitrary equations remain. Since we can apply them to any two of the twelve undetermined coefficients, I propose to apply them to two of the four quantities B, B', B'', B'''. I first proceed by computing the lines, going through the terms x^3, $x^2 y$, xy^2, y^3, x^2, xy, y^2, x, y, and the term with neither x nor y, successively. As soon as I reach the eigth line, I can eliminate all terms in this line where A, A', A'', A''' remain, as well as all those where any combination of any three among C, C', C'', and C''' remains. When computing the ninth line, I eliminate the terms where any pairwise, arbitrary combination of C, C', C'', C''' remains. When computing the tenth line, I omit all terms containing any of the quantities C, C', C'', C'''. In fact, as soon as I compute the third line, I eliminate all terms where $AA'A''A'''$ remains. When computing the fifth line, I eliminate all terms where any combination of three of these letters appears; when computing the fifth line, I eliminate all terms where any pairwise combination of these letters appears.

However, when we group these coefficients together, only the terms $AA'A''A'''$, $A'A''A'''$, $A''A'''$, A''', $BB'B''B'''$, $B'B''B'''$, $B''B'''$, B''', $CC'C''C'''$, $C'C''C'''$, $C''C'''$, C''' appear successively. Thus, we eliminate these products successively when proceeding with the computation of the lines, whose orders we have indicated above.

Performing these eliminations excludes a very large number of terms as we proceed through computations and leads us to the tenth line given by

$$
\begin{aligned}
& -(ab'c''f''').(ab'e''f''')ce'B''B''' \\
& -(bc'e''f''').[(ab'd''f''')bc'B''B''' - (ac'd''f''')ac'B''B'''] \\
& +(ac'e''f''').[(ab'c''f''')cd'B''B''' + (ab'e''f''')bc'B''B''' - (ac'e''f''')ac'B''B'''] \\
& +(ab'c''f''')^2 cf'B''B''' - (ab'c''d''').(bd'e''f''')ce'B''B''' \\
& +(ab'c''e'').(ad'e''f''')cd'B''B''' \\
& -(cd'e''f''').[(ab'd''e''')bc'B''B''' - (ac'd''e''')ac'B''B''' + (ab'c''d''')cd'B''B'''] \\
& -(ab'c''d''').(bc'd''f''')cf'B''B''' + (ab'c''e'').(ac'd''f''')cf'B''B'''.
\end{aligned}
$$

At this point, recall that the quantity $ce'B''B'''$, for example, is nothing but shorthand for

$$
(ce')B''B''' - (ce'')B'B''' + (ce''')B'B'' + (c'e'')BB''' - (c'e''')BB'' + (c''e''')BB'
$$

in the computations. The same is true of $bc'B''B'''$, which is shorthand for

$$(bc')B''B''' - (bc'')B'B''' + (bc''')B'B'' + (b'c'')BB''' - (b'c''')BB'' + (b''c''')BB',$$

and so on for the other expressions.

Assume, for example, that $B'' = 0$ and $B''' = 0$, and use these two equations to compute the quantity

$$(ce')B''B''' - (ce'')B'B''' + (ce''')B'B'' + (c'e'')BB''' - (c'e''')BB'' + (c''e''')BB'.$$

Representing the two equations $B'' = 0$, $B''' = 0$ as

$$0B + 0B' + 1B'' + 0B''' = 0,$$
$$\text{and} \quad 0B + 0B' + 0B'' + 1B''' = 0,$$

it is easy to conclude that $(ce'B''B''')$, which is also

$$(ce')B''B''' - (ce'')B'B''' + (ce''')B'B'' + (c'e'')BB''' - (c'e''')BB'' + (c''e''')BB',$$

successively becomes

$$(ce')B''' - (ce'')B' - (c'e''')B,$$
$$\text{and} \quad (ce').$$

Thus, assuming $B'' = 0$ and $B''' = 0$ leads to one of the condition equations

$$\left.\begin{array}{l}
-(ab'c''f''').(ab'e''f''').(ce') \\
-(bc'e''f''').[(ab'd''f''').(bc') - (ac'd''f''').(ac')] \\
+(ac'e''f''').[(ab'c''f''').(cd') + (ab'e''f''').(bc') - (ac'e''f''').(ac')] \\
+(ab'c''f''')^2.(cf') - (ab'c''d''').(bd'e''f''').(ce') \\
+(ab'c''e''').(ad'e''f''').(ce') \\
-(cd'e''f''').[(ab'd''e''').(bc') - (ac'd''e''').(ac') + (ab'c''d''').(cd')] \\
-(ab'c''d''').(bc'd''f''').(cf') + (ab'c''e''').(ac'd''f''').(cf').
\end{array}\right\} = 0.$$

Since we can also assume $B = 0$ and $B' = 0$, we can also get the second condition equation

$$\left.\begin{array}{l}
-(ab'c''f''').(ab'e''f''').(c''e''') \\
-(bc'e''f''').[(ab'd''f''').(b''c''') - (ac'd''f''').(a''c''')] \\
+(ac'e''f''').[(ab'c''f''').(c''d''') + (ab'c''f''').(b''c''') - (ac'e''f''').(a''c''')] \\
+(ab'c''f''')^2.(c''f''') - (ab'c''d''').(bd'e''f''').(c''e''') \\
+(ab'c''e''').(ad'e''f''').(c''e''')] \\
-(cd'e''f''').[(ab'd''e''').(b''c''') - (ac'd''e''').(a''c''') + (ab'c''d''').(c''d''')] \\
-(ab'c''d''').(bc'd''f''').(c''f''') + (ab'c''e''').(ac'd''f''').(c''f''').
\end{array}\right\} = 0.$$

It is easy to see that many more such equations can be obtained. However, they are not independent from these two. However, we believe deriving these equations is useful, and we now consider this problem.

About a much broader use of the arbitrary coefficients and their usefulness
to reach the condition equations with lowest literal dimension

(487.) We can use the so-called *useless* coefficients for any purpose, except that forbidden in (230 ff); thus it follows that, when looking for the final equation resulting from any form of polynomial multipliers, the last line must be the final equation whether the useless coefficients are determined before, during, or after the computations of the lines or whether they are not determined at all or are used in the form of arbitrary equations.

Since the final equation must in no way depend on these useless coefficients, it follows that the final equation will yield as many final equations as there are possible combinations of these useless coefficients, whether these coefficients are taken alone or in combinations of 2, 3, or, etc.

Indeed, since the last *line* must be zero irrespective of the values of the useless coefficients, each term in front of any combination of these coefficients must be zero.

(488.) Consider the case when the number of unknowns is equal to the number of equations. Assume we have proceeded with the computation of the *last line*, without assigning any value to the useless coefficients. Assume we then perform one more line computation using the total undetermined coefficient for each term of the final equation as an additional equation. Finally assume we use this new line as the coefficient of the term of the final equation that yielded this line: The resulting final equation can be decomposed in as many final equations as there are different combinations of the remaining undetermined coefficients. Moreover, these final equations must occur simultaneously, and thus they will differ only by a factor specific to each equation.

It is easy to realize that the computations involved in the process are much longer and complex than when the useless coefficients are arbitrarily determined; at the same time, however, we see that these computations bring all the information about the proposed equations in a single and unique equation.

(489.) Consider now the case when the number of proposed equations exceeds the number of unknowns. Assume we compute the last line without determining any of the useless coefficients. Then from (487) above, we will obtain as many final equations, that is, as many condition equations linking the known coefficients, as there are different combinations among the arbitrary coefficients of the polynomial multipliers.

However, these equations will not be the same and they do not have to occur simultaneously. They will also be more or less complex combinations of one another. In general, if the number of equations is n and the number of unknowns is p, $n - p$ of these equations will be independent from one another. The others will either be the same as some of these equations, or a combination of them, or a combination of products involving them. In other words, they will necessarily hold if the $n - p$ first equations hold.

(490.) Before casting more light on all this using examples, let us add two important remarks.

When reaching the last *line* after having used all equations provided by the sum equation, we do not always have to use all terms generated by the general expression.

Indeed, assume, for example, that the sum equation contains an arbitrary number of undetermined coefficients A, B C, D, etc., and that only three are known to be useless. The last *line* will contain all possible combinations of any three among A, B, C, D, etc. However, by virtue of the reasoning performed in (487), we can set the coefficient of any one out of this combination to zero only if the undetermined coefficients have been identified as useless. Following what was said in (230 ff) we enjoy much freedom in that regard, but this freedom remains limited. For example, assume A, B, C, D, E, etc. belong to the first, or highest, dimension of the sum equation, the second dimension, the third, etc. If the first dimension in the sum equation were to give no arbitrary equation, then there would be no reason to set the coefficient of any combination where A enters to zero.

Indeed, since A cannot be determined by any arbitrary equation, it cannot be part of any arbitrary equation. Thus, if, after forming these arbitrary equations, we perform the computation of the *lines*, any combination containing A will eventually vanish and not be part of the final line.

(491.) If instead we do not form arbitrary equations, we will have to exclude from the last *line* any combination containing even one undetermined coefficient which does not qualify as useless; the only informative equations are those that are obtained by setting to zero the total coefficient of a combination of variables which can all be chosen as useless.

Lack of proper care of this issue could lead to irrelevant condition equations.

(492.) For example, let p be the known coefficient of any combination of the three letters or coefficients C, D, E; that is, let $pCDE$ be one of the terms in the last line; we assume it satisfies the conditions elicited above. The reason why we can set up the equation $p = 0$ is that C, D and E are part of the useless coefficients; that is, we can assume $C = 0$, $D = 0$ or $E = 0$. Thus we can form the three arbitrary equations

$$\begin{aligned} 1C + 0D + 0E &= 0, \\ 0C + 1D + 0E &= 0, \\ 0C + 0D + 1E &= 0. \end{aligned}$$

If we use these three arbitrary equations to further the computation of the last *line*, we will see that all other terms vanish and only the term $pCDE$ will remain until the end, becoming successively pDE, pE, and finally p, leading to $p = 0$.

Assume we take instead a term such as $qABC$, in which only B and C could be used as useless coefficients. Then the three arbitrary equations would be $B = 0$, $C = 0$, $D = 0$, or

$$
\begin{aligned}
1B + 0C + 0D &= 0, \\
0B + 1C + 0D &= 0, \\
0B + 0C + 1D &= 0.
\end{aligned}
$$

Continuing the computation of the *lines* would successively transform $qABC$ into the quantities $-qAC$, $+qA$ and 0. Thus, the term $qABC$ would end up disappearing; however, if r were the coefficient of BCD, r would be the last result of the computation of these lines, leading to $r = 0$ and not $q = 0$.

(493.) There are a few cases where the condition equation with the lowest possible dimension is unique; in other terms, there is only one equation whose literal dimension is lowest, and all the others have higher degree. We have already encountered such cases in (462). We have then already shown that we must use polynomial multipliers whose form is immediately higher in order to reach the other condition equations.

This case occurs when the form of the simplest polynomial multipliers does not have any useless coefficient. Then we can only reach one condition equation. However, if we use polynomial multipliers whose form is immediately higher and compute the lines without assigning any specific value to the useless coefficients, the process will yield several condition equations, among which the simplest equation will also be present.

(494.) We now consider again the equations covered in (462) to clarify and confirm all that has been said. These equations have the form

$$
ax^2 + bx + c = 0.
$$

We have already seen that the simplest solution consists of the equation $(ab'c'') = 0$ and any one of the following three equations:

$$
\begin{aligned}
(ab').(bc') - (ac')^2 &= 0, \\
(ab'').(bc'') - (ac'')^2 &= 0, \\
(a'b'').(b'c'') - (a'c'')^2 &= 0.
\end{aligned}
$$

However, if we had used the polynomial multipliers, whose form immediatley follows the simplest possible form, we would have looked for the condition equations by assigning values to the two arbitrary coefficients, say by cancelling one polynomial multiplier. We would then have found the three previous condition equations only; it would have been rather difficult to extract the equation $(ab'c'') = 0$, although this equation follows from the three others.

(495.) If instead we keep polynomial multipliers of the form $Ax + B$ without assigning any specific value to the useless coefficients, we get a sum equation of the form

$$
\begin{array}{rcrcrcr}
Aa & x^3 & + & Ab & x^2 & + & Ac & x & + & Bc = 0. \\
 & & + & Ba & & & + & Bb & &
\end{array}
$$

Computing the lines that must lead to the final equation, we get

First line   ... $aA'A''.BB'B''$

Second line ... $ab'A''.BB'B'' + aA'A''.aB'B''$

Third line ... $(ab'c'')BB'B'' - ab'A''.bB'B'' + ac'A''.aB'B''$
$+ aA'A''.ab'B''$

Fourth line ... $(ab'c'')cB'B'' - ab'A''.bc'B'' - (ac')A''.ac'B''$
$- aA'A''(ab'c'').$

Moreover, if we remember that (1) $cB'B''$ is only shorthand for $cB'B'' - c'BB'' + c''BB'$, (2) $ab'A''$ is shorthand for $(ab')A'' - (ab'')A' + (a'b'')A$, and so on for the others, the general final equation becomes

$$(ab'c'').(cB'B'' - c'BB'' + c''BB')$$
$$-[(ab')A'' - (ab'')A' + (a'b'')A][(bc')B'' - (bc'')B' + (b'c'')B]$$
$$-[(ac')A' - (ac'')A' + (a'c'')A][(ac')B'' - (ac'')B' + (a'c'')B]$$
$$-(ab'c'').(aA'A'' - a'AA'' + a''AA') = 0.$$

From (462), we know the two arbitrary equations can be formed in any dimension of the sum equation; thus all combinations of the coefficients A, A', A'', B, B', B'' satisfy the required properties stated in (491). Bringing together the coefficients of each combination $B'B''$, BB'', BB', $A'A''$, AA'', AA', $A''B''$, $A''B'$, $A'B''$, etc., we obtain the following ten distinct equations

$$(ab'c'') = 0,$$
$$(ab').(bc') - (ac')^2 = 0,$$
$$(ab'').(bc'') - (ac'')^2 = 0,$$
$$(a'b'').(b'c'') - (a'c'')^2 = 0,$$
$$(ab').(bc'') - (ac').(ac'') = 0,$$
$$(ab').(b'c'') - (ac').(a'c'') = 0,$$
$$(ab'').(bc') - (ac'').(ac') = 0,$$
$$(ab'').(b'c'') - (ac'').(a'c'') = 0,$$
$$(a'b'').(bc') - (a'c'').(ac') = 0,$$
$$(a'b'').(bc'') - (a'c'').(ac'') = 0.$$

When two of them occur, so do the other eight.

(496.) Our second example is the three equations of the form

$$ax^3 + bx^2 + cx + d = 0.$$

Using the same multiplier polynomials as in (463), the last line is

$$[(ab'c'').(cd') - (ab'd'').(bd') + (ac'd'').(ad')]B''$$
$$-[(bc'd'').(ab') - (ac'd'').(ac') + (ab'd'').(ad')]A'',$$

which is shorthand for

$$[(ab'c'').(cd') - (ab'd'').(bd') + (ac'd'').(ad')]B''$$
$$-[(ab'c'').(cd'') - (ab'd'').(bd'') + (ac'd'').(ad'')]B'$$
$$+[(ab'c'').(c'd'') - (ab'd'').(b'd'') + (ac'd'').(a'd'')]B$$
$$-[(bc'd'').(ab') - (ac'd'').(ac') + (ab'd'').(ad')]A''$$
$$+[(bc'd'').(ab'') - (ac'd'').(ac'') + (ab'd'').(ad'')]A'$$
$$-[(bc'd'').(a'b'') - (ac'd'').(a'c'') + (ab'd'').(a'd'')]A.$$

Since each coefficient A'', A', etc.; B'', B', etc. qualifies as a useless coefficient, we can extract the following six equations from the above line:

$$
\begin{aligned}
(ab'c'').(cd') - (ab'd'').(bd') + (ac'd'').(ad') &= 0, \\
(ab'c'').(cd'') - (ab'd'').(bd'') + (ac'd'').(ad'') &= 0, \\
(ab'c'').(c'd'') - (ab'd'').(b'd'') + (ac'd'').(a'd'') &= 0, \\
(bc'd'').(ab') - (ac'd'').(ac') + (ab'd'').(ad') &= 0, \\
(bc'd'').(ab'') - (ac'd'').(ac'') + (ab'd'').(ad'') &= 0, \\
(bc'd'').(a'b'') - (ac'd'').(a'c'') + (ab'd'').(a'd'') &= 0.
\end{aligned}
$$

If any two of these equations hold, then so do the other four.

(497.) Assuming $d = 0$, $d' = 0$, $d'' = 0$, the three equations become

$$ ax^3 + bx^2 + cx = 0, \ a'x^3 + b'x^2 + c'x = 0, \ a''x^3 + b''x^2 + c''x = 0, $$

that is,

$$
\begin{aligned}
ax^2 + bx + c &= 0, \\
a'x^2 + b'x + c' &= 0, \\
a''x^2 + b''x + c'' &= 0.
\end{aligned}
$$

Thus the six equations we have just found must yield the ten equations found in (495). This is what happens indeed. Instead of writing $d = 0$, $d' = 0$, $d'' = 0$ directly, assume these quantities are infinitesimally small. The six equations become

$$
\begin{aligned}
(ab'c'').(cd') &= 0, \\
(ab'c'').(cd'') &= 0, \\
(ab'c'').(c'd'') &= 0, \\
(bc'd'').(ab') - (ac'd'').(ac') &= 0, \\
(bc'd'').(ab'') - (ac'd'').(ac'') &= 0, \\
(bc'd'').(a'b'') - (ac'd'').(a'c'') &= 0
\end{aligned}
$$

that is,

$$
\begin{aligned}
(ab'c'') &= 0, \\
(bc'd'').(ab') - (ac'd'').(ac') &= 0, \\
(bc'd'').(ab'') - (ac'd'').(ac'') &= 0, \\
(bc'd'').(a'b'') - (ac'd'').(a'c'') &= 0.
\end{aligned}
$$

Each of the last three equations generates three new equations by setting any two quantities among d, d', d'' to zero and dividing the resulting expression by the remaining nonzero quantity. The ten resulting equations are those found in (495).

(498.) A larger number of equations with more unknowns would essentially yield more complex computations, without shedding more light on our problem. Thus these examples will not be treated here.

(499.) However, we must point out the essential role of the useless coefficients to obtain the final equation with the lowest possible literal dimension.

Indeed, assume instead that we limit ourselves to a few of the equations obtained by forming arbitrary equations; we would then be led, for example, to the three equations

$$
\begin{aligned}
(ab').(bc') - (ac')^2 &= 0, \\
(ab'').(bc'') - (ac'')^2 &= 0, \\
(a'b'').(b'c'') - (a'c'')^2 &= 0.
\end{aligned}
$$

Considering the case considered in (462), it would be rather difficult to find the simplest equation $(ab'c'') = 0$ that we know is relevant.

If we had, instead, formed the arbitrary equations leading to the three other equations

$$(ab').(bc') - (ac')^2 = 0,$$
$$(ab').(bc'') - (ac').(ac'') = 0,$$
$$(ab').(b'c'') - (ac').(a'c'') = 0,$$

we could have found the equation $(ab'c'') = 0$ more easily. Multiply the first, second and third equations by a'', a' and a, respectively; then add the first and third products and subtract the second product. We then get $(ab').(ab'c'') - (ac').(ac'c'') = 0$, that is, $(ab').(ab'c'') = 0$, since $(ac'c'') = 0$. The equation $(ab').(ab'c'') = 0$ gives $(ab') = 0$, which does not solve the problem, since it does not contain all parameters of the problem and $(ab'c'') = 0$, which is the equation sought.

(500.) Although it is easier to reach the equation $(ab'c'') = 0$ using the last three equations, the process requires the use of a specific trick, which does not seem to extend to a set of general rules.

(501.) If we proceed with the computation of the *lines* without using the arbitrary equations, then we automatically get all condition equations. These equations may have different levels of complexity. Some give the equation(s) with lowest letter dimension, as seen in (495), either immediately or multiplied by a factor; others will wrap around this or these equations, possibly multipled by several factors, in such a way that the simplest equation becomes very difficult to recognize.

In short, denote by E, E', E'' etc. the equations with lowest literal dimension; the current method will yield equations such as $aE = 0$, $a'E' = 0$, $a''E'' = 0$, etc.; equations such as $bE + b'E' = 0$, $bE + b''E'' = 0$ or $bE + b'E' + b''E'' = 0$. It is easy to see that E, E', E'', etc. are not explicit in these expressions. Thus, if we limited ourselves to only a few condition equations, we would look in vain for these equations.

On the other hand, if we do not neglect any condition equation, we are guaranteed to find equations with a factor, and removing this factor immediately yields one of the equations with lowest literal dimension.

Consider, for example, the expression

$$(ab'c'').(cB'B'' - c'BB'' + c''BB')$$
$$-[(ab')A'' - (ab'')A' + (a'b'')A][(bc')B'' - (bc'')B' + (b'c'')B]$$
$$-[(ac')A'' - (ac'')A' + (a'c'')A][(ac')B'' - (ac'')B' + (a'c'')B]$$
$$-(ab'c'').(aA'A'' - a'AA'' + a''AA') :$$

From this expression we find: (1) $(ab'c'')c = 0$, $(ab'c'')c' = 0$, $(ab'c'')c'' = 0$, $(ab'c'')a = 0$, $(ab'c'')a' = 0$, $(ab'c'')a'' = 0$. All these equations give the equation with lowest literal dimension alone, multiplied by one factor; (2) the other equations we have seen above, but whose equation with lowest dimension can be extracted only by combining several of these equations together. These combinations are somewhat ad hoc and vary with the number of equations. When we do not omit any of the condition equations, we are always guaranteed that some of them will be of the form $aE = 0$, $a'E' = 0$, etc. The general process to get E will therefore be to look for the greatest common divider of these equations.

Admittedly, there is no general method to recognize whether any of these equations is of the simple form $aE = 0$ or the more convoluted form $bE + b'E' = 0$, etc. We have to take these equations in pairs and check whether they have a common divider; at least, this method is guaranteed to reach equations with the lowest literal dimension. If we limit ourselves to only a few condition equations, these might only be of the form $bE + b'E' = 0$, $bE + b'E' + b''E'' = 0$, etc. Then it seems to me that reaching the final equations with lowest literal dimension becomes a very arduous task.

About systems of n equations in p unknowns, where p < n

(502.) When the number of unknowns p is less than the number n of equations, whether the system of equations admits a solution depends upon $n - p$ condition equations, relating the coefficients of the proposed equations.

We need to use polynomial multipliers with lowest possible dimension to obtain condition equations whose literal dimension is as low as possible. We propose to show how to determine the lowest dimension of the polynomial multipliers for all cases.

(503.) Denote by s the sum of the exponents of the degrees of each proposed equation and by t, t', t'', t''', etc. the exponents of the degrees of the first, second, third, fourth, etc. equations. The form of the polynomial multipliers is the following:

First equation	...	$(x \dots p)^{s-t}$
Second equation	...	$(x \dots p)^{s-t'}$
Third equation	...	$(x \dots p)^{s-t''}$
Fourth equation	...	$(x \dots p)^{s-t'''}$

and so on.

The form of the sum equation is $(x \dots p)^s$. The difference between the total number of terms in the sum equation and the number of useful coefficients is

$$d^n N(x \dots p)^s \dots \left(\begin{array}{c} s \\ t, t', t'', t''', \text{ etc.} \end{array} \right).$$

Since $p < n$, this quantity is necessarily zero.

Remembering that the last term in the quantity

$$d^n N(x \dots p)^{s-q} \dots \left(\begin{array}{c} s \\ t, t', t'', t''', \text{ etc.} \end{array} \right)$$

is $\pm N(x \dots p)^{-q}$ and that this term is zero until $q = p$, we see that we also have

$$d^n N(x \dots p)^{s-p} \dots \left(\begin{array}{c} s - p \\ t, t', t'', t''', \text{ etc.} \end{array} \right) = 0;$$

consequently the form of the polynomial multipliers can be reduced to the following

First equation   ... $(x \ldots p)^{s-t-p}$,
Second equation   ... $(x \ldots p)^{s-t'-p}$,
Third equation   ... $(x \ldots p)^{s-t''-p}$,
Fourth equation   ... $(x \ldots p)^{s-t'''-p}$,

Consequently the sum equation is of the form $(x \ldots p)^{s-p}$.

(504.) However, this is not, by far, the simplest form; let us lift our standpoint in order to better see the path towards this simplest form.

1. When the number of unknowns equals the number of equations, the difference between the number of terms in the sum equation and the total number of useful coefficients is not zero, but a function of the known exponents of the proposed equations: This should now be obvious to the reader.

2. When the number of equations exceeds the number of unknowns by one unit, the expression of the difference between the number of terms in the sum equation and the total number of useful coefficients in the polynomial multipliers is zero; however, this is an aggregate result only. Assume we compute the difference between the number of terms in that dimension and the number of useful coefficients in the corresponding dimensions of the polynomial multipliers for each dimension in the sum equation. The sum of these quantities is guaranteed to be zero only when s spans the whole range from its maximum value to zero. This can be seen from what was said from (338) to (440).

(505.) When the difference between the number of equations and the number of unknowns is greater than one, the difference between the number of terms in the sum equation and the number of useful coefficients in the polynomial multipliers is zero in several possible ways. To be exact, if this difference is computed as a partial sum for each dimension starting from the highest, it will go through zero. However, as long as this partial sum is positive, we can suppress all the dimensions which have yielded these results in the sum equation, and all corresponding dimensions in the polynomial multipliers. Thus, we need to determine the time when these partial sums become zero for the time before last, moving from a positive to a negative value for the last time. This leads us to the lowest possible dimension.

(506.) Although this dimension can always be easily found numerically, it is much more difficult to find an algebraic expression for it. This is true for two reasons: (1) There are many cases arising from the different relative values of the exponents, resulting in as many different expressions, this we have already seen. (2) The algebraic expression for each of the results referred to above changes almost all the time, and thus so do the partial sums.

(507.) We will therefore limit our investigation to numerical examples. We will show that the algorithm remains the same in any case.

(508.) Assume that the proposed equations are all of first order; recall that

$$d^n [N(x\ldots p)^{s-p}]\ldots \left(\begin{array}{c} s-p \\ t,t',t'',\ \text{etc.} \end{array} \right) = N(x\ldots p)^{s-p} - nN(x\ldots p)^{s-t-p}$$
$$+ \frac{n.(n-1)}{2} N(x\ldots p)^{s-2t-p} - n.\frac{n-1}{2}.\frac{n-2}{3} N(x\ldots p)^{s-3t-p} + \text{etc.}$$

In agreement with several of our previous observations, we must reject terms where the exponent of $N(x\ldots p)$ becomes negative in this expression.

This being said, the expression for the difference between the number of terms in the highest dimension and the number of useful coefficients in the corresponding dimensions in the polynomial multipliers is

$$N(x\ldots p-1)^{s-p} - nN(x\ldots p-1)^{s-t-p} + n.\frac{n-1}{2} N(x\ldots p-1)^{s-2t-p}$$
$$-n.\frac{n-1}{2}.\frac{n-2}{3} N(x\ldots p-1)^{s-3t-p} + \text{etc.}$$

The following examples will provide insight into what happens when s successively takes all possible positive values from $s = nt$ to $s = p$, rejecting as appropriate the terms where the exponent of $N(x\ldots p-1)$ becomes negative.

(509.) Consider first six first-order equations and one unknown only. The expression for the difference between the number of terms in each dimension of the sum equation and the number of useful coefficients in the polynomial multipliers is the following sequence of expressions:

$$\begin{aligned}
&N(x\ldots 0)^5 - 6N(x\ldots 0)^4 + 15N(x\ldots 0)^3 - 20N(x\ldots 0)^2 \\
&+15N(x\ldots 0)^1 - 6N(x\ldots 0)^0 && \ldots\ = -1, \\
&N(x\ldots 0)^4 - 6N(x\ldots 0)^3 + 15N(x\ldots 0)^2 - 20N(x\ldots 0)^1 \\
&+15N(x\ldots 0)^0 && \ldots\ = +5, \\
&N(x\ldots 0)^3 - 6N(x\ldots 0)^2 + 15N(x\ldots 0)^1 - 20N(x\ldots 0)^0 && \ldots\ = -10, \\
&N(x\ldots 0)^2 - 6N(x\ldots 0)^1 + 15N(x\ldots 0)^0 && \ldots\ = +10, \\
&N(x\ldots 0)^1 - 6N(x\ldots 0)^0 && \ldots\ = -5, \\
&N(x\ldots 0)^0 && \ldots\ = +1,
\end{aligned}$$

We see that

1. The sum of the first two results is 4; thus the first two dimensions of the sum equation and the polynomial multipliers can be eliminated. Four arbitrary equations remain to be formed in the remaining sum equation.

2. The sum of the first four results is 4; thus we can eliminate the first four dimensions of the original sum equation, and the first four dimensions of the polynomial multipliers; four arbitrary equations will have to be formed in the sum equation.

3. Since the sum of the results in the lower dimensions is neither positive nor zero, except in the last dimension, it is not possible to lower the dimension of the sum equation and the polynomial multipliers.

The sum equation with the lowest dimension therefore has the form $(x\dots1)^1 = 0$, the polynomial multipliers have the form $(x\dots1)^0$, and four arbitrary equations must be formed in the sum equation.

Denoting the polynomial multipliers as A, A', A'', A''', A^{IV}, A^V, the last line or final equation is

$$ab'A''A'''A^{IV}A^V,$$

which is shorthand for

$$(ab')A''A'''A^{IV}A^V - (ab'')A'A'''A^{IV}A^V + (ab''')A'A''A^{IV}A^V - (ab^{IV})A'A''A'''A^V$$
$$+(ab^V)A'A''A'''A^{IV} + (a'b'')AA'''A^{IV}A^V - (a'b''')AA''A^{IV}A^V + (a'b^{IV})AA''A'''A^V$$
$$-(a'b^V)AA''A'''A^{IV} + (a''b''')AA'A^{IV}A^V - (a''b^{IV})AA'A'''A^V + (a''b^V)AA'A'''A^{IV}$$
$$+(a'''b^{IV})AA'A''A^V - (a'''b^V)AA'A''A^{IV} - (a^{IV}b^V)AA'A''A'''..$$

Since four arbitrary equations must be formed in the sum equation and these equations can involve any four out of six undetermined coefficients, we will obtain the following fifteen equations according to what was said in (487):

$$(ab') = 0,\ (ab'') = 0,\ (ab''') = 0,\ (ab^{IV}) = 0,\ (ab^V) = 0,$$
$$(a'b'') = 0,\ (a'b''') = 0,\ (a'b^{IV}) = 0,\ (a'b^V) = 0,\ (a''b'') = 0,$$
$$(a''b^{IV}) = 0,\ (a''b^V) = 0,\ (a'''b^{IV}) = 0,\ (a'''b^V) = 0,\ (a^{IV}b^V) = 0.$$

These are indeed all the equations that may be obtained by combining the proposed equations pairwise; no other simple combination can be built.

If five of these fifteen equations are satisfied, so are the remaining ten.

(510.) For our second example, assume five second-order equations with three unknowns. The difference between the number of terms in each dimension of the sum equation and the number of useful coefficients in the corresponding dimensions can be written sequentially as follows:

$$N(x\dots2)^7 - 5N(x\dots2)^5 + 10N(x\dots2)^3 - 10N(x\dots2)^1 \dots\ = +1,$$
$$N(x\dots2)^6 - 5N(x\dots2)^4 + 10N(x\dots2)^2 - 10N(x\dots2)^0 \dots\ = +3,$$
$$N(x\dots2)^5 - 5N(x\dots2)^3 + 10N(x\dots2)^1 \dots\ = +1,$$
$$N(x\dots2)^4 - 5N(x\dots2)^2 + 10N(x\dots2)^0 \dots\ = -5,$$
$$N(x\dots2)^3 - 5N(x\dots2)^1 \dots\ = -5,$$
$$N(x\dots2)^2 - 5N(x\dots2)^0 \dots\ = +1,$$
$$N(x\dots2)^1 \dots\ = +3,$$
$$N(x\dots2)^0 \dots\ = +1.$$

Thus we can eliminate the first four dimensions of the polynomial multipliers since the total number of terms in the first four dimensions of the sum equation is precisely the same as the number of useful coefficients in the first four dimensions of the polynomial multipliers. However, past this term, the form of the numbers that follow: -5, $+1$, $+3$, $+1$ becomes zero only at the end. Thus the form of the polynomial multipliers cannot be further reduced, and it therefore reduces to $(x\dots3)^1$.

(511.) In our third example we consider six second-order equations in three unknowns. The difference between the number of terms in each dimension of the sum equation and the number of useful coefficients of the polynomial multipliers is as follows:

$$
\begin{aligned}
N(x\ldots 2)^9 - 6N(x\ldots 2)^7 + 15N(x\ldots 2)^5 - 20N(x\ldots 2)^3 + 15N(x\ldots 2)^1 &= -1 \\
N(x\ldots 2)^8 - 6N(x\ldots 2)^6 + 15N(x\ldots 2)^4 - 20N(x\ldots 2)^2 + 15N(x\ldots 2)^0 &= -3 \\
N(x\ldots 2)^7 - 6N(x\ldots 2)^5 + 15N(x\ldots 2)^3 - 20N(x\ldots 2)^1 \ldots &= 0 \\
N(x\ldots 2)^6 - 6N(x\ldots 2)^4 + 15N(x\ldots 2)^2 - 20N(x\ldots 2)^0 \ldots &= +8 \\
N(x\ldots 2)^5 - 6N(x\ldots 2)^3 + 15N(x\ldots 2)^1 \ldots &= +6 \\
N(x\ldots 2)^4 - 6N(x\ldots 2)^2 + 15N(x\ldots 2)^0 \ldots &= -6 \\
N(x\ldots 2)^3 - 6N(x\ldots 2)^1 \ldots &= -8 \\
N(x\ldots 2)^2 - 6N(x\ldots 2)^0 \ldots &= +0 \\
N(x\ldots 2)^1 &= +3 \\
N(x\ldots 2)^0 \ldots &= +1.
\end{aligned}
$$

From these computations we can first eliminate the first four dimensions of the polynomial multipliers; then the sum equation has the form $(x\ldots 3)^5$ and four arbitrary equations remain since the sum of -1, -3, 0, $+8$ is $+4$, a positive number.

Continuing to compute the partial sum of these numbers leads one to realize that two more dimensions can be rejected in the polynomial multipliers, since the corresponding sum of -1, -3, 0, $+8$, $+6$, -6 is $+4$. Four arbitrary equations remain to be formed in the sum equation. However, past this term the dimension of the polynomial multipliers cannot be reduced anymore. Indeed, if we keep adding the numerical values, the sum becomes zero again only when reaching the last dimension.

(512.) We will not extend these examples any further; they can be chosen so as to involve more equations in more unknowns and higher degrees. We observe that, although we have assumed all the equations share the same degree, what we said in (502 ff) still holds when these equations have different degrees. The process to find whether the polynomial multipliers can be reduced remains exactly the same. Nevertheless, we need to say a word about the different terms composing the expressions resulting from computing the difference between the number of terms in each dimension of the sum equation and the number of useful coefficients in the corresponding dimensions of the polynomial multipliers.

(513.) This general expression is always

$$
d^n N(x\ldots p)^{s-p} \ldots \left(\begin{array}{c} s-p \\ t, t', t'', t''', \text{ etc.} \end{array}\right)
$$

once all terms where $N(x\ldots p)$ has a negative exponent are rejected. We can expand $d^n N(x\ldots p)^{s-p} \ldots \left(\begin{array}{c} s-p \\ t, t', t'', t''', \text{ etc.} \end{array}\right)$ by observing that

1. the first term only contains $N(x\ldots p)^{s-p}$.

2. the second term bears the negative sign and includes the expression $N(x\ldots p)$, whose exponent is all possible exponents resulting from the sum of $n-1$ of the quantities t, t', t'', etc., minus p.

3. the third term bears the $+$ sign and includes the expression $N(x\ldots p)$, whose exponent is all possible exponents resulting from all sums of $n-2$ of the quantities t, t', t'', etc., minus p.

4. the fourth term bears the $-$ sign and includes the expression $N(x\ldots p)$, whose exponent is all possible exponents resulting from all sums of $n-3$ of the quantities t, t', t'', etc., minus p; and so on.

For example, expanding $d^3 N(x\ldots p)^{t+t'+t''-p}$ results in

$$
\begin{array}{llll}
N(x\ldots p)^{t+t'+t''-p} & -N(x\ldots p)^{t+t'-p} & +N(x\ldots p)^{t-p} & -N(x\ldots p)^{-p} \\
 & -N(x\ldots p)^{t+t''-p} & +N(x\ldots p)^{t'-p} & \\
 & -N(x\ldots p)^{t'+t''-p} & +N(x\ldots p)^{t''-p}.
\end{array}
$$

However, we would omit the term $N(x\ldots p)^{-p}$ when using this expression to solve the problems of interest here.

(514.) Although it is easy to see what to do when the proposed equations do not share the same degree, we will give an example to illustrate the detail of the expressions arising during the computations.

Consider four equations in two unknowns, whose forms are $(x\ldots 2)^4 = 0$, $(x\ldots 2)^3 = 0$, $(x\ldots 2)^2 = 0$, $(x\ldots 2)^1 = 0$.

The expression for the difference between the total number of terms in the sum equation and the number of useful coefficients in the four polynomial multipliers, is

$$
d^4 N(x\ldots 2)^{10-2} \left(\begin{array}{c} 10-2 \\ 4,3,2,1 \end{array}\right).
$$

The expression for the same difference, but only considering the highest dimension in the sum equation is

$$
d^4 N(x\ldots 1)^{10-2} \ldots \left(\begin{array}{c} 10-2 \\ 4,3,2,1 \end{array}\right).
$$

Assume that we expand this expression according to what we said in (513) and that we reject all terms where $N(x\ldots 1)$ has a negative exponent; assume we then compute the expressions corresponding to the other dimensions of the sum equation. We get

$$
\begin{array}{lllll}
N(x\ldots 1)^8 & -N(x\ldots 1)^7 & +N(x\ldots 1)^5 & -N(x\ldots 1)^2 & +N(x\ldots 1)^{-2} \\
 & -N(x\ldots 1)^6 & +N(x\ldots 1)^4 & -N(x\ldots 1)^1 & \\
 & -N(x\ldots 1)^5 & +N(x\ldots 1)^3 & -N(x\ldots 1)^0 & \\
 & -N(x\ldots 1)^4 & +N(x\ldots 1)^3 & -N(x\ldots 1)^{-1} & \\
 & & +N(x\ldots 1)^2 & & \\
 & & +N(x\ldots 1)^1. & &
\end{array}
$$

Rejecting the terms $N(x\ldots 1)^{-1}$ and $N(x\ldots 1)^{-2}$, we get after reduction

$$
N(x\ldots 1)^8 - N(x\ldots 1)^7 - N(x\ldots 1)^6 + 2N(x\ldots 1)^3 - N(x\ldots 1)^0 = +1.
$$

For the following dimensions, we get

$$
\begin{aligned}
N(x\ldots1)^7 - N(x\ldots1)^6 - N(x\ldots1)^5 + 2N(x\ldots1)^2\ldots &= +1\\
N(x\ldots1)^6 - N(x\ldots1)^5 - N(x\ldots1)^4 + 2N(x\ldots1)^1\ldots &= 0\\
N(x\ldots1)^5 - N(x\ldots1)^4 - N(x\ldots1)^3 + 2N(x\ldots1)^0\ldots &= -1\\
N(x\ldots1)^4 - N(x\ldots1)^3 - N(x\ldots1)^2\ldots &= -2\\
N(x\ldots1)^3 - N(x\ldots1)^2 - N(x\ldots1)^1\ldots &= -1\\
N(x\ldots1)^2 - N(x\ldots1)^1 - N(x\ldots1)^0\ldots &= 0\\
N(x\ldots1)^1 - N(x\ldots1)^0\ldots &= +1\\
N(x\ldots1)^0\ldots &= +1.
\end{aligned}
$$

If we compute the partial sums $+1$, $+1+1$, $+1+1+0$, $+1+1+0-1$ for the first four dimensions, we see that they are all positive. These sums begin to be negative only at the fifth dimension. Hence, we can eliminate the four highest dimensions in the sum equation. Since the latter sum $+1+1+0-1$ equals $+1$, one arbitrary equation remains to be formed in the sum equation. Moreover, this equation arises from the higher dimensions and can therefore apply to any desired coefficient.

By forming no equation as in (487 ff), we can extract all necessary condition equations and those resulting from these from the last *line*.

Since the dimension of the sum equation is reduced to 4, we see that it will have the form $(x\ldots2)^4 = 0$ and that consequently the polynomial multipliers of the equations are as follows:

For the equation $(x\ldots2)^4 = 0$ ... $(x\ldots2)^0$,
For the equation $(x\ldots2)^3 = 0$ ... $(x\ldots2)^1$,
For the equation $(x\ldots2)^2 = 0$ ... $(x\ldots2)^2$,
For the equation $(x\ldots2)^1 = 0$ ... $(x\ldots2)^3$.

(515.) We said that there remained one arbitrary equation to create in the sum equation. According to what we have already said, we must remind the reader that the number of arbitrary equations found following the above process is not the total number of arbitrary equations. To determine how many more such equations exist, we must observe that the number of useful coefficients of the polynomial multipliers for the equations

$$
\begin{aligned}
(x\ldots p)^t = 0 &\quad \text{is} \quad d^{n-1}N(x\ldots p)^{s-t-p}\ldots\binom{s-t-p}{t',t'',t''',\text{ etc.}},\\
(x\ldots p)^{t'} = 0 &\quad \text{is} \quad d^{n-2}N(x\ldots p)^{s-t'-p}\ldots\binom{s-t'-p}{t'',t''',\text{ etc.}},\\
(x\ldots p)^{t''} = 0 &\quad \text{is} \quad d^{n-3}N(x\ldots p)^{s-t''-p}\ldots\binom{s-t''-p}{t''',\text{ etc.}},
\end{aligned}
$$

and so on.

Thus the number of arbitrary coefficients or equations is

First Polynomial

$$N(x \ldots p)^{s-t-p} - d^{n-1}N(x \ldots p)^{s-t-p} \ldots \left(\begin{array}{c} s-t-p \\ t', t'', t''', \text{ etc.} \end{array} \right),$$

Second Polynomial

$$N(x \ldots p)^{s-t'-p} - d^{n-2}N(x \ldots p)^{s-t'-p} \ldots \left(\begin{array}{c} s-t'-p \\ t'', t''', \text{ etc.} \end{array} \right),$$

Third Polynomial

$$N(x \ldots p)^{s-t''-p} - d^{n-3}N(x \ldots p)^{s-t''-p} \ldots \left(\begin{array}{c} s-t''-p \\ t''', \text{ etc.} \end{array} \right),$$

and so on.

(516.) From (234 ff), it is also important to know how to distinguish the number of arbitrary coefficients or equations in each dimension. We have the following expressions:

First polynomial

$$N(x \ldots p - 1)^{s-t-p} - d^{n-1}N(x \ldots p - 1)^{s-t-p} \ldots \left(\begin{array}{c} s-t-p \\ t', t'', t''', \text{ etc.} \end{array} \right),$$

Second Polynomial

$$N(x \ldots p - 1)^{s-t'-p} - d^{n-2}N(x \ldots p - 1)^{s-t'-p} \ldots \left(\begin{array}{c} s-t'-p \\ t'', t''', \text{ etc.} \end{array} \right),$$

Third Polynomial

$$N(x \ldots p - 1)^{s-t''-p} - d^{n-3}N(x \ldots p - 1)^{s-t''-p} \ldots \left(\begin{array}{c} s-t''-p \\ t''', \text{ etc.} \end{array} \right),$$

and so on.

(517.) Proceeding this way for the previous example, we find four more arbitrary equations, beyond the one indicated by the numerical computations.

One equation comes from the polynomial multiplier of the equation $(x \ldots 2)^3 = 0$; the arbitrary equation can be formed in any desired dimension of the sum equation.

Three coefficients come from the polynomial multiplier of the equation $(x \ldots 2)^2 = 0$; two of them can be used to form arbitrary equations in any dimension of the sum equation. The third coefficient can lead to arbitrary equations in any dimension except the first.

When not all proposed equations are necessary to obtain the condition equation with lowest literal dimension

(518.) We can use the previous method to recognize the case when a subset of the proposed equations can yield a condition equation with lower literal dimension than when combining all equations together. Here is how.

The form of the polynomial multipliers can be determined after we have determined the lowest possible dimension of the sum equation. All the polynomial multipliers whose resulting form has negative exponent indicate that the corresponding equation can be removed when computing the final condition equations with lowest literal dimension.

For example, consider four equations of the form

$$(x \ldots 1)^2 = 0, \ (x \ldots 1)^1 = 0, \ (x \ldots 1)^1 = 0, \ (x \ldots 1)^1 = 0.$$

The degree of some of the condition equations must not exceed 2. However, the degree of condition equations resulting from combining the four proposed equations must be at least 4.

The current method is able to identify these facts. Indeed, the successive expressions for the difference between the number of terms in each dimension in the sum equation and the number of useful coefficients in the corresponding dimensions of the polynomial multipliers are as follows:

$$N(x \ldots 0)^4 - 3N(x \ldots 0)^3 + 2N(x \ldots 0)^2 + 2N(x \ldots 0)^1 - 3N(x \ldots 0)^0 \quad \ldots = -1,$$
$$N(x \ldots 0)^3 - 3N(x \ldots 0)^2 + 2N(x \ldots 0)^1 + 2N(x \ldots 0)^0 \qquad\qquad \ldots = +2,$$
$$N(x \ldots 0)^2 - 3N(x \ldots 0)^1 + 2N(x \ldots 0)^0 \qquad\qquad\qquad\quad \ldots = 0,$$
$$N(x \ldots 0)^1 - 3N(x \ldots 0)^0 \qquad\qquad\qquad\qquad\qquad\qquad \ldots = -2,$$
$$N(x \ldots 0)^0 \qquad\qquad\qquad\qquad\qquad\qquad\qquad\qquad\qquad\quad \ldots = +1.$$

The sum $-1 + 2 + 0$ of the first three results is positive, thus the sum equation can be reduced to the form $(x \ldots 1)^1 = 0$, with one remaining arbitrary equation. Extracting the form of the polynomial multipliers from this form yields the polynomial multiplier $(x \ldots 1)^{-1}$ for the equation $(x \ldots 1)^2 = 0$. This therefore indicates that this equation does not enter in the expression of the sum equation with lowest dimension.

(519.) However, we still have to look further than the equations with lowest possible dimension. In the current situation, for example, we could find three condition equations whose literal dimension is 2. However, these three condition equations are not sufficient to solve our problem completely, since they are not independent.

In order to find the other condition equations, we must rely on the multiplier form immediately beyond the one we have used so far, and we must increase the dimension of the polynomial multipliers up to the moment when the exponents of all polynomial multipliers are greater than or equal to zero. Then we will obtain the one or more condition equations with lowest possible degree, resulting from the combination of all proposed equations.

(520.) Moreover, if we proceed according to what was said in (487 ff), the latter form for the polynomial multipliers provides all condition equations and, consequently, it also provides those condition equations resulting from polynomial multipliers of lower degrees; however, these equations will arise with a factor. At any rate, we must use polynomial multipliers such that their exponents are all positive to obtain all condition equations that answer our problem; in some cases (see (493)), we will even have to use the form immediately above that one.

Let us return to the above example. The sum of the results is $-1 + 2 + 0$ and therefore a positive quantity; it is then possible to reduce the dimension of the sum equation down to 1. However, I must use a sum equation with the form $(x \ldots 1)^2$ if I want to obtain all condition equations. In fact, this sum

equation provides me with all the equations that are provided by the form
$(x \ldots 1)^1$. Although these equations appear more complicated than they are,
nothing prevents us from finding them.

*About the way to find, given a set of equations, whether some of them
necessarily follow from the others*

(521.) When we eliminate p unknowns from a set of n equations, the
question of whether these equations can all be simultaneously satisfied is
recast to whether $n - p$ condition equations involving the coefficients of
these equations are satisfied.

We have seen, however, that the number of equations is almost always
much larger than $n - p$; thus several of them are a necessary consequence
of the others. Choosing $n - p$ equations at random may not always bring a
complete solution to the problem: If q equations are implied by the $n-p-q$
others, the complete set of equations contains no more information than
these $n - p - q$ equations, and this is not sufficient.

(522.) In some cases, checking whether a given equation follows from the
others can be done by inspection. However, in the vast majority of cases,
this task is very complex. Likewise, there are a few cases where we can check
whether an equation follows from the others faster and easier than with the
method we are about to propose. However, for many more cases, it is very
hard to find and use tricks to speed up this task.

(523.) Consider, for example, the three equations $(ab') = 0$, $(ab'') = 0$,
$(a'b'') = 0$, the condition equations provided by the equations

$$ax + b = 0,$$
$$a'x + b' = 0,$$
$$a''x + b'' = 0.$$

These equations depend on two condition equations only; thus one of the
three above equations must follow from the other two. Indeed I can use
simple computations and theorems similar to the ones we have taught in (215
ff): If I multiply the first equation by a'', the second by a' and I subtract
the second product from the first, I get $(ab')a'' - (ab'')a' = 0$. From (219)
we have $(ab')a'' - (ab'')a' + (a'b'')a = 0$; thus $(a'b'')a = 0$, which implies
$(a'b'') = 0$. Thus the equation $(a'b'') = 0$ follows from the two equations
$(ab') = 0$ and $(ab'') = 0$.

(524.) What we just found very easily may not be so easy to find when
considering three equations

$$(ab').(bc') - (ac')^2 = 0,$$
$$(ab'').(bc'') - (ac'')^2 = 0,$$
$$(a'b'').(b'c'') - (a'c'')^2 = 0,$$

arising from the three equations

$$ax^2 + bx + c = 0,$$
$$a'x^2 + b'x + c' = 0,$$
$$a''x^2 + b''x + c'' = 0.$$

It is not easy to see what functions should multiply two of these equations to yield the third one, as the sum or difference of these two products; even if we knew it, we would still lack the theorem of the form given in (215) which would make it possible to find the third equation buried in the result of the computations.

(525.) Thus we will not rely on specific tricks that may apply to the equations under certain circumstances, to check whether among many proposed equations, one results from the others. Rather, we will develop a general methodology.

We first observe that an equation can be (1) the necessary consequence of another; (2) the necessary consequence of two others, without resulting from any one of these separately. For example, the equation $a'x + b' = 0$ follows necessarily from the two equations $ax + b = 0$, $(a + ma')x + b + mb' = 0$, although it follows from neither the equation $ax + b = 0$ nor the equation $(a + ma')x + b + mb' = 0$ taken alone; (3) the necessary consequence of three other equations, although it does not follow from any of these equations taken alone or pairwise, and so on.

(526.) We can use the following process to determine whether one equation follows from another.

Pick a letter common to the two equations and consider this letter as an unknown: Eliminate this unknown using the two equations, using the rules we have given until now. The result of the elimination is a trivial identity; that is, it becomes zero all by itself. Equivalently, when computing the *lines*, the last line will be zero.

For example, the equation $max^2 + mbx + mc = 0$ necessarily follows from the equation $ax^2 + bx + c = 0$, because if I eliminate x, I reach

$$(mab - mab).(mbc - mbc) - (mac - mac)^2 = 0,$$

that is, $0 = 0$.

(527.) We can use the following process to determine whether one equation follows from two others.

Choose two letters common to the three equations or such that if one of the letters enters into two equations, the other enters into another two equations. Looking at these letters as two unknowns, proceed with their elimination using the three equations. Computing the *lines* leads to zero as the value of the last line; that is, the last line will be zero by itself.

(528.) We can use the following process to determine whether an equation follows from three others.

Choose three letters common to the four equations or such that the value of any letter cannot be determined independently from the other two. Looking at these letters as three unknowns, proceed with their elimination using the four equations. If one of the equations follows from the other three, this elimination process will lead to zero.

(529.) Here is the general process to determine whether an equation follows from $n-1$ others; choose $n-1$ letters common to the n equations, or at least such that none can be determined independently from the $n-2$ others. Looking at these $n-1$ letters as $n-1$ unknowns, eliminate them using the n equations: The result of this elimination is identically zero if one of the equations results from the $n-1$ others.

We can use this process to determine, for example, whether any one of the three equations

$$(ab').(bc') - (ac')^2 = 0,$$
$$(ab'').(bc'') - (ac'')^2 = 0,$$
$$(a'b'').(b'c'') - (a'c'')^2 = 0$$

follows from the other two (which we already know is true). I begin by picking a and a' as the unknowns. Eliminating a and a' using the rules given previously and the three equations, I reach an equation where all terms vanish independent of the values of a, a', a''; b, b', b''; c, c', c''.

About equations that only partially follow from the others

(530.) The equations under consideration in (521 ff) are those which express nothing else than what the $n-1$ others already express; thus these $n-1$ equations carry as much information as the n equations do.

There are also equations such that the following happens: When $n-1$ of these equations are satisfied, the nth is also satisfied; however, the $n-1$ equations do not carry as much information as the n equations.

Consider for example the two equations

$$max^2 + mbx + nb = 0,$$
$$+ na$$

$$ma'x^2 + mb'x + nb' = 0;$$
$$+ na'$$

If we replace x by the quantity $-\frac{n}{m}$, we will see both equations are satisfied. Thus one of the equations follows necessarily from the other.

Indeed, if we eliminate x using these two equations, we find $m^2n^2(ab' - a'b)^2 - m^2n^2(ab' - a'b)^2 = 0$, that is $0 = 0$.

We would be mistaken, however, if we were led to conclude that one of the two proposed equations summarizes the entire problem.

Indeed, one of the equations satisfies the other only through one of its factors, $mx + n$.

If we solve the second equation for example, we will find it can be decomposed in two factors, $a'x + b'$ and $mx + n$. Solving the first equation, we also find it can be decomposed in two factors $ax + b$ and $mx + n$.

Thus the question can be seen as the two pairs of equations

$$mx + n = 0, \quad mx + n = 0,$$
$$\& \quad ax + b = 0, \quad a'x + b' = 0.$$

The first pair of equations would obviously belong to the type of equations studied in (521 ff), if it were alone. However, the second pair of equations leads to the condition equation $ab' - a'b = 0$, or $(ab') = 0$.

Thus, although eliminating x from the two equations yields a trivial identity, it does not mean one of the two proposed equations is enough to summarize the problem at hand.

(531.) Thus, we may conclude that a trivial identity does not necessarily mean that $n-1$ out of n equations hold the keys to the complete solution to the problem; and, that the proposed method in (521 ff) does not indicate whether one equation completely follows from the $n-1$ others; rather this method indicates whether this equation partially follows from the others.

This is indeed true, unless we simplify the equations as much as possible before using the method presented in (521 ff); that is, unless we remove their common divider.

If instead we make sure to remove the common divider, then a trivial identity can occur only if one equation completely follows from the others.

By the way, when equations only partially follow from each other, they do not always lead to a trivial equation. This depends upon which quantity is chosen to be the unknown.

For example, consider the two equations

$$
\begin{aligned}
amxy + nax + mby + nb &= 0, \\
a'mxy + na'x + mb'y + nb' &= 0.
\end{aligned}
$$

If we choose x as the unknown and eliminate it, we are not led to a trivial equation. The opposite happens when y is chosen to be the unknown. This is because the common divider of these two equations is $my + n$ and does not contain x, but only y.

Now would be the time to discuss how the methods we have given in this book can be used to look for the common divider of polynomials. However, this application is not difficult, and we will instead pursue the subjects that remain to be treated.

Reflections on the successive elimination method

(532.) We believe the reader can now build a fair opinion of the state of matters relative to elimination, before we introduced the methods described in this work.

Up to now, all known elimination methods relied on successive elimination. Consider three equations in three unknowns. The process consisted of recasting these equations as equations in one unknown, then to eliminate this unknown using the three equations, leading to two condition equations in the coefficients of the initial equations. Rewriting these equations as equations in one of the two remaining unknowns and performing another elimination then leads to one equation in one unknown only.

All known general elimination methods reduce to that process:

(533.) We now examine whether this process can (1) lead to the true final equation, that is, the one with lowest possible degree, and (2) lead us instead to equations that are more complex than necessary, with no hope to find the complicating factor, (3) possibly lead us to nothing at all.

If we decide to combine the equations two-by-two to obtain the two condition equations, the resulting condition equations would necessarily be much more complex than needed, and it would not be possible simplify them by extracting a factor.

If we used more than two equations to obtain simpler condition equations, this would provide only partial relief from the difficulties we just pointed out. The last final equation would still be too complex. The case of three second-order equations in one unknown, which is related to three second-order equations in three unknowns, provides us with a simple example: We know from (462) that the two simplest condition equations are $(ab'c'') = 0$ and $(ab').(bc') - (ac')^2 = 0$. However, a, b, c; a', b', c'; and a'', b'', c'' have degrees 0, 1 and 2, respectively. Thus, the first of the final equations is third order, and the second is fourth order. Thus the final equation would be of order 12, whereas we know it is only order 8.

Also, by recasting all equations as equations in one unknown and using the condition equations with lowest dimension, we do not have any indication about whether some of these condition equations naturally follow from the others. If this had been the case, we would not have found the final equation, but rather a trivial identity after all terms cancel each other.

We can therefore conclude that the successive elimination method may not only unavoidably lead to a final equation with many factors, but also lead to no final equation at all.

About equations whose form is arbitrary, regular or irregular.
Determination of the degree of the final equation in all cases

(534.) We say that the form of an equation is regular if the number of its terms can be written as a finite algebraic expression. Conversely, we say the form of an equation is irregular if the number of its terms cannot be written as a finite algebraic expression, either because the equation can only be written in exhaustive form or because we do not know the rule on the main exponents of the unknowns, and we therefore do not know the number of terms either.

(535.) We first discuss the equations whose number of equations is equal to the number of unknowns; the case where the number of equations exceeds the number of unknowns will then only require a short discussion.

We have described how to compute the general algebraic form for the degree of the final equation when dealing with an infinite variety of regular equation forms. What we have then said is more than enough to determine the same degree for an infinite number of other forms.

Since it is impossible to obtain an algebraic expression for the number of terms of an arbitrary equation or polynomial, the same is true for the degree of the general final equation.

However, it is always possible to obtain numerical expressions for the degree of the final equation resulting from an arbitrary number of equations, no matter how irregular they are.

This last subject will complete our study of the degree of the final equation for an arbitrary number of equations; it is all the more useful that the method we propose can also lead to algebraic expressions whenever these exist, provided the form of the polynomial multipliers is identified directly.

(536.) When dealing with incomplete equations of various order, we have seen in (168 ff) that the first and simplest form of the polynomial multipliers is not appropriate to obtain the algebraic expression for the degree of the final equation, except in certain cases. Depending upon the relative values of the known exponents in the form of the proposed equations, the polynomial multipliers must be incomplete polynomials whose order may vary.

If we do not use the most appropriate polynomial multiplier form, it will not lead to an exact differential, because it does not give the right number of terms that may be cancelled in each polynomial multiplier; the resulting expression regarding the difference between the number of terms to be cancelled in the sum equation and the number of useful coefficients in the polynomial multipliers is therefore untrue. Thus it is obvious that the undetermined exponents of the polynomial multiplers do not vanish from the algebraic expression of the degree of the final equation.

For example, assume the form of the polynomial multipliers is such that cancelling undetermined coefficients in the sum equation results in the vanishing of even more terms in that equation. It would then be obvious that we would have made a mistake by assuming the number of useful coefficients in the polynomial multipliers must exceed the number of terms in the sum equation by one, since some of these terms vanish with fewer coefficients. The form of the polynomial multipliers then becomes corrupted and the degree of the final equation would not be true.

(537.) We must therefore make sure that cancelling some of the coefficients in these polynomial multipliers does not result in a larger number of terms vanishing in the sum equation, if we want to get the true form of the polynomial multipliers.

(538.) Let p be the number of these coefficients and n the number of terms in the sum equation that may disappear when these coefficients are cancelled. The sum equation will therefore provide n equations containing only p undetermined coefficients. Assuming all these coefficients are zero, n equations in the sum equation would already be satisfied, and the form of the polynomial multipliers remains as before, minus the terms whose coefficients have just been assumed to be zero. This form of the polynomial multipliers is, however, not the true one yet.

We examine again whether there still exists n' partial equations in the sum equation containing only $p' < n'$ undetermined coefficients; if the answer is positive, we must proceed as above, until the number of coefficients remaining in the sum equation is larger than the total number of terms in the sum equation, minus those containing only the unknown to be present in the final equation. At the same time, these coefficients must be so coupled that none can be determined independently from the others.

(539.) We must of course pay attention to the number of arbitrary equations, which is obtained by counting the number of useless coefficients in the polynomial multipliers. For example, assuming all terms in the polynomial multipliers have been properly substituted, we should then determine the q terms in the sum equation which, when combined with q' arbitrary equations, leads to n equations where n is greater than the p coefficients these equations contain. Alternatively, we can first cancel all useless coefficients in the polynomial multipliers and then examine whether the sum equation contains groups of equations whose number is larger than the number of unknowns these groups contain.

(540.) If the number of equations generated by the sum equation is higher than the number of remaining coefficients, after performing all of the above, then the form of the multipliers is too low. There is a general and trustworthy method to avoid this problem; here it is.

(541.) We pick complete polynomials as polynomial multipliers, such that the total dimension of the sum equation is the same as if the proposed equations were all complete equations. Then we are guaranteed that the unknown polynomial multipliers can be part of these more general polynomial multipliers. The method we proposed above can be used to remove the superfluous terms, without removing any useful ones. If instead we picked a lower dimensional form or an incomplete form, it could happen that we perform illegal operations. Starting with complete polynomials instead guarantees that no illegal operation is performed.

(542.) Once we are able to have more undetermined coefficients than there are terms to be cancelled in the sum equation, we need to determine the allowed uses for the superfluous and therefore arbitrary coefficients.

We will denote as *general arbitrary coefficients* those that can always be cancelled in the polynomial multipliers using the proposed equations. If we use complete polynomials as polynomial multipliers, the number of general arbitrary coefficients is the same as if the proposed equations were complete.

We will denote as *specific arbitrary coefficients* those which turn out to be arbitrary because the sum equation is lacking the terms it should have if all equations were complete.

(543.) This being said, we can determine the number of specific arbitrary coefficients or the number of specific arbitrary equations as follows: We first determine the difference between the number of terms to be cancelled in the sum equation and the number of terms to be cancelled in the sum equation if the equations had been complete. We also compute how many of the additional terms belong to each dimension. There are as many specific arbitrary coefficients as there are such terms.

We then look for the terms in the sum equation whose number of arbitrary coefficients is less than the number of these terms. Let n be the number of these terms and p the number of these coefficients; assuming these coefficients equal zero, n equations are satisfied; thus $n - p$ terms are cancelled, beyond the number of coefficients that was used; thus there are $n - p$ more specific arbitrary coefficients in the sum equation.

This process is continued until there are no more terms to be cancelled that contain more coefficients than there are equations to find them. Accounting for the number of specific arbitrary equations provided by this process as it proceeds, we can use them to lower the equation whenever possible. However, lowering the degree of the equation must not be arbitrarily imposed, but rather must follow from using the arbitrary equations. We now illustrate this by means of examples.

(544.) From (285) we have computed the final equation resulting from two equations of the form $(x \ldots 2)^2 = 0$. Assume the first-dimensional terms are missing from these two equations, that is, the equations are of the form

$$ax^2 + byx + cy^2 + f = 0.$$

I now form the product equation that follows, to know whether the degree of the equation can be lowered or whether the polynomial multipliers can be simpler than the general form:

$$
\begin{array}{llllllllll}
& Aa & x^4 & + & Ab & x^3y & + & Ac & x^2y^2 & + & Bc & xy^3 & + & Cc & y^4 = 0. \\
& & & + & Ba & & + & Bb & & + & Cb & & & & \\
& & & & & & + & Ca & & & & & & & \\
\\
+ & Da & x^3 & + & Db & x^2y & + & Dc & xy^2 & + & Ec & y^3 & & & \\
& & & + & Ea & & + & Eb & & & & & & & \\
\\
+ & Fa & x^2 & + & Fb & xy & + & Fc & y^2 & & & & & & \\
+ & fA & & + & fB & & + & fC & & & & & & & \\
\\
+ & fD & x & & + & fE & y & & & & & & & & \\
\\
+ & fF & & & & & & & & & & & & &
\end{array}
$$

This is a complete equation, and I conclude there is no specific arbitrary coefficient, but only one general arbitrary coefficient. We know this coefficient can be chosen in any desired dimension.

However, when observing the terms y^4, x^2y, xy^2, y^3 and y, I see that the resulting equations only contain six undetermined coefficients C, C'; D, D'; E, E'. Combined with the general arbitrary equation, we have six equations containing these six coefficients only; we can therefore assume

$$C = 0, \ C' = 0, \ D = 0, \ , D' = 0, \ E = 0, \ E' = 0.$$

The form of the polynomial multipliers can therefore be simpler than the general form. It is

$$
\begin{array}{lllll}
Ax^2 & + & Bxy & \ldots & A'x^2 & + & B'xy \\
+ & F & & & + & F'
\end{array}
$$

such that the sum equation reduces to

$$
\begin{aligned}
Aax^4 &+ Abx^3y + Acx^2y^2 + Bcxy^3 = 0 \\
&+ Ba \quad\; + Bb
\end{aligned}
$$

$$
\begin{aligned}
+\; Fax^2 &+ Fbxy + Fcy^2 \\
+\; fA &+ fB
\end{aligned}
$$

$$
+\; fF
$$

and leads to the final equation

$$
(bc')\left\{ \begin{array}{c} (ac')^2 \\ -(ab').(bc') \end{array} \; x^4 \; + \; \begin{array}{c} (bc').(bf') \\ -\;2(ac').(cf') \end{array} \; x^2 \; + \; (cf')^2 \right\} = 0.
$$

The factor (bc') indicates a case when the degree of the equation can be lowered.

(545.) Assume the proposed equations are of the form

$$
ax^2 + bxy + f = 0.
$$

Ignoring the form that must be taken by the polynomial multipliers and the final equation, let us use the same polynomial multiplier form as if these equations were complete.

I examine the following form of the sum equation to determine the true form of the polynomial multipliers and the true degree of the final equation:

$$
\begin{aligned}
Aa \;\; x^4 &+ Ab \;\; x^3y + Bb \;\; x^2y^2 + Cb \;\; xy^3 = 0. \\
&+ Ba
\end{aligned}
$$

$$
\begin{aligned}
+\; Da \;\; x^3 &+ Db \;\; x^2y + Eb \;\; xy^2 \\
&+ Ea
\end{aligned}
$$

$$
\begin{aligned}
+\; Fa \;\; x^2 &+ Fb \;\; xy + fC \;\; y^2 \\
+\; fA &+ fB
\end{aligned}
$$

$$
+\; fD \;\; x \;\; + fE \;\; y
$$

$$
+\; fF
$$

Considering the highest dimension, I see there is one fewer term to be cancelled than if the proposed equations were complete; thus there are two arbitrary equations in this dimension: one general arbitrary dimension and one specific arbitrary dimension.

Likewise, the third dimension has one fewer term to be cancelled than if the equations were complete. Thus this dimension provides a specific arbitrary equation.

This being said, the three equations provided by cancelling the terms x^3y, x^2y^2, xy^3, and the two arbitrary equations in the same dimension contain a total of six unknowns; moreover, the equation provided by the term in y^2

does not contain any other unknowns. Thus we have six equations in six unknowns and each coefficient is zero; we therefore have

$$A = 0, \ A' = 0, \ B = 0, \ B' = 0, \ , C = 0, \ C' = 0.$$

The sum equation thus reduces to

$$\begin{aligned} Da \ \ x^3 \ \ &+ \ \ Db \ \ x^2 y \ \ + \ \ Eb \ \ xy^2 = 0. \\ &+ \ \ Ea \end{aligned}$$

$$+ \ \ Fa \ \ x^2 \ \ + \ \ Fb \ \ xy$$

$$+ \ \ fD \ \ x \ \ + \ \ fE \ \ y$$

$$+ \ \ fF$$

However, the two equations provided by cancelling the terms $x^2 y$, xy^2, and the arbitrary equation in this dimension contain four unknowns; together with the equation provided by the y term, it gives four equations in four unknowns; thus we conclude $D = 0$, $D' = 0$, $E = 0$, $E' = 0$.

The sum equation thus reduces to the form

$$Fax^2 + Fbxy = 0$$
$$+fF.$$

The polynomial multipliers are therefore simply F and F'. The final equation reduces to a second-order equation; this is obvious from the proposed equations.

(546.) Assume we have two equations of the form $(x^2, y^1)^3 = 0$.

From (62), we know the degree of the final equation must be 4, and we have shown in (317) how to find the simplest polynomial multipliers.

However, let us behave as if we knew neither the degree of the final equation nor the form of the polynomial multipliers.

I therefore use two complete polynomials with degree $3 \times 3 - 3$, that is, with degree 6. The sum equation therefore has the form $(x^8, y^7)^9 = 0$.

Thus, there are two fewer terms to be cancelled in the highest dimension than if the proposed equations were complete; and there is one such term in the dimension 8. Thus, there are three specific arbitrary coefficients standing beyond the general arbitrary coefficients; two of these belong to the dimension 9, and one of these coefficients belongs to the dimension 8.

This being said, I notice there are three terms in y^7, and only two coefficients can be provided by the polynomial multipliers to cancel these terms. Thus if we assume these two coefficients are set to zero, three terms in the sum equations are cancelled, thereby yielding one more arbitrary equation in the sum equation. This arbitrary equation can be used in any dimension of the sum equation.

Thus the polynomial multipliers reduce to the form $(x^6, y^5)^6$, and the sum equation reduces to the form $(x^8, y^6)^9 = 0$.

Looking at this new form, I notice the sum equation has four terms in y^6, which can be cancelled by using the only four useful coefficients provided by the polynomial multipliers; each of these coefficients can therefore be assumed to be zero.

The form of the polynomial multipliers thus reduces to $(x^6, y^4)^6$ and that of the sum equation reduces to $(x^8, y^5)^9 = 0$.

In this form, the sum equation has five terms in y^5, which can be cancelled by the six coefficients provided by the two polynomial multipliers. These six coefficients reduce to five, because one of them is in fact a general arbitrary coefficient. Thus there are only as many coefficients as there are equations, and we can assume these coefficients are zero. The form of the polynomial multipliers thus reduces to $(x^6, y^3)^6$, and that of the sum equation reduces to $(x^8, y^4)^9 = 0$.

Considering the new form for the sum equation, there are six terms in y^4, which can be cancelled by eight coefficients from the polynomial multipliers. However, two of these coefficients are general arbitrary coefficients; thus we can still set all these coefficients to zero. The form of the polynomial multipliers therefore reduces to $(x^6, y^2)^6$ and that of the sum equation is $(x^8, y^3)^9 = 0$.

Following the same argument, the form of the polynomial multipliers reduces to $(x^6, y^0)^6$ or $(x \ldots 1)^6$. The form of the sum equation reduces to $(x^8, y^1)^9 = 0$ and then no general arbitrary coefficient remains.

Since four specific arbitrary coefficients remain, we must now use them. Assume first we only use one arbitrary equation in the highest dimension; together with the equation from the term $x^8 y$, this equation makes the number of equations equal to the number of undetermined coefficients they contain. Consequently these two coefficients are 0 each, the form of the polynomial multipliers is $(x \ldots 1)^5$, and the form of the sum equation is $(x^7, y^1)^8 = 0$.

Following a similar argument, we see that we can use one of the three remaining specific arbitrary equations to transform the form of the polynomial multipliers into $(x \ldots 1)^4$ and that of the sum equation into $(x^6, y^1)^7 = 0$. Likewise, if we use one of the two remaining specific arbitrary equations in the new sum equations, the form of the polynomial multipliers becomes $(x \ldots 1)^3$ and that of the sum equation becomes $(x^5, y^1)^6$. Finally, using the last specific arbitrary equation, the form of the polynomial multipliers becomes $(x \ldots 1)^2$ and the form of the sum equation becomes $(x^4, y^1)^5 = 0$; from this we conclude that, indeed, the final equation is fourth order only; and the simplest polynomial multipliers are the same as those resulting from what was said in (317).

(547.) Consider now three equations of the form $[x, (y, z)^1]^2 = 0$. We have already discussed these equations more than once, but we now behave as if we knew neither the degree of the final equation nor the form of the polynomial multipliers.

Consider three complete polynomial multipliers with degree $2 \times 2 \times 2 - 2$, that is, 6.

The sum equation is of the form $[x, (y, z)^7]^8 = 0$. Thus all nine terms where y and z reach the dimension 8 are missing. We therefore have nine specific arbitrary coefficients. Since these belong to the highest dimension, they can be used for any other dimension.

Consider the terms where y and z reach the dimension 7. They enter the sum equation through the terms in the polynomial multipliers where y and z reach the dimension 6; thus cancelling the sixteen terms in y and z that reach the dimension 7 depends upon twenty-one coefficients; however, nine of them are specific arbitrary coefficients, as we just said. If we use only five of them together with the sixteen equations provided by the terms in y and z reaching dimension 7, we can assume all these coefficients are zero. Consequently the form of the sum equation is $[x, (y, z)^6]^8 = 0$, and the form of the polynomial multipliers is $[x, (y, z)^5]^6$. Four specific arbitrary coefficients remain.

We now examine the terms where y and z reach the dimension 6. We see that, except for the general arbitrary coefficients, the polynomial multipliers can provide only as many coefficients as there are terms to be cancelled; thus we can set to zero all coefficients of the terms where y and z reach dimension 5 in the polynomial multipliers. The form of the polynomial multipliers thus reduces to $[x, (y, z)^4]^6$, and the form of the sum equation reduces to $[x, (y, z)^5]^8 = 0$.

Likewise, we look at the terms in the sum equation where y and z reach dimension 5 together, then all terms where y and z reach dimension 3, then those where they reach dimension 2. This process leads to the following successive forms for the polynomial multipliers:

$$[x, (y, z)^3]^6; \ [x, (y, z)^2]^6; \ [x, (y, z)^1]^6; \ [x, (y, z)^0]^6,$$

that is, $(x \ldots 1)^6$.

At this point there remain four arbitrary equations. Assume we only use one of them in the highest dimension of the sum equation $[x, (y, z)^1]^8 = 0$. Together with the two equations provided by the terms $x^7 y$ and $x^7 z$, we obtain as many equations as there are coefficients. Each of these coefficients is therefore zero and the term x^8 vanishes automatically.

The sum equation thus reduces to $[x, (y, z)^1]^7 = 0$, and the polynomial mulipliers reduce to $(x \ldots 1)^5$.

Using the three specific arbitrary equations, one in each dimension 7, 6, and 5 in the sum equation, we find that the sum equation successively reduces to

$$[x, (y, z)^1]^6 = 0, \ [x, (y, z)^1]^5 = 0, \ [x, (y, z)^1]^4 = 0.$$

The latter form is the most reduced, since no arbitrary coefficient remains.

Thus the lowest degree of the final equation is 4, and the simplest form for the polynomial multipliers is $(x \ldots 1)^2$; this is consistent with the results in (320) and (329).

Remark

(548.) We have said in (234) that the values of the arbitrary coefficients can be freely chosen, but not completely so. For example, there is a maximum number of such coefficients in each dimension of the sum equation. We have already taught how to determine this number.

We have also said that, provided we do not exceed the maximum number of arbitrary equations in each dimension, we can choose the terms in each dimension to which the equation should apply.

This is true in general, when the polynomials expressing the number of terms that may be cancelled can provide all terms contained in the polynomial multipliers, when these polynomials are multiplied by the proposed equations. When the form of the polynomial multipliers is unknown and we use an arbitrary form, such as complete polynomials instead, then we cannot freely assign the general arbitrary equations to any term, regardless of its dimension. However, it is always possible to identify which terms cannot benefit from these arbitrary equations.

Indeed, assume we want to determine whether an arbitrary coefficient in the polynomial multipliers can be considered as a general arbitrary coefficient. To that end, we must consider the equations that can contribute to cancelling the terms in this polynomial multiplier. Can these equations, when multiplied by a complete polynomial, result in a polynomial whose degree is the same as that of the polynomial multiplier of interest? If the term of interest is part of those arising from these products, then this term can be considered to be a general arbitrary term. Otherwise it cannot.

Thus, in the last example, the terms where y and z reach the dimension 6 in the complete polynomial multipliers cannot be considered to be general arbitrary coefficients. Indeed, no such terms can arise from the multiplication of the proposed equations by complete polynomials of degree 4.

Follow-up on the same subject

(549.) Now consider two fourth-order equations, whose terms with order less than 3 are all missing. We represent these polynomials as $(x \ldots 2)_3^4 = 0$. We propose to determine the degree of the final equation and the simplest possible form of the polynomial multipliers.

I first use two polynomials of the form $(x \ldots 2)^{4 \times 4 - 4} = (x \ldots 2)^{12}$ as polynomial multipliers; the number of general arbitrary coefficients is then $N(x \ldots 2)^8$ and the form of the sum equation is $(x \ldots 2)_3^{16} = 0$.

Since the terms with dimension less than 3 are missing from this equation, there are three fewer terms to be cancelled than if the proposed equations were complete; thus there are three specific arbitrary coefficients.

However, the third dimension provides the terms x^2y, xy^2 and y^3, which must be cancelled; we only have two coefficients in the dimension 0 of the two polynomial multipliers to cancel these three terms. Assuming these two coefficients to be equal to zero, we have one more arbitrary coefficient or equation. Thus, using polynomial multipliers of the form $(x \ldots 2)_1^{12}$ leads to a sum equation of the form $(x \ldots 2)_4^{16} = 0$, with four specific arbitrary equations and $N(x \ldots 2)^8$ general arbitrary equations.

The fourth dimension of the sum equation provides four terms to be cancelled. The first dimension of the polynomial multipliers provides only four coefficients for that purpose. Each of these coefficients is therefore zero.

Consequently the form of the polynomial multipliers becomes $(x \ldots 2)_2^{12}$ and the form of the sum equation becomes $(x \ldots 2)_5^{16} = 0$, with the same number of general and specific arbitrary equations.

The fifth dimension in the sum equation provides five terms to be cancelled. The second dimension of the polynomial multipliers provides six coefficients to that end; assuming we use one of the four specific arbitrary equations, we have six equations in six unknowns and each of the six coefficients is zero. Consequently, the form of the polynomial multipliers becomes $(x \ldots 2)_3^{12}$, and the form of the sum equation is $(x \ldots 2)_6^{16} = 0$, along with three specific arbitrary equations only and $N(x \ldots 2)^8$ general arbitrary equations.

Considering this new form, the sixth dimension of the sum equation contains six terms that must be cancelled. The third dimension of the polynomial multipliers provides eight coefficients to that end. The 0th dimension of the polynomial $(x \ldots 2)^8$ (which expresses the number of the general arbitrary coefficients) provides one such coefficient in that dimension. Thus we can count on only seven coefficents, provided by the third dimension of the polynomial multipliers. However, if we use one of the three remaining specific arbitrary equations, we can set all coefficients in the third dimension of the polynomial multipliers to zero; consequently, the form of the polynomial multipliers reduces to $(x \ldots 2)_4^{12}$, and the form of the sum equation reduces to $(x \ldots 2)_7^{16} = 0$, with two specific arbitrary coefficients and $N(x \ldots 2)_1^8$ general arbitrary coefficients.

At this point, the dimension 7 in the sum equation contains seven terms to be cancelled. The fourth dimension of the polynomial multipliers provides ten coefficients, two of which are useless, by virtue of the first dimension of the polynomial $(x \ldots 2)_1^8$. Hence, we still have eight useful coefficients. Using one of the two remaining specific arbitrary coefficients, only seven coefficients remain, that is, as many as there are terms to be cancelled. Thus each coefficient in the fourth dimension of the polynomial multipliers can be assumed to be zero; consequently the form of the polynomial multipliers reduces to $(x \ldots 2)_5^{12}$, and the form of the sum equation reduces to $(x \ldots 2)_8^{16} = 0$. In addition, there are $N(x \ldots 2)_2^8$ general arbitrary coefficients and one specific arbitrary coefficient.

Considering this new form, the eighth dimension of the sum equation contains eight terms to be cancelled. The fifth dimension of the polynomial multipliers provides twelve coefficients, three of which are useless, by virtue of the second dimension of the polynomial muliplier. Thus nine coefficients remain. Using the remaining specific arbitrary equation, only eight of these coefficients are really useful, as many as the number of terms to be cancelled. Thus each coefficient in the fifth dimension of the polynomial multipliers can be assumed to be zero. Therefore the form of the polynomial multipliers reduces to $(x \ldots 2)_6^{12}$, and the form of the sum equation reduces to $(x \ldots 2)_9^{16} = 0$. $N(x \ldots 2)_3^8$ general arbitrary coefficients remain. The polynomial multipliers cannot be reduced any further, and the same is true of the sum equation. Indeed, examining the ninth dimension of the sum equation as we have done before indicates that nine terms must be cancelled. On the other hand, the sixth dimension of the polynomial multipliers gives fourteen coefficients, three of which are useless by virtue of the third dimension of the polynomial $(x \ldots 2)^8_3$. Since no specific arbitrary equation remains, the number of useful coefficients exceeds the number of terms to cancel and we cannot assume these coefficients to be zero anymore.

Therefore, the final equation still has degree 16; that is, it has sixteen roots; however, nine of these roots are zero. Thus this equation is to be solved as a seventh-order equation, although its degree is actually 16.

(550.) Using the same argument with three equations of the form $(x \ldots 3)^3_2$, we see the sum equation reduces to the form $(x \ldots 3)^{27}_8 = 0$; the polynomial multipliers reduce to the form $(x \ldots 3)^{24}_6$, whose number of general arbitrary coefficients is $3N(x \ldots 3)^{21}_4 - N(x \ldots 3)^{18}_2$ and whose number of specific arbitrary coefficients is zero.

Therefore, the form of the final equation is $(x \ldots 1)^{27}_8 = 0$. Its degree is 27, and eight of its roots are zero.

(551.) Consider the general case of n equations of the form

$$(x \ldots n)^t_t = 0, \ (x \ldots n)^{t'}_{t'} = 0, \ (x \ldots n)^{t''}_{t''} = 0, \ (x \ldots n)^{t'''}_{t'''} = 0, \ \text{etc.}$$

The sum equation always reduces to the form $(x \ldots n)^{tt't''t''',\text{etc.}}_{tt't''t''',\text{etc.}} = 0$. The form of the polynomial multipliers is

First equation   ... $(x \ldots n)^{tt't''t''',\text{etc.}-t}_{tt't''t''',\text{etc.}-t} = 0$

Second equation   ... $(x \ldots n)^{tt't''t''',\text{etc.}-t'}_{tt't''t''',\text{etc.}-t'} = 0$

Third equation   ... $(x \ldots n)^{tt't''t''',\text{etc.}-t''}_{tt't''t''',\text{etc.}-t''} = 0$

Fourth equation   ... $(x \ldots n)^{tt't''t''',\text{etc.}-t'''}_{tt't''t''',\text{etc.}-t'''} = 0,$

and so on. The number of general arbitrary coefficients is very complex to write but easy to find from all we have said.

(552.) So far we have dealt with specific examples of low-order equations. Our goal was to keep the reader's attention focused on the process rather than the complexity of equations; this process is, however, the same irrespective of the degree of the equations and the number of unknowns.

When the equations are regular, this process can always be generalized. This does not require knowing the final equation to determine the general algebraic form of its degree. This is the way, for example, which allows us to determine the most appropriate form of the polynomial multipliers for incomplete equations with arbitrary orders (181 ff). The reason why the form we have used in (181 ff) is not appropriate in general is because this form has terms that generate other terms in the sum equation and these terms are more numerous than the number of equations available to cancel them. Thus it means there really are more arbitrary coefficients than was really accounted for originally. Only by accounting for the true number of arbitrary coefficients can we find the true form of the polynomial multipliers and the true expression for the degree of the final equation. In order for us to leave no uncertainty aside, we must first use complete polynomial multipliers, whose degree is the same as if the equations were complete. We then have to go through all terms that can yield specific arbitrary equations because they can cancel each other; we must also list the number of terms that would appear if the sum equation were complete. Then putting together the number of specific arbitrary equations with the number of general arbitrary equations, the difference between the total number of coefficients in the polynomial multipliers and the total number of general and specific arbitrary equations will always be sufficient to cancel the terms that must disappear in the sum equation, that is, to give the final equation.

However, when the equations are regular, we can always determine all these numbers algebraically and consequently obtain an algebraic expression for the degree of the final equation.

If these equations have an irregular form, then we cannot determine the algebraic expression for this degree; we can, however, obtain a numerical expression for it. Looking for the number of specific algebraic expressions most often requires inspecting the sum equation. As for the general arbitrary equations, we can always obtain their number, since it is the same as if the proposed equations were complete.

(553.) Thus we can determine the degree of the final equation and the simplest polynomial multipliers for any set of regular or irregular algebraic equations.

(554.) All that we said in (534 ff) applies similarly to equations whose number exceeds the number of unknowns; however, the degree of the polynomial multipliers is not the product of the exponents of all proposed equations minus the exponent of the equation corresponding to this polynomial multiplier anymore. Instead, the degree must be determined from what was said from (338) to (518), as if the proposed equations were complete. We then determine the simplest possible form for these polynomial multipliers, from what was said in (534 ff).

Consider three equations of the form

$$ax^3 + bx^2y + cxy^2 + dy^3 + cxy = 0.$$

I argue as follows: If these equations were complete, the polynomial multipliers would be of the form $(x \ldots 2)^4$ and the number of general arbitrary coefficients would be $3N(x \ldots 2)^1 = 9$. Six of these coefficients can be used in any dimension of the sum equation.

For simplicity, assume first that the dimension 2 in the proposed equations is complete; that is, we deal with these equations as if they were of the form $(x \ldots 2)_2^3 = 0$.

The form of the sum equation is therefore $(x \ldots 2)_2^7 = 0$. In this equation, three terms are missing, compared with the case of complete equations. Therefore, we begin with three specific arbitrary equations.

Looking at the dimension 2 in the sum equation, we see three terms to cancel. This task can be performed using the three coefficients provided by the dimension 0 of the polynomial multipliers. Since there are as many coefficients as there are equations, we can assume that each one of them is zero. Consequently, the form of the polynomial multipliers reduces to $(x \ldots 2)_1^4$, and the form of the sum equation reduces to $(x \ldots 2)_3^7 = 0$. In addition, there are nine general arbitrary coefficients and three specific arbitrary coefficients.

The dimension 3 of the new sum equation contains four terms to be cancelled. The dimension 1 of the three polynomial multipliers provides six coefficients; thus, if we use two of the three specific arbitrary equations, we can assume each coefficient in the dimension 1 of the polynomial multipliers to be zero. The form of these polynomials is therefore $(x \ldots 2)_2^4$; the form of the sum equation is $(x \ldots 2)_4^7 = 0$, with nine general arbitrary coefficients and one specific arbitrary coefficient.

The dimension 4 in the sum equation contains five terms to be cancelled; however, the dimension 2 of the polynomial multipliers gives nine coefficients; three of them are useless due to the dimension 0 of the three polynomials $(x \ldots 2)^1$ that give the number of general arbitrary coefficients; thus six coefficients remain; using the remaining specific arbitrary equation, we end up with as many equations as there are useful coefficients in the dimension 2 of the polynomial multipliers. Thus we can assume these coefficients to be zero; the form of the polynomial multipliers reduces to $(x \ldots 2)_3^4$, and the form of the sum equation reduces to $(x \ldots 2)_5^7 = 0$, along with six general arbitrary coefficients.

At this point, there are six terms to cancel in the dimension 5 of the sum equation. The dimension 3 of the polynomial multipliers provides twelve coefficients; six of them are useless, as given by the dimension 1 of the polynomials expressing the number of useless coefficients. Thus there are as many useful coefficients as the the number of terms to cancel. Therefore, each coefficient in the dimension 3 of the polynomial multipliers can be assumed to be zero. Consequently the form of the polynomial multipliers reduces to $(x \ldots 2)_4^4$, and the form of the sum equation reduces to $(x \ldots 2)_6^7 = 0$, with no general or specific arbitrary coefficient.

This would be the simplest form of the polynomial multipliers if the dimension 2 of the proposed equations were complete; however, examining the missing terms from the equations shows that the terms x^6 and y^6 are missing from the sum equation. Thus, there are two specific arbitrary equations. We can put them to best use by cancelling a few more terms in the polynomial multipliers, if possible. For example, cancelling the term y^7 in the sum equation leads to an equation which leads to setting to zero the coefficient of y^4 in each polynomial multiplier, when it is combined with the two specific arbitrary equations.

However, the coefficients entering y^7 are the same as those entering xy^5; thus this term disappears when the term y^4 in the polynomial multipliers vanishes, and one specific arbitrary equation remains.

This equation can be used by noting that the coefficients entering the term x^7 are the same as those entering the term $x^5 y$. Thus, combining the two equations provided by canceling the terms x^7 and $x^5 y$ with the remaining specific arbitrary equation allows us to cancel the term x^4 in the polynomial multipliers.

Thus, given three equations of the form

$$ax^3 + bx^2 y + cxy^2 + dy^3 + exy = 0,$$

the form of the three simplest polynomial multipliers is

$$Ax^3 y + Bx^2 y^2 + Cxy^3,$$

with no general or specific arbitrary coefficient.

The final equation is easy to compute from there.

About equations whose number is smaller than the number of unknowns they contain: New observations about the factors of the final equation

(555.) When the number of unknowns exceeds the number of equations, the problem reduces to finding an equation that contains only $n - p + 1$ unknowns, where n is the number of unknowns and p is the number of equations.

Three methods are available to solve this problem:

1. We can rewrite the equations as equations containing only $p - 1$ unknowns. Consider, for example, two equations of the form

$$
\begin{aligned}
& ax^2 \;+\; bxy \;+\; cxz \;+\; dy^2 \;+\; eyz \;+\; fz^2 = 0. \\
& +\; gx \;+\; hy \;+\; kz \\
& +\; l
\end{aligned}
$$

   We ask for the equation in y and z. I introduce

$$a = A, \; by + cz + g = B, \text{ and } dy^2 + eyx + fz^2 + hy + kz + l = C$$

   and I recast the two proposed equations as

$$Ax^2 + Bx + C = 0.$$

   I can then eliminate x using the methods given in (338 ff).

2. We can rewrite the proposed equations as equations in p unknowns. Considering the same example, I introduce

$$a = A, \; b = B, \; d = C, \; cz + g = D, \; ez + h = E, \; fz^2 + kz + l = F,$$

   and I write the two proposed equations as

$$
\begin{aligned}
& Ax^2 \;+\; Bxy \;+\; Cy^2 = 0. \\
& +\; Dx \;+\; Ey \\
& +\; F
\end{aligned}
$$

   I can then compute the equation in y using the method proposed in (285); replacing the A, B, C, D, E, F by their values in the final equation, we get the equation in y and z.

3. Finally, the proposed equations can be used *as is* to proceed with eliminating $p - 1$ unknowns, using polynomial multipliers containing all n unknowns.

(556.) Among the three methods presented above, the first one is undoubtedly the fastest and it leads to the simplest relation linking the $n - p + 1$ unknowns of interest. However, this method may hide certain factors, which can bring additional information about the proposed equations. When these factors are set to zero, they may sometimes lead to equations that indeed occur. In that case, the "simplest relation" linking the $n - p + 1$ unknowns is in fact not as simple as it could be; it may also contain irrelevant solutions.

Therefore, this first method may be the fastest, but not the most reliable, as long as we cannot guarantee that no relation exists among the coefficients that could lead to a final equation with lower degree.

(557.) The second method follows the first in terms of computational speed. In addition, it can reveal some of the relations among the coefficients which, whenever they occur, lead to lowering the degree of the final equation; however, it does not indicate all such relations when there are more unknowns than equations. Let us clarify this through the simple example of two equations of the form

$$ax^2 \;+\; bxy \;+\; cxz \;+\; dy^2 \;+\; eyz \;+\; fz^2 = 0.$$
$$+\; gx \;+\; hy \;+\; kz$$
$$+\; l$$

Writing these two equations as

$$Ax^2 + Bx + C = 0,$$

we get an equation in A, B, C; A', B', C' by eliminating x. Substituting the values of A, B, C; A', B', C' in y and z leads to a fourth-degree equation in y and z.

This is the simplest relation involving y and z if the coefficients of the proposed equation are arbitrary and unrelated or if there exists no value of z that is independent of x and y, or no value of y that is independent of x and z, such that the two proposed equations are satisfied.

If any of these cases were to arise, the final equation in y and z would not be of degree 4; we have already seen this in (290). However, we see that the current process does not indicate this possibility.

On the other hand, assume we write the proposed equations as

$$Ax^2 \;+\; Bxy \;+\; Cy^2 = 0$$
$$+\; Dx \;+\; Ey$$
$$+\; F$$

and that we compute the final equation in y by eliminating x, following the method given in (285). Then the final equation in y has degree 4 in z after substitution of the values of A, A'; B, B', etc. in z; however, this equation has degree 6 relative to z. This final equation is the same as the one given by the previous method, but it has a factor that we have studied in (290). This factor is a second-order equation in z. We have already shown that, when we assume this factor to be zero, the degree of the final equation can be lowered. This second process is longer than the first but is able to indicate the cases when the degree of the final equation can be lowered.

(558.) This second process is not able to detect all possible cases of this type. Other cases may be detected or not, depending upon which unknowns are chosen to be embedded inside the coefficients of the reduced equations. By using different such variables, we would eventually be able to detect more cases when the degree of the final equation can be lowered. However, there still are more cases that this process is unable to indicate.

(559.) Indeed, consider first the cases that may be detected by changing the variables embedded inside the coefficients of the reduced equations. For example, assume that we eliminate x and z instead of x and y; using the second process, we would have reached an equation where z reaches degree 4 and y reaches the degree 6; the resulting second-degree factor would have indicated two values of y for which the two proposed equations are satisfied independently of x and z, similar to the factor found in the first case, which indicates two values of z for which the two equations are satisfied independently of x and y. At the same time, if this factor were zero independently from z, it would indicate a case when the degree of the final equation in y and z can be lowered; or it would indicate that, if z is given any of these two values, the equation in y can be lowered. If the difference between the number of unknowns and the number of equations is larger, we see many more cases arise that the first process is unable to reveal and that the second process can reveal only by applying it to all the forms that may arise from the proposed equations via this process.

Because of the requirement to use this process many times in order to find these different cases, we must conclude the process is not general enough and that it may even leave more cases aside. And this is indeed what happens:

We have seen in (285) that eliminating y from two generic equations of the form $(x \ldots 2)^2 = 0$ leads to a factor which is a function of all coefficients in the two equations. When this factor becomes zero, the degree of the final equation can be lowered. Likewise, eliminating x from two equations of the form $(x \ldots 3)^2 = 0$ and, in general, eliminating $n - 1$ unknowns in n generic equations in p unknowns where p is greater than n, yields a factor which is a function of all coefficients of these equations; when this factor is zero, it indicates that the degree of the final equation can be lowered. Thus in general, we must deal with these equations in all their generality to make sure none of the possible cases is inadvertently discarded.

(560.) Thus we must use polynomial multipliers that each contain all the unknowns of interest to determine all that must be known about the final equation resulting from n equations in p unknowns, where p is greater or less than n. Any other process can yield only a subset of the information contained in these equations.

(561.) Consider the n general equations

$$(u\ldots p)^t = 0, \ (u\ldots p)^{t'} = 0, \ (u\ldots p)^{t''} = 0, \text{ etc.,}$$

where $n < p$. We multiply the first equation by the undetermined polynomial $(u\ldots p)^{T-t}$, the second by the polynomial $(u\ldots p)^{T-t'}$, the third by the polynomial $(u\ldots p)^{T-t''}$, etc. We create the sum equation by summing these products. We then set the coefficients of the terms whose unknowns must not appear in the final equation to zero.

We then form as many arbitrary equations in the sum equation as the number of terms that can be cancelled in the first polynomial using the $n-1$ last equations, plus the number of terms that can be cancelled in the second polynomial multiplier using the $n-2$ last equations, plus the number of terms that can be cancelled in the third polynomial multiplier using the $n-3$ last equations, etc. We then proceed with the computation of the coefficients and consequently with the computation of the final equation, following the same method as described previously.

However, the value of T is not known, and we must make a few obesrvations.

The difference between the number of terms in the sum equation and the number of useful coefficients of the polynomial multipliers is in general

$$d^n N(u\ldots p)^T \ldots \left(\begin{array}{c} T \\ t, t', t'', \text{etc.} \end{array} \right).$$

The number of terms in the final equation is $N(u\ldots p - n + 1)^T$. Thus we must have

$$N(u\ldots p - n + 1)^T > d^n N(u\ldots p)^T \ldots \left(\begin{array}{c} T \\ t, t', t'', \text{etc.} \end{array} \right),$$

and T must satisfy this condition.

However, this condition only gives a lower bound for T and no upper bound.

Moreover, the case $n < p$ is different from the case $n = p$. In the latter case, any value of T greater than $tt't''\ldots$ would bring in factors of the final equation indicating only specific solutions or simpler cases either in the final equation or in the polynomial multipliers. However, the degree of the final equation would not increase.

The same factors would be functions of the unknowns in the final equations, if the same process were applied to the case $n < p$, after recasting the equations as equations in n unknowns. It should therefore be no surprise that the degree of the final equation increases with the degree of the multipliers in the case $n < p$. This happens because there are infintely many specific cases expressed by the original equations, and all these cases must be contained in the final equation. This necessarily results in increasing the degree of this equation.

Since these specific solutions are not necessarily correlated, some of them can be expressed by equations of a given degree and others can be expressed by equations with a higher degree, and they do not necessarily belong to the same question. Only certain solutions will be part of all possible final equations: These solutions can be found in the final equation found by rewriting all equations as equations in n unknowns. Moreover, all superfluous factors must be eliminated from this final equation. The other particular solutions and the symptoms to find an even simpler final equation are found by trying various values of T. This is why the degree of the final equation is not determined.

Anyway, the simplest way to reach the final equation for generic equations is to pick the smallest value of T immediately greater than $tt't'' \ldots$, which also satisfies the condition

$$N(u \ldots p - n + 1)^T - d^n[N(u \ldots p)^T] \ldots \left(\begin{array}{c} T \\ t, t', t'', \text{etc.} \end{array} \right) > 0.$$

We then leave a few arbitrary coefficients to find the general final equation from this equation. Following our remarks in (487 ff), several final equations appear, all of which are multiples of the general final equation. We can therefore obtain the final equation by computing their common divider.

THE END